# Detection-Oriented Derivatization Techniques in Liquid Chromatography

# CHROMATOGRAPHIC SCIENCE

*A Series of Monographs*

Editor: JACK CAZES
*Sanki Laboratories, Inc.*
*Sharon Hill, Pennsylvania*

Volume 1: Dynamics of Chromatography
*J. Calvin Giddings*

Volume 2: Gas Chromatographic Analysis of Drugs and Pesticides
*Benjamin J. Gudzinowicz*

Volume 3: Principles of Adsorption Chromatography: The Separation of Nonionic Organic Compounds (out of print)
*Lloyd R. Snyder*

Volume 4: Multicomponent Chromatography: Theory of Interference (out of print)
*Friedrich Helfferich and Gerhard Klein*

Volume 5: Quantitative Analysis by Gas Chromatography
*Joseph Novák*

Volume 6: High-Speed Liquid Chromatography
*Peter M. Rajcsanyi and Elisabeth Rajcsanyi*

Volume 7: Fundamentals of Integrated GC-MS (in three parts)
*Benjamin J. Gudzinowicz, Michael J. Gudzinowicz, and Horace F. Martin*

Volume 8: Liquid Chromatography of Polymers and Related Materials
*Jack Cazes*

Volume 9: GLC and HPLC Determination of Therapeutic Agents (in three parts)
*Part 1 edited by Kiyoshi Tsuji and Walter Morozowich*
*Parts 2 and 3 edited by Kiyoshi Tsuji*

Volume 10: Biological/Biomedical Applications of Liquid Chromatography
*Edited by Gerald L. Hawk*

Volume 11: Chromatography in Petroleum Analysis
*Edited by Klaus H. Altgelt and T. H. Gouw*

Volume 12: Biological/Biomedical Applications of Liquid Chromatography II
*Edited by Gerald L. Hawk*

Volume 13: Liquid Chromatography of Polymers and Related Materials II
*Edited by Jack Cazes and Xavier Delamare*

Volume 14: Introduction to Analytical Gas Chromatography: History, Principles, and Practice
*John A. Perry*

Volume 15: Applications of Glass Capillary Gas Chromatography
*Edited by Walter G. Jennings*

Volume 16: Steroid Analysis by HPLC: Recent Applications
*Edited by Marie P. Kautsky*

Volume 17: Thin-Layer Chromatography: Techniques and Applications
*Bernard Fried and Joseph Sherma*

Volume 18: Biological/Biomedical Applications of Liquid Chromatography III
*Edited by Gerald L. Hawk*

Volume 19: Liquid Chromatography of Polymers and Related Materials III
*Edited by Jack Cazes*

Volume 20: Biological/Biomedical Applications of Liquid Chromatography IV
*Edited by Gerald L. Hawk*

Volume 21: Chromatographic Separation and Extraction with Foamed Plastics and Rubbers
*G. J. Moody and J. D. R. Thomas*

Volume 22: Analytical Pyrolysis: A Comprehensive Guide
*William J. Irwin*

Volume 23: Liquid Chromatography Detectors
*Edited by Thomas M. Vickrey*

Volume 24: High-Performance Liquid Chromatography in Forensic Chemistry
*Edited by Ira S. Lurie and John D. Wittwer, Jr.*

Volume 25: Steric Exclusion Liquid Chromatography of Polymers
*Edited by Josef Janča*

Volume 26: HPLC Analysis of Biological Compounds: A Laboratory Guide
*William S. Hancock and James T. Sparrow*

Volume 27: Affinity Chromatography: Template Chromatography of Nucleic Acids and Proteins
*Herbert Schott*

Volume 28: HPLC in Nucleic Acid Research: Methods and Applications
*Edited by Phyllis R. Brown*

Volume 29: Pyrolysis and GC in Polymer Analysis
*Edited by S. A. Liebman and E. J. Levy*

Volume 30: Modern Chromatographic Analysis of the Vitamins
*Edited by Andre P. De Leenheer, Willy E. Lambert, and Marcel G. M. De Ruyter*

Volume 31: Ion-Pair Chromatography
*Edited by Milton T. W. Hearn*

Volume 32: Therapeutic Drug Monitoring and Toxicology by Liquid Chromatography
*Edited by Steven H. Y. Wong*

Volume 33: Affinity Chromatography: Practical and Theoretical Aspects
*Peter Mohr and Klaus Pommerening*

Volume 34: Reaction Detection in Liquid Chromatography
*Edited by Ira S. Krull*

Volume 35: Thin-Layer Chromatography: Techniques and Applications, Second Edition, Revised and Expanded
*Bernard Fried and Joseph Sherma*

Volume 36: Quantitative Thin-Layer Chromatography and Its Industrial Applications
*Edited by Laszlo R. Treiber*

Volume 37: Ion Chromatography
*Edited by James G. Tarter*

Volume 38: Chromatographic Theory and Basic Principles
*Edited by Jan Åke Jönsson*

Volume 39: Field-Flow Fractionation: Analysis of Macromolecules and Particles
*Josef Janča*

Volume 40: Chromatographic Chiral Separations
*Edited by Morris Zief and Laura J. Crane*

Volume 41: Quantitative Analysis by Gas Chromatography, Second Edition, Revised and Expanded
*Josef Novak*

Volume 42: Flow Perturbation Gas Chromatography
*N. A. Katsanos*

Volume 43: Ion-Exchange Chromatography of Proteins
*Shuichi Yamamoto, Kazuhiro Nakanishi, and Ryuichi Matsuno*

Volume 44: Countercurrent Chromatography: Theory and Practice
*Edited by N. Bhushan Mandava and Yoichiro Ito*

Volume 45: Microbore Column Chromatography: A Unified Approach to Chromatography
*Edited by Frank J. Yang*

Volume 46: Preparative-Scale Chromatography
*Edited by Eli Grushka*

Volume 47: Packings and Stationary Phases in Chromatographic Techniques
*Edited by Klaus K. Unger*

Volume 48: Detection-Oriented Derivatization Techniques in Liquid Chromatography
*Edited by Henk Lingeman and Willy J. M. Underberg*

Volume 49: Chromatographic Analysis of Pharmaceuticals
*Edited by John A. Adamovics*

Volume 50: Multidimensional Chromatography: Techniques and Applications
*Edited by Hernan Cortes*

Volume 51: HPLC of Biological Macromolecules: Methods and Applications
*Edited by Karen M. Gooding and Fred E. Regnier*

Volume 52: Modern Thin-Layer Chromatography
*Edited by Nelu Grinberg*

*Additional Volumes in Preparation*

Chromatographic Analysis of Alkaloids
*Milan Popl, Jan Fahnrich, and Vlastimil Tatar*

# Detection-Oriented Derivatization Techniques in Liquid Chromatography

*edited by*

**HENK LINGEMAN**
Free University
Amsterdam, The Netherlands

**WILLY J. M. UNDERBERG**
State University of Utrecht
Utrecht, The Netherlands

**MARCEL DEKKER, INC.** **New York and Basel**

Library of Congress Cataloging-in-Publication Data

Detection-oriented derivatization techniques in liquid chromatography / edited by Henk Lingeman, Willy J. M. Underberg.
p. cm. -- (Chromatographic science ; v.)
Includes bibliographical references.
ISBN 0-8247-8287-9 (alk. paper)
1. Liquid chromatography. 2. Derivatization. I. Lingeman, Henk. II. Underberg, Willy J. M. III. Series.
QD79.C454D45 1990
543'.0894--dc20 89-23752
CIP

This book is printed on acid-free paper.



MARCEL DEKKER, INC.
270 Madison Avenue, New York, New York 10016

Current printing (last digit):
10 9 8 7 6 5 4 3 2 1

PRINTED IN THE UNITED STATES OF AMERICA

# Preface

This volume comprehensively covers the principles, methods, kinetics, and applications of derivatization reactions in liquid chromatography. Our primary aim is to provide all the information necessary to develop an analytical procedure, starting from sample pretreatment and proceeding through detection. Toward this end, each chapter has a number of references and examples from the pharmaceutical, environmental, and bioanalytical fields.

The first four chapters of this book provide the theoretical background and practical details to aid selection of an optimum derivatization procedure to enhance the detectability of the compound(s) under investigation. Hence, the principles of the detection mode, the influence of the labels and the applied solvent on the detectability, the design and practical aspects (including prechromatographic and postchromatographic possibilities), and the physicochemical properties of the detectability introducing or enhancing reagents are reviewed. These detection-oriented chapters contain up-to-date surveys, presented as tables, of the available reagents and applications, with emphasis on the quantitative determination of drugs and drug metabolites in pharmaceutical formulations and biological fluids. Consideration is given to trends and future prospects.

The first chapter of the book describes the situations where derivatization is appropriate, the improvement of selectivity or chromatographic behavior, the choice of prechromatographic or postchromatographic derivatization techniques with the concomitant advantages and disadvantages, the necessary equipment, the conditions for optimum derivatization reactions, the choice of suitable derivatization reagents, and some indications to optimize a prechromatographic derivatization procedure.

In Chapter 2, the reaction mechanisms, kinetics, reaction conditions (e.g., solvents, temperature), influence of type of reagent, and functional groups of the analytes are discussed. The relevant reactions considered are alkylation, acylation, arylation, silylation, esterification, and cyclization.

Chapter 3, on sample pretreatment, discusses various methods for the isolation of analytes from matrices and, in particular, a biological matrix. These include liquid-solid isolation, deproteination, dehydration, liquid-liquid extraction, and ion-pair extraction. Particular attention is paid to the effective link-up of the derivatization procedure to the sample pretreatment and the chromatographic separation steps.

In the following chapter, the design and principles of the different types of liquid-phase and solid-phase reactions or extraction detectors (e.g., capillary, bed, segmented stream) are described, and the problems associated with band broadening, pressure drop, and phase separation with the different reaction detectors are discussed. Miniaturization and the use of narrow-bore columns are also considered.

Chapter 5 provides an overview of the separation of enantiomers after conversion to their respective diastereomers. Special attention is given in Chapter 6 to the use of enzymatic derivatization and, in particular, to postchromatographic enzyme immunoassay procedures to increase the selectivity of the analytical procedures.

Chapters 7-10 emphasize detection-oriented derivatization procedures that can be used to enhance to detectability of a wide variety of solutes using absorbance, amperometric, and chemiluminescence or fluorescence detection.

We wish to express our gratitude to the contributing authors from the different fields of derivatization in chromatography for their work and suggestions to improve the quality of this monograph. Preparation of a book of this nature is impossible without the support and cooperation of many people. We are particularly grateful to H. J. E. M. Reeuwijk, C. M. B. van den Beld, and R. A. M. van der Hoeven from the Division of Analytical Chemistry of the Center for Bio-Pharmaceutical Sciences, Leiden University, and P. W. van Dorp van Vliet from the Pharmaceutical Laboratory, State University of Utrecht for the drawings. Last, but by no means least, we thank our wives, Anne-Marie and Urmila, for their understanding and patience during the many evenings we spent on the preparation of this work.

Henk Lingeman
Willy J. M. Underberg

# Contents

# Abbreviations

| | |
|---|---|
| AASP | Advanced Automated Sample Processor |
| Ab | Antibody |
| ACDA | 2-amino-1-cyclopentene-1-dithiocarboxylic acid |
| AChE | Acetylcholine esterase |
| Ag | Antigen |
| AD | Amperometric detection |
| APM | N-(4-Anilinophenyl)maleimide |
| ASTED | Automated Sequential Trace Enrichment of Dialysates |
| ATP | Adenosine-5'-triphosphate |
| AUX | Auxiliary electrode |
| BMP | 2-Bromo-1-methylpyridinium iodide |
| BOC | tert.-Butyloxycarbonyl |
| BW | Band width |
| c | Concentration (mol/l) |
| CDI | N,N'-Carbonyldiimidazole |
| CE | Collection efficiency |
| CFA | Continuous-flow analysis |
| ChO | Choline oxidase |
| CI | Chemical ionization |
| CIF | Conventional-induced fluorescence |

| | |
|---|---|
| CF | Continuous flow |
| ClMIS | N-Chloromethylisatine |
| ClMNPI | N-Chloromethyl-4-nitrophthalimide |
| ClMPI | N-Chloromethyl-4-phthalimide |
| CPG | Controlled pore glass |
| CS | Column-switching |
| CV | Current-voltage |
| $D_x$ | Distribution ratio |
| D | Dielectric constant |
| $\delta$ | Solubility parameter |
| DABS-Cl | 4-N,N'-dimethylazobenzenesulfonyl chloride |
| DAS | 9,10-Dimethoxyanthracene-2-sulphonate |
| DCC | Dicyclohexylcarbodiimide |
| DCI | Direct chemical ionization |
| DDTC | Diethylammonium·diethyldithiocarbamate |
| DMAPI | p-N,N-dimethylaminophenylisothiocyanate |
| DNBB | 3,5-Dinitrobenzoylbromide |
| DNBCl | 3,5-Dinitrobenzoyl chloride |
| DNCP | 2-Chloro-3,5-dinitropyridine |
| DNPH | 2,4-Dinitrophenylhydrazine |
| Dns-Cl | 5-Dimethylamino-1-naphthalenesulfonyl chloride |
| DPNT | 3,6-Dinitrophthalic acid anhydride |
| DTC | Dithiocarbamate |
| DTNA | 6,6'-Dithiodinicotinic acid |
| DTNB | 5,5'-Dithio-2-nitrobenzoic acid |
| $\varepsilon$ | Molar absorptivity |
| $E_{1/2}$ | Half-wave potential |
| EC | Electrochemical |
| EDCI | 1-Ethyl-3-(dimethylaminopropyl)carbodiimide |
| EDTA | Ethylenetetraacetic acid |
| EIA | Enzyme immunoassay |
| EMG | Exponentially Modified Gauss |
| $\phi$ | Phase ratio |
| $\Phi_f$ | Quantum yield of fluorescence |

| | |
|---|---|
| FAB | Fast atom bombardment |
| FD | Fluorescence detection |
| FDNB | 1-Fluoro-2,4-dinitrobenzene |
| FEDA | N-(Fluorescein)-ethylenediamine |
| FIA | Flow-injection analysis |
| FITC | Fluoresceinisothiocyanate |
| GITC | 2,3,4,6-tetra-O-acetyl-β-D-glucopyranosylisothiocyanate |
| G(L)C | Gas (liquid) chromatography |
| GP | Gel permeation |
| h | Constant of Planck |
| HEPT | Height equivalent of theoretical plates |
| HMP | Hexamethylphosphoramide |
| HPAC | High-performance affinity chromatography |
| HPIC | High-performance immuno affinity chromatography |
| HPLC | High-performance liquid chromatography/<br>High-pressure liquid chromatography |
| HSD | 3-α-Hydroxysteroid dehydrogenase |
| $I_a$ | Intensity absorbed light |
| $I_f$ | Intensity fluorescence light |
| $I_o$ | Intensity primary light |
| IE | Ion-exchange |
| IEP | Immobilized enzyme probe |
| IFS | Intrinsic fluorescence sensitivity |
| IMEAC | Immobilized enzyme affinity column |
| IMER | Immobilized enzyme reactor |
| IP | Ion-pair |
| k | Reaction rate |
| K | Distribution coefficient |
| k' | Capacity value/ratio |
| $K_{ex}$ | Equilibrium constant |
| $K_M$ | Michaelis constant |
| l | path length |
| L, $L^-$ | Leaving group |
| LC | Liquid chromatography |
| LIF | Laser-induced fluorescence |

| | |
|---|---|
| LLC | Liquid-liquid chromatography |
| LLE | Liquid-liquid extraction |
| LSC | Liquid-solid chromatography |
| MMC | Mitomycin C |
| MS | Mass spectrometry |
| N | Number of theoretical plates (efficiency) |
| n | Refractive index |
| $\eta$ | Viscosity |
| NITC | Naphthylisothiocyanate |
| NP | Normal phase |
| NAD(PH) | Nicotinamide adenine dinucleotide (phosphate) |
| NBr | Naphthacylbromide |
| Nu, $Nu^-$ | Nucleophile |
| ODS | Octadecylsilica |
| OPA | o-Phthaldialdehyde |
| OTR | Open-tubular reactor |
| P' | Solvent polarity value |
| PBr | Phenacylbromide |
| PBR | Packed-bed reactor |
| Peri | 1,2,3-Perinaphthindantrione |
| PITC | Phenylisothiocyanate |
| pNBB | p-Nitrophenylbromide |
| PNBDI | p-Nitrobenzyl-N,N'-diisopropylisourea |
| pNPB | p-Nitrophenacylbromide |
| POD | Peroxidase |
| PROSPEKT | Programmable On-line Solid Phase Extraction Technique |
| PTFE | Polytetrafluorethylene |
| PTH | Phenylthiohydantoin |
| $Q_{10}$ | Temperature coefficient |
| R | Alkyl group / Gas constant |
| RI | Refractive index |
| $R_s$ | Resolution |
| REF | Reference electrode |
| RP | Reversed phase |

| | |
|---|---|
| SCE | Saturated calomel electrode |
| $S_N$ | Nucleophilic substitution |
| SDS | Sodium dodecylsulfate |
| SE | Shielding efficiency |
| SFC | Supercritical chromatography |
| SPI | Solid-phase isolation |
| t | Time |
| $t_o$ | Hold-up time of non-retained solute |
| $t_r$ | Retention time of solute |
| TBA | Tetrabutylammonium |
| THxA | Tetrahexylammonium |
| TLC | Thin-layer chromatography |
| TNB | 5-Thio-2-nitrobenzoic acid |
| TNBS | 2,4,6-Trinitrobenzenesulfonic acid |
| TPA | Tetrapentylammonium |
| TSP | Thermospray |
| UF | Ultrafiltration |
| UV | Ultra-violet |
| UV-Vis | Ultra-violet-Visible |
| v | Reaction velocity |
| w/v | Weight/volume |
| WE | Working electrode |
| X, $X^-$ | Anion |
| Y, $Y^-$ | Nucleophile |

# Contributors

*Jos H. Beijnen, Ph.D.* Hospital Pharmicist, Pharmacy, Slotervaart Hospital, The Netherlands Cancer Institute, Amsterdam, The Netherlands

*Udo A. Th. Brinkman* Prof. Dr., Department of Analytical Chemistry, Free University, Amsterdam, The Netherlands

*G. J. de Jong* Department of Analytical Chemistry, Free University, Amsterdam, The Netherlands

*R. W. Frei* Department of Analytical Chemistry, Free University, Amsterdam, The Netherlands

*Adrianus N. Fruijtier, M.D.* Junior Scientist, Department of Pharmaceutical Analysis, Faculty of Pharmacy, State University of Utrecht, Utrecht, The Netherlands

*Sándor Görög, Ph.D., D.Sc.* Member, Hungarian Academy of Science, Professor, and Director of Synthetic Research, Scientific Department, Chemical Works of Gedeon Richter, Ltd., Budapest, Hungary

*Junichi Goto, Ph.D.* Research Associate, Department of Analytical Chemistry, Pharmaceutical Institute, Tohoku University, Aobayama, Sendai, Japan

*Kazuhiro Imai, Ph.D.** Associate Professor, Faculty of Pharmaceutical Science, University of Tokyo, Bunkyo-ku, Tokyo, Japan

---

**Present affiliation*: Associate Professor, Branch Hospital Pharmacy, University of Tokyo, Bunkyo-ku, Tokyo, Japan

*Henk Lingeman** Research Scientist, Division of Analytical Chemistry, Center for Bio-Pharmaceutical Sciences, Leiden University, Leiden, The Netherlands

*J. A. Lisman* Department of Chemical Pharmacy, Pharmaceutical Laboratory, State University of Utrecht, Utrecht, The Netherlands

*Jan A. P. Meulendijk, LL.B., LL.M.*† Research Scientist, Department of Chemical Pharmacy, Pharmaceutical Laboratory, State University of Utrecht, Utrecht, The Netherlands

*Ubbo R. Tjaden* Department of Analytical Chemistry, Center for Bio-Pharmaceutical Science, Leiden University, Leiden, The Netherlands

*Willy J. M. Underberg* Associate Professor, Department of Pharmaceutical Analysis, Faculty of Pharmacy, State University of Utrecht, Utrecht, The Netherlands

*Petra C. van Krimpen, Ph.D.*‡ Research Assistant, Department of Chemical Pharmacy, Faculty of Pharmacy, State University of Utrecht, Utrecht, The Netherlands

*Marian A. J. van Opstal* Department of Pharmaceutical Analysis, State University of Utrecht, Utrecht, The Netherlands

*Present affiliations*:

*Research Scientist, Department of Analytical Chemistry, Free University, Amsterdam, The Netherlands

†Licensing Manager, Organon International B. V., Den Bosch, The Netherlands

‡Methodology Manager, Unilever Research Laboratorium, Vlaardingen, The Netherlands

# Detection-Oriented Derivatization Techniques in Liquid Chromatography

# 1

# Derivatization in Liquid Chromatography: Introduction

HENK LINGEMAN* *Leiden University, Leiden, The Netherlands*

WILLY J. M. UNDERBERG *State University of Utrecht, Utrecht, The Netherlands*

## INTRODUCTION

Accurate, sensitive, and selective bioanalytical determination methods for drugs and drug products (e.g., degradation products and metabolites) are of major importance in a number of pharmaceutical research fields: therapeutic drug monitoring, metabolic profiling, toxicology, drug pharmacology and pharmacokinetics, and stability [1]. The concentration of drugs and metabolites in whole blood, plasma, urine, saliva, or tissue homogenates is in many cases in the nanogram (ng/ml) or even the picogram (pg/ml) range, necessitating the use of selective and sensitive separation and detection methods after isolation and cleaning of the analyte from excess of endogenous and exogenous components present in the biological matrix.

Chromatographic separation techniques such as high-performance liquid chromatography (HPLC or LC) and gas chromatography (GC) in combination with selective and sensitive detection methods have gained an enormous popularity in the analysis of compounds of pharmaceutical and environmental interest [2].

The chemical structure of the analyte and the physicochemical properties of the matrix together with the desired selectivity and sensitivity determines the best suitable isolation and quantitation

**Present affiliation*: Free University, Amsterdam, The Netherlands

method. The required clean-up procedure strongly depends on the chosen chromatographic separation-detection combination. In the case of a very selective detection technique for the analyte under investigation, isolation from the matrix may be superfluous. However, even with the variety of currently available separation-detection systems it is not always possible to achieve the desired detection limit with the necessary precision without a chemical manipulation, *derivatization*, of the drug [3-7].

However, *derivatization is a necessary evil.* In developing accurate and precise methods the number of manipulations should be kept as low as possible. Every derivatization procedure consists of several manipulations which are potential error sources. Therefore, the analytical chemist must be sure, for instance, that no impurities are present in the solvents and the reagents. Blank derivatization procedures are an essential part in the development of every derivatization procedure. Furthermore, a great number of derivatives are not very stable, e.g., hydrolysis under the influence of moisture as well as other degradative processes may occur.

The main reasons for derivatization before or after the chromatographic separation are:

- Improvement of the detectability of the analyte
- Introduction or improvement of the resolution of closely related compounds during chromatography
- Establishment of analyte identity
- Improvement of the selectivity for the solute in a complex matrix
- Improvement of the chromatographic behavior of the drug under investigation
- Prevention of decomposition during chromatography and/or detection of the drug
- Achievement of better chemical and/or physical properties (e.g., volatility of polar compounds, solubility) of the solute to be chromatographed
- Introduction of an additional sample clean-up step

Derivatization reactions necessary to improve the chromatographic properties of analytes must be performed before the chromatographic separation (prechromatographic), but structure manipulation reactions to improve the sensitivity and/or the selectivity can be performed either before or after (postchromatographic) the separation. Prechromatographic derivatization techniques are far more often applied than postchromatographic methods because of their simplicity in comparison with the latter.

In general, derivatization reactions can be performed off-line with the chromatographic separation and on-line with the chromatography. The result is that four different derivatization principles can be distinguished (Scheme 1).

| | | | |
|---|---|---|---|
| Derivatization | Prechromatographic | Off-line | (A) |
| | | On-line | (B) |
| | Postchromatographic | Off-line | (C) |
| | | On-line | (D) |

Scheme 1

The two most widely applied derivatization procedures are the prechromatographic off-line and the postchromatographic on-line principles. Both procedures have their own advantages and disadvantages as discussed below. The postchromatographic off-line mode (D) is becoming more important after the introduction of the solid phase isolation cartridges (see Chap. 3). The principle is that a chromatographic system is used to clean up the samples (e.g., biological samples), followed by a derivatization step and a second separation step in order to quantitate the resulting derivatives. The prechromatographic on-line procedure (B) is also more widely used during the last few years because of the use of sample processors, which allow the automation of mixing solvents and analysis (see Chap. 3).

In choosing the proper derivatization technique for a certain problem, one of the questions is whether to perform the derivatization before or after the chromatographic separation.

### Prechromatographic Derivatization

Prechromatographic derivatization is the most widely used derivatization technique with many advantages and only a few disadvantages [4,6,8,9]. The major advantages are:

- There are no restrictions on the reaction kinetics, provided that the reaction goes to completion within a reasonable time, yielding one derivatization product for the compound in question without the formation of side products.
- There is a free choice in varying the conditions in order to optimize the reaction time and reaction yield.
- The solvent in which the prechromatographic reaction takes place need not be compatible with the chromatographic system (e.g., the mobile phase of the LC system).
- Chromatographic properties (e.g., capacity factors, resolution) can be optimized.
- Side products formed during the derivatization, degradation products of the reagent, and the excess of the reagent will be separable from the derivative, either on the chromatographic system or by a prechromatographic clean-up step.

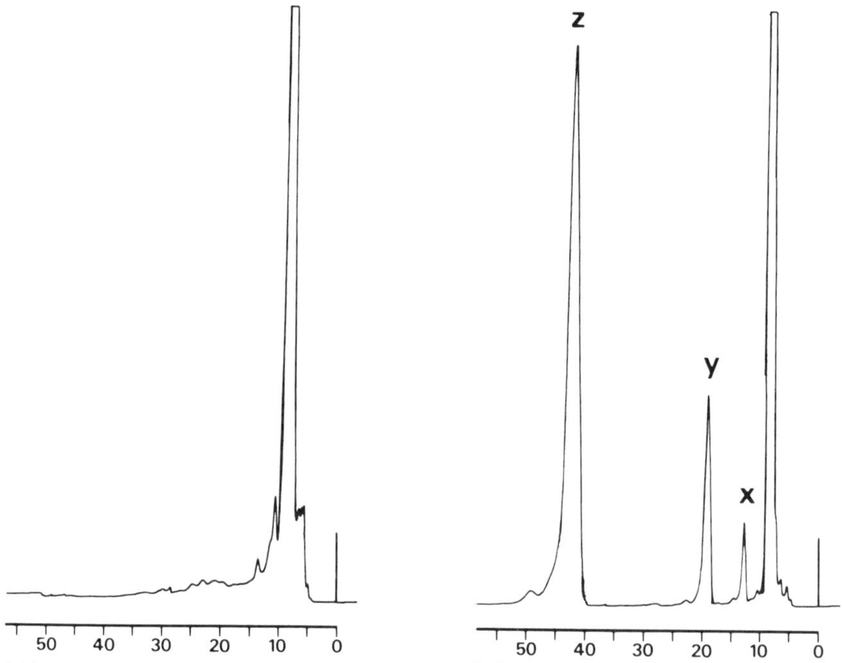


Figure 1.1 Chromatograms obtained after the derivatization of α-naphthyl-β-D-glucuronide, β-estradiol-17-β-D-glucuronide, and mentholglucuronide with *N*-(1-naphthyl)ethylenediamine (NED) according to the 2-bromo-1-methylpyridinium iodide method. (A) blank chromatogram. (B) derivatization of 10 μg of the three glucuronides. x, amide of mentholglucuronide and NED. y, amide of estradiolglucuronide and NED. z, amide of mentholglucuronide and NED. (Reprinted with permission from Lingeman, H., U. R. Tjaden, C. M. B. van den Beld, and J. van der Greef, *J. Pharm. Biomed. Anal.* [12].)

The main disadvantages of the prechromatographic derivatizations are sometimes:

There is formation of side products and artifacts, which may interfere in the chromatographic analysis or in the reproducibility of the derivatization reaction.

Each sample needs individual handling (e.g., laborious, time consuming).

There is a need for an internal standard in case the derivatization reaction is incomplete.

Target compounds may be more alike after derivatization.

One of the most frequently encountered problems in derivatization procedures is the occurrence of interfering peaks in the chromatograms due to degradation products or impurities of reagents and/or solvents. If an LC separation is used in combination with a prechromatographic derivatization step, the possibility exists that the derivative will be unstable in the mobile phase and will be partly degraded during the chromatographic process or during the pretreatment [10]. In these cases a postchromatographic derivatization method may be advantageous.

Ideally the components of the derivatization mixture do not change the chromatographic system through chemical reactions, precipitation, or demixing of the eluent upon injection or otherwise, and do not interfere with the detection of the derivatized compounds. If these conditions are fulfilled, aliquots of the derivatization mixture can be directly injected into the chromatographic system (Fig. 1.1). However, it is frequently necessary to remove the excess reagent or the solvent or other components of the reaction mixture prior to the chromatographic analysis.

Evaporation of the reagent mixture under a stream of dry nitrogen, either at room temperature or under heating, is a simple and convenient way of removing the solvents and, if volatile, the reagent. It is also a concentration step [11].

Prechromatographic derivatization can also be accompanied by liquid-liquid extraction or other prechromatographic separation step. Excess reagent is sometimes removed by allowing it to react with an excess of another compound; neither that compound nor its derivative must interfere with the analysis.

Although automated postchromatographic labeling is more frequently used, automated prechromatographic labeling methods have also been described and will become increasingly popular in the next years by the introduction of sample processors and robotic systems (see Chap. 3).

### Postchromatographic Derivatization

Considerable efforts have been put into the development of on-line postchromatographic derivatization methods and reaction detectors (see Chap. 4). Postchromatographic reactions take place in a part of the chromatographic system, called the reactor, between the column and the detector [2,13-17].

Postchromatographic derivatization techniques have some important advantages over prechromatographic methods:

The reaction must be reproducible without the necessity of being quantitative or the formation of only one derivative.

- The underivatized compounds are separated and eluted from the column and can therefore be detected with other, nondestructive detection methods before being derivatized.
- Limited sample preparation.
- Signal improvement by multiderivatization of the analyte.
- Increased accuracy and reproducibility by automation of the derivatization procedure, which is easier than in prechromatographic derivatization.

Obvious disadvantages of postchromatographic derivatization methods are:

- Restricted freedom in the choice of reaction conditions
- Band broadening in the reactor with a resulting loss of chromatographic resolution
- The possible interference by excess reagent or reagent degradation products in the detection of the derivatized compounds
- The need for instrument modifications to suit the postchromatographic reaction

The reactor consists of one or more units (made of inert material), depending on the chemical operations to be carried out. Low-pressure peristaltic pumps and low-volume mixing tees are used to add reagent solutions, solvents, or air to the column effluent. The design of postchromatographic reaction detectors and the theory concerning the effects influencing the sensitivity and resolution in these systems have been discussed in some recent reviews [2,14] and will be discussed extensively in Chap. 4.

### Prerequisites for Ideal Labeling Procedure

In choosing a labeling procedure, the prerequisites for an ideal derivatization technique are:

- The derivatization reagent is selective for a single functional group.
- The compound of interest is rapidly and quantitatively derivatized under mild conditions with a minimum of manipulations, yielding a single product.
- The label introduces a high degree of sensitivity.
- The reagent solution used for the derivatization is stable over prolonged periods.
- The derivatization reagent and its degradation products formed during the reaction are either nondetectable or well separable from the derivatized compounds of interest.
- The reagent is nontoxic.

The following prerequisites are only valid in the case of prechromatographic derivatization:

The derivatives are stable and possess favorable chromatographic properties.

The formation of relatively nonpolar derivatives is advantageous to achieve a successful isolation and concentration by extraction with organic solvents or by means of a solid phase isolation.

When developing a derivatization procedure for a certain compound, the influences of all reaction conditions (choice of solvent, concentrations of reactants and catalyst(s), temperatures, and reaction time) must be thoroughly investigated to obtain the optimal reaction conditions. The identity of the reaction product(s) must be determined. The reproducibility of the labeling procedure should be thoroughly tested, especially when the yield of the reaction is less than 100%.

The most important limitation of many of the published derivatization methods [8,13,18] is the lack of selectivity for a certain functional group, meaning that, especially in a biological matrix, accompanying components (e.g., amino acids, fatty acids) will be derivatized together with the analyte. Another serious problem is the fact that most derivatization reactions require nonpolar reagents, which involves the use of organic solvents that are not directly compatible with aqueous biological materials [2]. Furthermore, the choice of labels is limited because a good reactivity of both the functional group of the analyte and the label is required. A solution to this problem is the use of indirect derivatization techniques [1]. These methods, where the solute is first activated by a suitable activator, offer the possibility of using different types of derivatization reagents for the same functional group.

On-column derivatization applying solid phase isolation (SPI) cartridges is becoming quite popular nowadays [19]. There are several possibilities; the derivatization reagent can be immobilized on the sorbent or the analyte can be immobilized on the sorbent. The reaction is performed by loading the impregnated column with a solution of the sample or the reagent, and after some time (completion of the derivatization reaction) valves are switched and the excess of the reagents is washed off the precolumn whereafter the derivative is eluted in the analytical column.

### Labeling Versus Nonlabeling Reaction

The reasons for the introduction of a derivatization reaction in an analytical procedure were given earlier, but the definition of derivatization as a chemical manipulation of the analyte is not quite satisfactory. Derivatization reactions can be divided in labeling and nonlabeling reactions. In the nonlabeling reactions the analyte is converted to an appropriate derivative by means of several manipulation mechanisms that are not based, in general, on the attachment of another molecule to the analyte. Nonlabeling reactions include the following principles [2,20]:

Heat treatment reactions
Acid-base manipulations
Oxidation-reduction reactions
Photochemical reactions
Solvolytic reactions
Ion pair formation reactions
Micellar formation reactions

In labeling procedures, however, the analyte reacts with another organic molecule, e.g., a chromophore, a fluorophore, a radiophore, or an electrophore, resulting in a derivative with improved chromatographic properties or better detectability.

### Derivatization in Gas Chromatography Versus Derivatization in Liquid Chromatography

The use of analytical derivatization in GC analysis may be desirable for various reasons. The most important ones are the necessity to increase the volatility of polar compounds; to prevent irreversible adsorption, caused by hydrogen bonding between polar functional groups of the compounds under investigation and the free silanol groups of the column packing material or the glassware; to improve the separation between compounds; to avoid decomposition during chromatography; or to decrease the excessive volatility of compounds of very low molecular weight.

Because these problems are also occurring in LC, the large number of derivatization techniques that is available for GC (acylation, silylation, alkylation) can also be used in LC. In many cases the same derivatization reagents can be applied. However, in order to increase the detectability of the analytes, the same procedures can be applied also, but different reagents should be applied. Among the various derivatization principles in GC, the alkylation techniques are the most prominent procedures to transfer to LC, especially the methods using alkyl halides as the alkylating agents. The reason is that the reactions with alkyl halides combine a number of desirable properties compared with the other alkylation reactions, such as the following: (a) A wide range of derivatives can be prepared. (b) The reaction is usually fairly rapid under mild conditions. (c) Direct injection of the reaction mixture is frequently possible. (d) Side reactions are rare and usually one product is formed. (e) The derivatives are stable, and (f) The reagents used are of rather low toxicity. The major disadvantage of the alkylation procedure, limited selectivity, can be eliminated by combining the procedure with a highly selective sample pretreatment procedure or by applying, for instance, the extractive alkylation reaction.

### Physicochemical Properties of Analyte and Matrix

The direct analysis of a number of analytes with the help of chromatographic separation-detection combinations is often encountered with problems. These problems are mainly due to extreme polarity or extreme nonpolarity of the analytes, and weak intrinsic chromophoric properties of the solute or coeluting compounds.

#### *Physical Aspects*

The physical properties of the majority of the analytes are strongly influenced by the formation of hydrogen bonds (H bonds) with other potential H-bonding compounds. The carboxylic acid H bonds are, for instance, much stronger than the corresponding H bonds in the analogous alcohols and amines due to the high degree of polarization of the first. Moreover, carboxylic acids and to some extent alcohols and phenols are able to form additional H bonds by the negative oxygen of the carbonyl dipole resulting in cyclic dimers in the solid as well as in the liquid phases of the carboxylic acid [21]. H-bonding effects are also responsible for the high solubility of carboxylic acids and lower alcohols in polar (proton-donating) solvents; except for the carboxylic acids with long saturated or unsaturated chains or large aromatic systems, the solubility in very aprotic solvents will be limited. Compounds forming strong H bonds will have relatively high boiling points in comparison with the corresponding alcohols and chlorides.

The intrinsic possibility of an analyte to form H bonds is a very important factor in LC because the majority of the stationary phases is still based on polar materials such as silica and aluminum oxide which also have the possibility of H-bond formation. However, in quite a number of chromatographic processes this H-bond formation is an unwanted retention influencing parameter.

Except for the possibility of H-bond formation between the analyte and the stationary phase, the H-bonding capabilities of polar compounds that are present in the matrix (e.g., amino acids, peptides, vitamins, hormones) may interfere with the sample pretreatment and the LC procedures. An example is the bonding of a number of analytes to serum proteins which sometimes makes an accurate quantitation of these analytes very difficult.

Except for the H-bonding interactions between different types of molecules in the sample pretreatment and in chromatography, a number of other interactions are important (Table 1.1).

Table 1.1 Interaction Forces in Liquid Chromatography

| Interaction | kcal/mol |
|---|---|
| Covalent | 100-300 |
| Ionic | 50-200 |
| H bond | 5-10 |
| Dipole-dipole | 3-10 |
| Induced dipole-dipole | 2-6 |
| Dispersion | 1-5 |

The intrinsic chromophoric properties of the majority of functional groups (e.g., carboxyl, hydroxyl, amine, fenol, thiol) are weak. These groups show absorption bands in aqueous solutions at wavelengths below 220 nm. Due to conjugation of such functions with unsaturated bonds the molar absorptivity may be raised by approximately three decades. The most sensational increase in molar absorptivity is observed if aromatic ring structures are introduced. The introduction of aromatic rings or highly conjugated systems is even necessary for an analyte to possess fluorescence properties, while the presence of functional groups (electron donating or electron withdrawing groups) normally only enhances or diminishes the fluorescence. However, to allow amperometric detection of an analyte, the presence of one or more oxidable or reducible functional groups (e.g., carbonyl, nitro, amine, fenol, thiol) is a necessity. To allow chemiluminescence detection the analyte must possess, for instance, a carboxylic acid or an amine function to allow derivatization of the solute in order to apply the luminol/isoluminol or an oxidable (reducible) function to apply the lucigenin or peroxyoxalate chemiluminescence systems.

### *Chemical Aspects*

Compounds with functional groups possessing free electrons can be considered as acids (proton-donating compounds) or bases (proton-accepting compounds). For instance, the carboxylic acid function is a moderately weak acidic function, with a $pK_a$ (in water) of about 4.5-5.0. The acidic function of the carboxylic acid is much stronger than those of corresponding alcohols (e.g., ethanol, $pK_a$ 18), and phenols ($pK_a$ 10), caused by the high stabilization energy of the carboxylate anion.

The most important alkaline reaction group is the amine function with $pK_a$ values of about 8-10. For instance, the $pK_a$ of the primary amine function of amphetamine is 9.9 and of dopamine 8.8, the

$pK_a$ of the secondary amine function of phenylefrine is 10.1 and of ketamine 7.5, and the $pK_a$ of the tertiary amine of cocaine is 8.4. However, this $pK_a$ value can be shifted dramatically if aromatic functions are introduced adjacent to the amine function. The $pK_a$ of the primary amine function in aniline is shifted to about 4.6 and in benzocaine to 2.5. Electron-withdrawing functions neighboring an amine fuction can change the basic character of the group into a neutral function (e.g., primary and secondary amides) or even into an acidic function (e.g., sulfonamides, barbiturates).

Carboxylic acids with relatively low $pK_a$ values are normally very reactive. The most important reactions are cleavage of the O–H bond, or sometimes cleavage of the N–H or S–H bond, and dissociation reactions. However, nucleophilic attacks at the carbonyl function followed by a cleavage of a C–O bond (e.g., esterification, acid-chloride formation, amidification) are other important reactions of carboxylic acids and of alcohols and amines. Other possible reactions are decarboxylation or dehydroxylation and substitution reactions at, for instance, the α-carbon atom of a carboxylic acid function [21].

Extensive information on possible reaction pathways is presented in Chap. 2.

## CHROMATOGRAPHIC SYSTEMS USED IN COMBINATION WITH DERIVATIZATION REACTIONS

Improvement of detection limits or, rather, determination limits is one of the most important reasons to introduce derivatization techniques within the analytical assay. However, one has to realize that a number of chromatographic parameters (e.g., column length, column diameter, particle size, mobile phase) strongly influence the detection sensitivity. Therefore, before applying a derivatization technique, these parameters should be optimized [22,23].

The various modes of chromatography are among the most frequently applied separation methods in many laboratories. Components of mixtures brought into the chromatographic system are separated because of their different distribution behavior between a mobile phase, which is moving through an open bed or a column, containing a stationary phase [13,24-27]. Chromatographic systems can be classified by the aggregation state of the mobile phase, which is either a gas (gas chromatography; GC), a supercritical fluid (SFC), or a liquid (liquid chromatography; LC). The stationary phase can either be a liquid or a solid. Frequently applied systems are GLC, liquid-liquid chromatography (LLC), and liquid-solid chromatography (LSC).

In comparison with TLC [28] and GC, LC* is a young branch of chromatography. It is a still-developing technique, not in the least

*In this book, instead of the abbreviation HPLC (high-performance liquid chromatography), the abbreviation LC is used unless stated otherwise.

because of continuing advances in instrumentation and column technology. The combination of high-efficiency columns and selective and very sensitive on-line detection methods, e.g., electrochemical (conventional or laser) fluorescence, chemiluminescence [29], makes LC a powerful analytical tool.

Contrarily to GC, LC is not restricted to more or less volatile compounds but can also be used for the analysis of very polar compounds and macromolecules. GC and LC are particularly well suited for the analysis of micro quantities of drugs and drug metabolites.

In selecting a column for the separation of an analyte or a group of analytes, the sorbent as well as the column itself (material, length, diameter) are important. Nowadays most of the applied columns are made of stainless steel, but glass and plastic columns are also available. The column wall must be chemically inert to avoid irreversible adsorption of the analytes. Advantages of glass columns over stainless steel columns are the high wall smoothness, the inertness of the inner wall surface, and the possibility of visual inspection of the sorbent. Furthermore, the column dimensions are very important. Columns with a smaller internal diameter will show less dilution of the samples and so greater sensitivity in comparison with the larger diameter columns. Related parameters are the column fittings, which must be zero dead volume to avoid additional band broadening.

### *Microbore Chromatography*

Applying conventional LC, using columns with internal diameters of 3-5 mm, packed with 3- to 10-μm particles, about 5000-20,000 plates can be obtained. In order to obtain 100,000 or more plates, microbore columns with internal diameters less than 1 mm should be applied. Recently, the possibilities of applying narrowbore packed columns or open-tubular capillary columns in combination with derivatization techniques were described [30]. The most important advantage of microbore columns is the fact that these systems are ideally suited for analytes in very limited amounts of sample. The drawbacks of these systems are the lack of sufficiently low-volume cells and the limited sample capacity [31], which is especially troublesome in bioanalysis when small amounts of the analytes should be determined in relatively large and dirty sample volumes. However, the efficiency of these microbore columns is much higher in comparison with the conventional systems, which may be advantageous when very complex mixtures must be analyzed; thus separations can be executed much faster and much lower flow rates can be applied [31]. One of the most severe restrictions of microbore columns is the limited sample capacity of these systems, which means that these columns can only be applied in derivatization studies and in bioanalysis in combination with preconcentration columns.

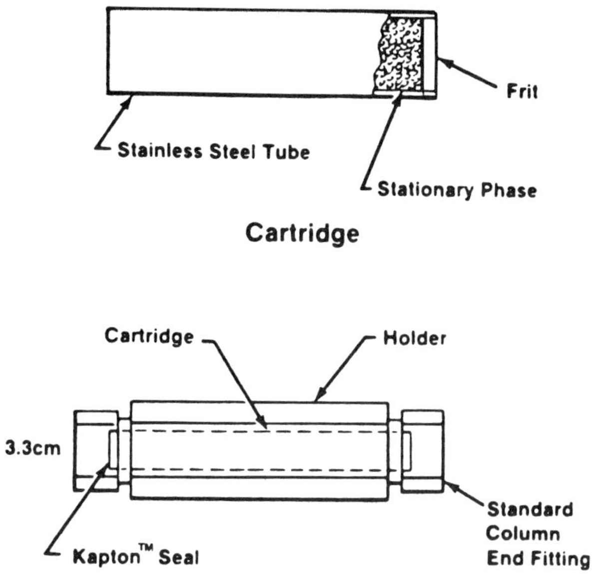


Figure 1.2 Chromatographic cartridge systems. (Reprinted with permission from Gant, J. R., W. A. America, M. W. Dong, and P. A. Perrone, Perkin-Elmer Liquid Chromatography Technical Bulletin 3 (1986).)

*Modular Chromatographic Systems*

Instead of the conventional columns, today modular cartridge systems are available [32]. These systems have a number of advantages:

- High flexibility.
- Simple handling.
- Great number of stationary phases available.
- Column exchange possible within a few seconds.
- Multicolumn chromatography (several cartridges can be placed within one cartridge holder).
- Guard columns can be placed inside the same cartridge holder.
- Low-cost systems.

In cartridge systems (Fig. 1.2) the stationary phase encased in a stainless steel or glass tube is pressed with frits into both ends of the tube to retain the packing material. Without the use of special tools the cartridge can be placed in a cartridge holder and the end fittings attached to the LC system. For some separation problems higher efficiencies than can be achieved with one single column are needed. In these cases two or more columns, with the same or different adsorbents, can be combined in one cartridge holder (Fig. 2). Furthermore, the features (efficiency, peak shape, lifetime) of the cartridge systems are comparable with those of the nonexchangeable columns.

## Stationary Phases

Considering the packing material itself, the nature of the particle surface (e.g., irregular, spherical), the particle size, the pore structure, and the pore diameter are important. Most LC sorbents nowadays are based on the polar silica material. In order to choose the appropriate sorbent for a certain separation, the characteristics of the analytes as well as those of the sorbents should be known.

The separation of solutes in LC systems is based on differences in one or more of the following properties: adsorption onto solid surfaces, partition in systems of two immiscible phases, interaction (in the case of ionic solutes) with surface ion exchange sites, or molecular size.

The following modes of LC systems can be distinguished: chromatography on nonmodified silica or other polar stationary phases with comparatively nonpolar mobile phases (normal phase; NP), chromatography on silica-based phases modified with polar/hydrophilic functional groups, chromatography with nonpolar stationary phases and polar mobile phases (reversed phase; RP), ion exchange chromatography (IE) on macromolecular organic materials containing ionogenic functional groups or nonmodified polar stationary phases with ionogenic surface groups, and ion pair (IP) chromatography. IP chromatography can be performed in various ways: RP adsorption and liquid-liquid partition, in the NP as well as in the RP mode.

Gel permeation chromatography (GPC) (also known as size exclusion chromatography), hydrophobic interaction chromatography, and affinity chromatography are not discussed in this chapter because the majority of the compounds to be separated have molecular weights in the range of 100-1000. However, some information about the application of these techniques, which are particularly suitable for the separation of large (bio)molecules, is discussed in Chap. 3. For the separation of the high molecular weight biopolymers (e.g., proteins), wide-pore (275 Å) materials can be applied using the separation principles mentioned.

Taking into account the molecular weights and the solubility features of the analytes and the analyte derivatives, a general classification can be made for the chromatographic analysis of these compounds (Table 1.2). RP materials are the most frequently applied stationary phases for the separation of low molecular weight uncharged compounds. The advantages of these materials are the short system equilibration times, high stability, excellent reproducibility,

Table 1.2 Chromatographic Systems for Separation of Low Molecular Weight Compounds

| | Ionic (acidic, basic, amphoteric) | | | Nonionic | |
|---|---|---|---|---|---|
| | | Analytes Soluble in Aqueous Solvents | | | |
| Type of chromatography | RP-IS[a] | RP-IP | IE | RP | Partition |
| Stationary phase | Apolar | Apolar | Polar[b] | Apolar | Hydrophili |
| Mobile phase | Methanol/ aq. buffer | Methanol/ aq. buffer + IP agent[c] | Aq. buffer | Methanol/ water | Methanol, water |
| | | Analytes Soluble in Organic Solvents | | | |
| Type of chromatography | RP-IS[a] | Partition | RP-IP | RP | Adsorption (partition) |
| Stationary phase | Apolar | Hydrophilic | Apolar | Apolar | Polar (hydrophilic) |
| Mobile phase | Methanol/ aq. buffer | Methanol, water, + buffer | Methanol/ aq. buffer + IP agent[c] | Methanol/ water | Organic solvents |

[a]IS, ion suppression.

[b]Ionized functions must be present.

[c]IP agent is ion-pairing reagent.

Note: Instead of methanol, acetonitrile may also be read in this table.

good peak symmetry, and wide application range because they can be applied to polar as well as nonpolar compounds.

To avoid irreversible adsorption of the analytes on chemically bonded silica phases, end capping (a silylation reaction) of the unreacted silanol functions is necessary to achieve optimal chromatographic separations.

The difference between the different $C_{18}$ materials, for instance, are related to particle size, surface area, and pore size. Smaller particles result in higher chromatographic efficiencies. The interrelationship between pore size, pore volume, and surface area is very critical and complex. For normal LC separations 50- to 200-Å particles are applied. The most important advantages of the spherical materials over the irregular materials are the more homogeneous partition of the particles, resulting in improved peak shapes, and the smaller back pressure which allows higher flow rates [33].

Whenever silica-based materials are used, the pH of the eluent must be between 2 and 8. At pH values over 8 silica can dissolve and below 2 the bonded phase can be stripped off the column.

The use of NP chromatography may be advantageous when fluorescence detection is used, but the limited reproducibility and system stability and the bad peak shapes (tailing effects) which are frequently obtained with uncharged polar compounds with this type of stationary phases (e.g., silica gel, aluminum oxide) outweigh the advantage of the higher intrinsic fluorescence sensitivity (IFS) of many compounds in this type of system.

Sample retention on silica is based on hydrogen bonding, dipole-dipole, and dipole-induced dipole interaction with the silanol (SiOH) groups. Since water deactivates this material, mixtures of organic solvents are used as mobile phase, but the addition of small, exactly known amounts of water, acetic acid, or ammonium hydroxide results in a controlled deactivation of the system.

The polar and hydrophilic stationary phases (e.g., diol, cyanopropyl, aminopropyl) are very suitable for the separation of some classes of nonionic compounds but cannot be used easily for the separation of charged compounds. The advantage of the application of the diol phase (COHCOH) over the use of bare silica, e.g., for the separation of carbohydrates, is the fact that the first sorbent is less sensitive to water deactivation.

One of the major problems in LC with silica-based phases is the bad peak symmetry which is sometimes obtained with many ionogenic compounds, resulting in difficulties in the resolution and quantitation of these analytes. Possible solutions are the use of ion-suppressing techniques, which can only be used in a limited pH range in order to avoid column damaging, or the use of tailing-suppressing agents, which also shortens the column life. Another possibility is the addition of an ion-pairing reagent to the eluent, but these systems have the disadvantage that they are very complex and may need long system equilibration times; furthermore, in most cases the peak shape is not significantly improved.

An obvious way to perform the analysis of charged compounds is through the use of cation or anion exchange materials; silica or IE resins modified with quaternary ammonium groups (anion exchanger) or sulfonic acid groups (cation exchanger) or the weak IE qualities of the silica or aluminum oxide (alumina) material itself can be used.

The described systems are applicable to the analysis of derivatization mixtures containing neutral and acidic compounds, but for mixtures containing neutral and basic compounds these separation systems, except for the cyanopropyl phases, are not always satisfactory.

One solution for the separation of an excess of a basic derivatization reagent (amine) and the derivative (neutral compound) in the derivatization mixture could be the use of the "buffered" silica technique [34]. With this system ionic compounds are separated via a partition mechanism and nonionic compounds by means of an adsorption process, but again the lack of stability of these systems prevents their practical use. Column-switching techniques involving utilization of two columns or more are increasingly used for the separation and identification of complex materials [35]. A nonmodified silica column in the IE mode followed by a RP analytical column has been used for the analysis of mixtures containing excess basic and neutral components. Both columns are eluted with the same isocratic methanol-aqueous buffer mobile phase [36]. This two-column device affords the separation of the protonated basic compounds from the neutral compounds on the first column and quantitation of the neutral derivatives on the second column.

It is important to make the right choice in choosing a chromatographic system. For instance, the anthrylmethyl ester derivatives of a number of fatty acids [37,38] cannot be separated with a number of $C_{18}$ phases despite the test of a large number of solvents. Applying the less hydrophobic $C_8$ material results in much improvement in the resolution between the derivatives.

In choosing the chromatographic matrices, a number of general prerequisites (indicated by asterisks) are valid for all stationary phases:

Close packing and minimum solvent swelling/shrinking
*(semi)rigid structures
Close packing and high efficiencies
*uniform particle size and shape
High resolution and efficiency
*relatively small particles and narrow particle size distribution

Table 1.3 Specifications for Chromatographic Eluents

| Solvent | $n^{20}_{D}$ | BP | DV | ST | DC | DP | P' | TW 20% | TW 80% |
|---|---|---|---|---|---|---|---|---|---|
| *n*-Heptane | 1.388 | 98.4 | 0.40 | 20.4 | | | | 200 | 220 |
| *n*-Hexane | 1.375 | 68.9 | 0.31 | 18.4 | 1.9 | 0 | 0 | 200 | 220 |
| *n*-Pentane | 1.358 | 36.1 | 0.22 | 16.0 | | | | | |
| Cyclohexane | 1.427 | 80.7 | 0.94 | 25.5 | 2.0 | 0 | 0 | 220 | 240 |
| Isooctane | 1.397 | 125.7 | | | | | | 210 | 235 |
| Carbon disulfide | 1.627 | 46.3 | 0.36 | | 2.6 | 0 | 1.0 | | |
| Carbon tetrachloride | 1.460 | 76.5 | 0.94 | 27.0 | 2.2 | 0 | 1.7 | 265 | 280 |
| Trichloroethylene | 1.481 | 86.9 | 0.57 | | 3.4 | | | | |
| Xylene | 1.500 | 138.5 | 0.68 | | | | 2.4 | | |
| Toluene | 1.496 | 110.6 | 0.57 | 28.1 | 2.4 | 0.4 | 2.3 | 290 | 310 |
| Benzene | 1.501 | 80.1 | 0.63 | 28.9 | 2.3 | 0 | 3.0 | | |
| Chloroform | 1.446 | 61.7 | 0.56 | 27.1 | 4.8 | 1.0 | 4.4 | 250 | 260 |
| Dichloroethane | 1.445 | 83.5 | 0.80 | 24.1 | 10.6 | | 3.7 | 230 | 245 |
| Dichloromethane | 1.424 | 40.0 | 0.43 | 26.5 | 9.1 | 1.6 | 3.4 | 235 | 245 |
| Diisopropylether | 1.368 | 68.0 | 0.35 | 32.0 | 3.9 | 1.3 | 2.2 | | |
| *tert*-Butanol | 1.386 | 82.6 | 2.82 | | 12.2 | 1.7 | 3.9 | | |
| Diethyl ether | 1.354 | 34.6 | 0.24 | 17.0 | 4.2 | 1.3 | 2.9 | | |
| Isobutanol | 1.398 | 107.7 | 3.71 | | 18.2 | | | | |
| Acetonitrile | 1.344 | 81.6 | 0.39 | 29.3 | 37.5 | 3.5 | 6.2 | | 205 |

| | | | | | | | | | |
|---|---|---|---|---|---|---|---|---|---|
| Isobutylmethylketone | 1.396 | 115.9 | 0.59 | 22.7 | | | | | |
| Isopropylalcohol | 1.378 | 82.4 | 2.27 | 21.7 | 18.3 | 1.7 | 3.9 | | |
| Ethylmethylketone | 1.379 | 79.6 | 0.43 | 24.6 | 18.5 | 1.9 | 4.5 | | |
| Ethyl acetate | 1.372 | 77.1 | 0.44 | 23.9 | 6.0 | 1.8 | 4.3 | 260 | 275 |
| Acetone | 1.359 | 56.2 | 0.32 | 23.7 | 20.7 | 2.9 | 5.4 | 340 | 350 |
| Ethanol | 1.361 | 78.5 | 1.14 | 22.8 | 24.3 | 1.7 | 5.2 | 210 | 240 |
| Dioxane | 1.422 | 101.0 | 1.21 | 33.7 | 2.2 | 0.4 | 4.8 | 220 | 270 |
| Tetrahydrofuran | 1.405 | 66.0 | 0.47 | | 7.4 | 1.6 | 4.2 | 245 | 270 |
| Methanol | 1.329 | 64.9 | 0.52 | 22.6 | 32.6 | 1.7 | 6.6 | 210 | 235 |
| Pyridine | 1.509 | 115.3 | 0.92 | | 12.3 | 2.2 | 5.3 | | |
| Water | 1.333 | 100.0 | 0.95 | 72.7 | 80.2 | | | | |

Symbols and abbreviations: $n^{20}D$, refractive index; BP, boiling point in °C; DV, dynamic viscosity in mPa/sec at 22°C; ST, surface tension in mN/m at 20°C; DC, dielectric constant at 20°C; DP, dipole moment according to Debye; P', retention parameter based on polarity and selectivity according to Snyder; TW, absorbance transmittance (%) and corresponding wavelength (nm).

High flow rates and extended lifetimes
*mechanical strength

### Mobile Phases

To allow successful LC separations the properties of the mobile phase must be known to the chromatographer, meaning that in order to avoid baseline drift, back pressure changes during gradient analysis, and allow detection at low absorbing wavelengths or to achieve an optimal signal-to-noise ratio, solvents of high purity should be used and that a number of physical and chromatographic parameters (e.g., 20% transmittance value, boiling point, viscosity, refractive index, dipole moment, polarity index) should be controlled carefully (Table 1.3) [39-43].

#### *Optimization Solvent Systems*

The development of a separation system for a certain problem is normally a continuing and laborious task [44]. In RP analysis, for instance, the first step will generally be the establishment of the percentage of organic modifier that must be added to the eluent in order to obtain acceptable capacity ratios (k'), using octyl- or octadecyl-modified silica as the stationary phase. This can be achieved by trial and error or by an initial gradient run [45]. In the next step several features can be optimized such as:

Stationary phase materials (e.g., phases with different hydrophobicity, or the same hydrophobicity but from different suppliers)
Organic modifiers (e.g., mixtures of methanol, acetonitrile, tetrahydrofuran, and water)
Eluent additives (e.g., pH, ionic strength, ion-pairing reagents)

The testing of different octadecyl-modified silica phases may be useful when non-end-capped materials are used which possess large differences in the free (nonderivatized) silanol functions [46]. Although optimizing the eluent by using a large number of different cosolvent combinations and concentrations is very time consuming, it is one of the most powerful optimization procedures [47]. Optimization of pH, ionic strength, and the addition of ion-pairing reagents is in principle only useful for ionic and ionizable analytes [48].

Although optimization of mobile phase systems can be performed manually or in automated procedures, most of the automated procedures require special hardware (LC equipment) and software (computer programs) investments [49,50]. However, the application of computer-based automated optimization procedures is, especially in bioanalysis, only useful if very complex mixtures of analytes are to be separated.

For the majority of separations (one to five analytes in a relatively complex matrix) manual optimization of the mobile phase is easier and faster, e.g., by applying one run using a solvent with low solvent strength (e.g., water in RP chromatography) and a run using a solvent with a high solvent strength (e.g., methanol or acetonitrile in RP chromatography). In case of excessive k' values of the analytes, an eluent mixture with a relatively high and an eluent mixture with a relatively low solvent strength can be tested. It is advisable to test methanol (proton-donating properties) as well as acetonitrile (proton-accepting properties). The required selectivity in bioanalysis can be obtained by manipulating the pH of the eluent or by the addition of an ion-pairing reagent for a selective retainment of either the analytes or the interfering endogenous compounds.

One of the possibilities for the optimization of the mobile phase in RP chromatography in a more or less optimized procedure is described below [51]:

One stationary phase ($C_{18}$ or $C_8$)
One buffer system (tetramethylammonium hydroxide)
Three pH values (2, 5, 8)
Two ionic strengths (1.25 mM and 2.50 mM)
Two binary eluents using two different compositions (methanol/tetrahydrofuran, acetonitrile/tetrahydrofuran)
Two solvent strengths (chosen as a result of the preliminary experiments)

In this procedure a series of 72 experiments is needed. Although with this scheme nearly all problems can be solved, it is almost never necessary to execute all the experiments to solve a specific problem.

### *Gradient Elution Systems*

The application of gradient techniques (programming of the mobile phase composition during the chromatographic process) is especially useful when complex samples (e.g., amino acid or organic acid profiling) or analytes in a complex matrix (e.g., urine) should be analyzed [36]. In the case of urine samples, for instance, the sugars, amino acids, and so forth, can be washed off the column using a low-strength solvent, and by gradually increasing the solvent strength the elution of the analytes can be optimized.

For the analysis of a number of solutes in complex matrices, such as the analysis of pesticides in environmental samples or the analysis of amino acids or fatty acids in biological fluids, a combination of a pre- or postchromatographic derivatization with gradient elution may be advantageous [52,53]. In general, it will be difficult to achieve good reproducibility if postchromatographic derivatization is combined with gradient elution. The main difficulties in this case are due to the fact that the continuous change in mobile phase composition may influence the reaction kinetics and detection conditions. Therefore, discontinuous or step gradients can be applied to solving these problems [36,53,54].

Furthermore, gradient systems can be combined with, for instance, RP-IP chromatography using electrochemical or fluorescence detection if the proper precautions are taken [55], such as the application of dual-electrode detection to meet the problem of baseline shifting in amperometric detection.

#### *Selectivity Manipulation*

Considering the mobile phase there are many ways to manipulate selectivity and retention in LC: pH, ionic strength, solvent composition (modifier), ion-pairing agents, etc. [56]. In addition to these chemical manipulation techniques, there are some important physicochemical parameters that influence chromatographic performance.

Viscosity has two effects on LC performance. The higher the viscosity, the greater the pressure drop and the lower the efficiency of the system.

The refractive index (RI) is not only important if the RI detector is applied, but also whenever absorbance or fluorescence detectors are applied because changes in the RI cause changes in the transmittance and, thus, in the noise and drift of the system. The 20% transmittance values (Table 1.3) give information of the lowest possible detection wavelengths.

The solvent polarity values are very useful; however, they do not accurately describe the eluting strength for all analytes. The explanation is that solvent strength is a combination of dispersion, orientation, and hydrogen-bonding interactions, meaning that two solvents with equal polarity can have different chromatographic selectivities [57-59].

### Application of Multicolumn Technology

#### *Column-Switching Techniques*

Simple column-switching (CS) systems (see Chap. 3) applying only one additional (six-port) valve and an autoinjector can be used

- To inject a sample on one column while a separation is performed on another column (advantage: increased throughput)
- To collect a sample in (sealed) vials (advantage: sample protection, possibility of reinjection)
- To concentrate samples on a precolumn (advantage: increased reproducibility, less laborious, increased lifetime analytical column)

To clean up samples on a precolumn (advantage: direct injection of nonpretreated samples, decreased number of manipulations
To select two analytical columns in series (advantage: alternative for gradient elution, increased selectivity)
To apply guard or scavenger columns (advantage: increased lifetime columns and pumps)

Furthermore, it is obvious that most of these techniques will result in an increased lifetime of the analytical column and an increased reproducibility of the whole procedure. The use of two additional six-port valves or one 10-port [59] valve will even increase the possibilities for the separation of complex mixtures of analytes or the analysis of analytes in a complex matrix. Because most of these procedures are identical with the clean-up of samples, the cutting techniques and the back and forward flushing systems are discussed in Chap. 3.

*Precolumn Technology—Concentration Columns*

Especially in environmental and bioanalysis, preconcentration of the samples is often necessary to allow sensitive detection of the analytes. An elegant solution is the application of on-line or off-line solid phase isolation columns or cartridges.

The first sample concentration procedure, applying chromatographic columns, was based on the adsorption of relatively nonpolar compounds on octadecyl-modified materials [32,60]. The basic principle of this procedure is that the analytes are transferred from an aqueous matrix onto the hydrophobic surface of the sorbent. The analytes will be immobilized on this surface until the elution strength of the eluents is increased, meaning that the solutes are concentrated in a very small volume at the beginning of the column. By using a solvent with a high solvent strength the solutes are subsequently eluted from the sorbent, with very little band broadening [52].

The on-line application of preconcentration columns is a very accurate and reproducible method for sample concentration, especially when stainless steel columns (2 × 4.6 mm ID or 10.0 × 2.0 mm ID) are used. These columns can be manually dry-packed with the chosen sorbent. In Fig. 1.3 a scheme is given which can be used in a simple preconcentration unit. Figure 1.3A represents the filling of the sample loop and the equilibration of the precolumn as well as the analytical column. Figure 1.3B is obtained after switching of the first valve and in this position the sample is loaded onto the precolumn. After sufficient cleaning and/or concentration of the sample, the second valve is switched (Fig. 1.3C) and the sample is eluted to the analytical column [61-64]. Instead of two six-port valves one 10-port valve can be applied to simplify the system [59].

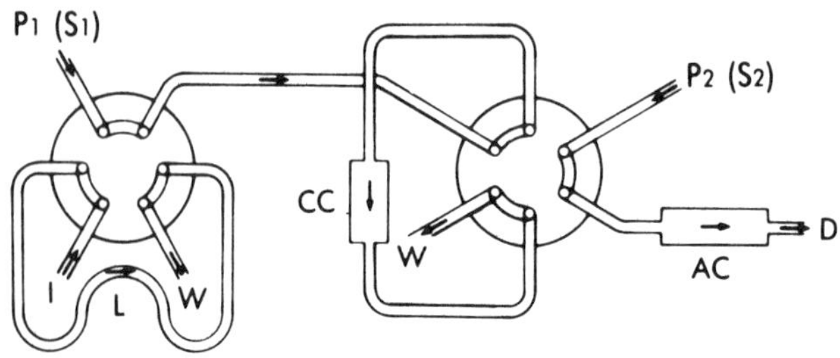


**Situation A:** Load Loop

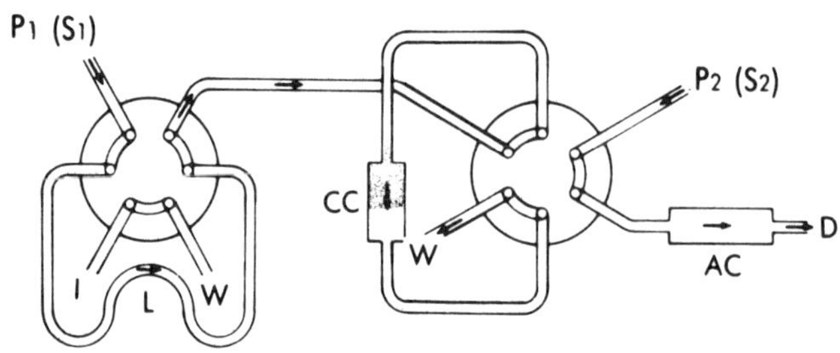


**Situation B:** Load CC

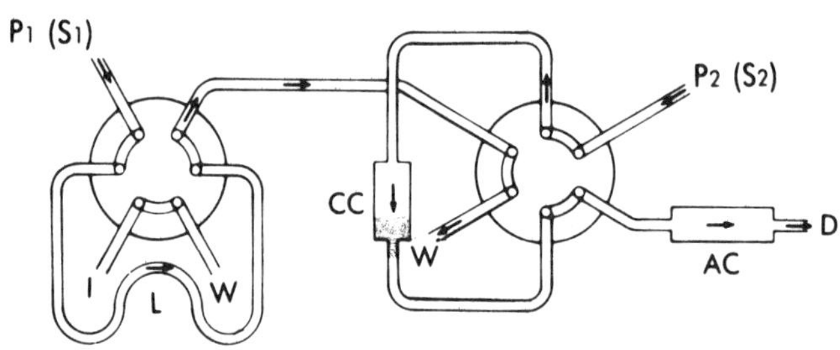


**Situation C:** Load AC

**Figure 1.3** Preconcentration column-switching system. Situation A, filing of sample loop. Situation B, loading of precolumn. Situation C, loading of analytical column. (Reprinted with permission from Chrompack International BV, Middleburg, The Netherlands.)

Although the use of on-line precolumn technology has a number of advantages over the use of off-line precolumn technology, the application of the nonselective hydrophobic sorbents do not offer always enough selectivity, especially in bioanalysis. Therefore, these sorbents should be applied in combination with multicolumn techniques and selective detection systems or they should be replaced by more selective sorbents (see Chap. 3).

Recent developments in precolumn technology include the systems with automated on-line cartridge exchange because it is not always possible to reuse the precolumns (Fig. 1.3) [65]. For instance, whenever a nonselective hydrophobic precolumn (e.g., PRP-1, XAD-2, $C_{18}$) is applied in combination with a more selective interacting precolumn (e.g., IE, metal-loaded) it may be advisable to exchange one or both cartridges after every analysis to avoid time-consuming reequilibration of the columns or to limit the number of manipulations in order to increase the reproducibility of the whole analytical procedure [66,67].

The programmable on-line solid phase extraction technique [68] allows automated sampling on disposable cartridges which can be purged with one or more solvents for sample pretreatment and/or concentration of the samples, and the purified and concentrated samples can subsequently be eluted on-line to the analytical column. Because the system consists of three six-port switching valves and the cartridge reservoir can be filled alternately with the nonselective and selective interacting precolumn, both columns can be replaced after every analysis. Solvation, equilibration, washing, and elution of the columns are performed as described before.

### *Precolumn Technology—Protector Columns*

The effective lifetime of a column is influenced by a number of parameters, such as particle size and shape, porosity of the sorbent, hardware of the columns, and by the mobile phase and the sample. However, nowadays the mobile phase is the main reason for decreased column lifetime. To increase the lifetime of the analytical column, which is especially important if sorbents are used with small particle sizes, scavenger columns and guard columns can be applied [69-72]. The use of this type of protector column is especially important if gradient elution is applied and if the analytes are present in a complex matrix (e.g., biological samples, environmental samples).

### *Precolumn Technology—Scavenger Columns*

Scavenger columns are placed between the pump and the injector to protect the analytical column from solid particles originating from, for instance, pump seals or the mobile phase. Especially with the increasing popularity of 3-μm particles the use of scavenger columns is essential [73]. The growth of bacteria in buffer-containing mobile phases is unavoidable and disastrous for analytical columns [74]. The solution is therefore the use of a scavenger column packed with the same sorbent as the analytical column, but normally with a somewhat larger particle size. Because the sample does not pass the scavenger column, the dimensions of this column may be chosen freely.

*Precolumn Technology—Guard Columns*

Guard columns are placed between the injector and the analytical column to protect the latter from interfering compounds in the sample and solid particles from the injector [75]. However, contrarily to the scavenger column, the dimensions of the guard columns are critical because the sample passes this precolumn. Therefore, the guard columns should have an internal volume as low as possible. In guard columns pellicular (about 40-50 μm) and microparticulate (about 5-10 μm) sorbents can be applied, but normally the pellicular materials are applied, having a number of advantages over the microparticulate sorbents: there is only a slight pressure increase, they can be easily dry-packed, and the sorbent of the guard column may be different from the analytical column. The disadvantages of the pellicular phases are some additional band broadening and the relatively low capacity [24,76,77]. If the proper guard column is chosen, the loss in plate height caused by the introduction of the guard column is less than 5%, and the peak symmetry and the k' values are almost the same as in a system without a guard column. The ratio between the volume of the guard column and the analytical column may not exceed 1:25 in order to avoid excessive band broadening.

*Precolumn Technology—Future Trends*

The present trends in precolumn technology are:

- The use of columns with dimensions of 2-10 mm × 2-4.6 mm internal diameter
- The use of sorbents with a relatively large particle size of 25-60 μm
- The use of stainless steel sieves instead of porous frits, to achieve as little back pressure as possible
- That if nonpolar packing materials are used, the less polar possible adsorbent is applied (see Chap. 3)
- That forward flushing is preferred over back flushing
- That the breakthrough volumes are carefully checked

The disadvantage of back flushing (see Chap. 3) is that all the dispersed particles (adsorbed on the frits) will be eluted immediately from the column. Back flushing is only necessary when long precolumns are applied.

However, if several precolumns are combined the sequence of the precolumns is essential. If, for instance, in the analysis of primary amines in dirty wastewater, a combination of a $C_{18}$, a PRP-1, and a cation exchange column is used, the last column in the series must be the IE column because this column will retain completely all the amines. Regarding the two hydrophobic precolumns, the less hydrophobic $C_{18}$ column must be placed in front of the more hydrophobic PRP-1 column.

For the selective isolation of thiols and phenols on precolumns, metal-loaded sorbents can be used. Desorption of the solute is subsequently performed, after washing the column to remove hydrophobic bonded interfering compounds, with an excess of another chelating agent. The advantage of these ligand exchange columns is the selective isolation of thiols and phenols, but because of the strong interaction between the metal-loaded phase and the analyte, quantitative desorption is difficult to achieve.

Nowadays it is even possible to combine precolumns with microbore columns [78]. Additional band broadening can be avoided by inserting a precolumn within the axis of a six-port switching valve.

## BASIC EQUIPMENT AND PROCEDURES IN DERIVATIZATION STUDIES

### Centrifugation

Centrifuges are an important piece of equipment in (bio)analytical derivatization studies, in the sample pretreatment as well as in the derivatization step itself. Centrifuges can be equipped with a number of different rotors (swing-out, fixed-angle, vertical, horizontal, continuous flow) with their own characteristics [79]. The speed of the centrifuge and the construction of the rotor are important parameters in choosing the proper centrifuge for a certain application.

The swing-out or the swinging bucket rotors are about the most popular centrifuges in analytical chemistry and are used, for instance, when fragile suspensions must be separated because precipitation with this type of rotor takes place along the length of the tube resulting in limited contact with the tube wall. These swinging bucket rotors are well suited to those cases in which an exact volume must be analyzed. The disadvantage is that they are relatively slow.

In the fixed-angle rotors, which are relatively fast, the precipitation path is short, but because the precipitate may be dispersed onto the tube wall, isolation of the supernatant may be difficult. Fixed-angle rotors are used for the separation of the protein-free and protein-bounded fractions of analytes [80]. Small (2000-3000 g) fixed-angle rotor centrifuges are very well suited for the preparation of differential precipitates and for the separation of nonmiscible solvents.

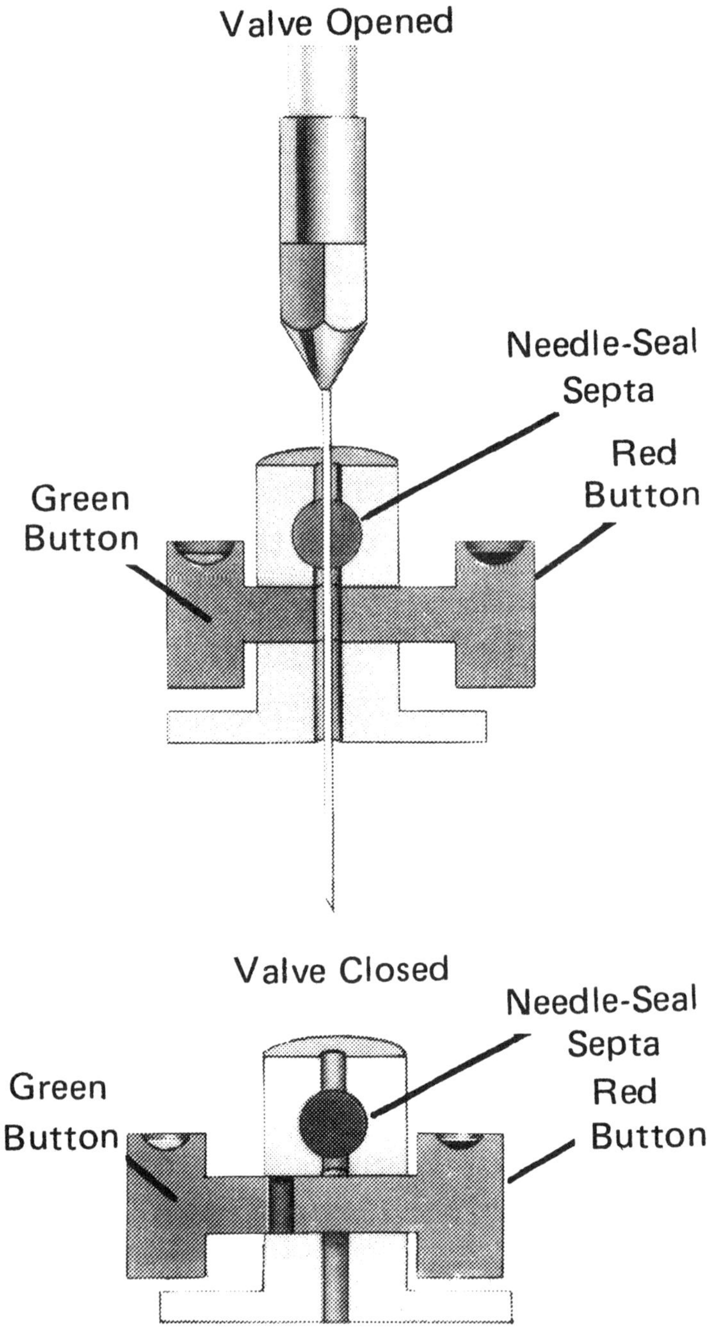


Figure 1.4 Push-button valves.

Vertical rotor centrifuges can be applied whenever during the increasing or decreasing of the centrifuge speed reorientation of the separated layers must be avoided.

### Reagent and Reaction Vials

The choice of reagent or reaction vials is very important. For instance, the surface of the vials (e.g., polymer-based vials) may contain phthalates or plasticizers which can dissolve in certain solvents and may interfere with the chromatographic analysis or the identification of the products. For instance, the phthalates which dissolve in dichloromethane are sometimes the reason for a series of unexpected peaks in the mass spectra of the derivative, and plasticizers may interfere with the fluorescence detection of the analytes.

Heating the vials is in some cases limited. Polypropylene vials can be heated up to 400 K while Teflon vials allow temperatures up to 500 K.

The color of the vial wall is also important when photolabile compounds are under investigation. In these circumstances brown vials are generally applied, although in some special cases green, blue, or red vials are more appropriate. However, the dyes in the colored vials sometimes interfere with the analysis (e.g., heavy metals).

The type of caps on the vials should be chosen in such a way that no disintegration by the solvent in the vial occurs, that the vials can be closed tightly to eliminate contamination of the vials with airborne particles, and that degradation of the samples by moisture is avoided. Furthermore, the method for capping the vials must be chosen in such a way that even after many times of opening and closing the vials, the caps are still reliable. A good system uses the principle of push-button valves (Fig. 1.4). This capping system can replace the septum stoppers and screw caps and allow leak-tight closures.

A few other important parameters are the shape, the material, and the volume of the reaction or reagent vial. The vials can be made of glass (e.g., borosilicate, soda glass) or of a polymer (e.g., polyethylene, polypropylene, Teflon). The use of soda glass vials is not recommended because of leaching problems. The choice of the material strongly depends on the physicochemical properties of the analyte. Because of irreversible adsorption of a number of compounds onto glass surfaces, nowadays the polymer-based vials are widely applied. The volume of the vial should be about the same as the total volume of the sample and small volumes (less than 1 ml) should be processed with an internal cone (Fig. 1.5).

Reaction vials, with an internal volume of 100 μl to 1 ml, possessing an internal cone, are very suitable for the isolation of residues

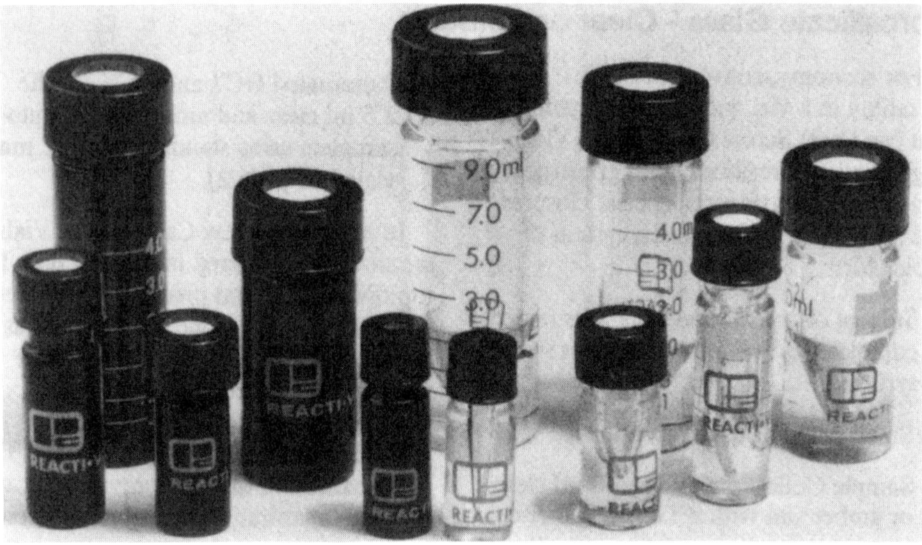


Figure 1.5 Reaction vials. (A) Overview of reaction vials with internal volumes of 100 μl to 10 ml with an internal cone. (B) Triangular magnetic stirrer for reaction vial. (Reprinted with permission from Pierce Eurochemie BV, Oud-Beijerland, Netherlands [81].)

or fractions, for derivatization reactions, for sample retrieval, for protection against moisture and airborne particles, and for sample storage [74]. Reaction vials with an internal cone are useful whenever small amounts of sample with a syringe must be removed or for concentration of the samples. Depending on the analyte, the chosen vials are made of clear or colored glass or a polymer.

To allow microderivatizations in these vials, triangular magnetic stirrers can be used to allow appropriate mixing of the samples during the derivatization (Fig. 1.5). Furthermore, this type of vial can be used for precipitation reactions, for centrifugation, and for solvent separation.

For the storage of reagents and standards under complete seal with instant access with a syringe, or to perform a derivatization reaction, 1- to 50-ml glass (e.g., borosilicate) vials can be applied.

For the handling of troublesome reagents or analytes the use of Teflon vials is recommended because they do not contain any plasticizers or organic acids. Furthermore, these vials can be used at temperatures from 50 K up to 500 K and no active sites on

Figure 1.5 (continued)

the surface are present which could adsorb polar compounds. Therefore, these vials can be used for the handling of strong acids and bases and other aggressive reagents.

### Evaporation and Heating

Evaporation and heating are two different processes which can be combined easily in one system [81]. An evaporation/heating device consists of an aluminum block with holes to allow the insertion of the reaction vials (Fig. 1.6). To extend the possibilities of the block, holes with relatively large diameters should be used, and by means of thin aluminum plates with different hole diameters, all types of reaction vials can be inserted.

To execute the derivatization procedure at elevated temperatures in these blocks, the aluminum block should be fixed on a magnetic stirring device and a temperature control unit should be present to allow temperatures up to 400 K with an accuracy of about 1 K. Mixing of the samples during the derivatization can be achieved in reaction vials with an internal cone with the corresponding triangular magnetic stirrers.

The reaction can be accelerated by using the concentration-evaporation technique. On top of the block a plate is fixed contain-

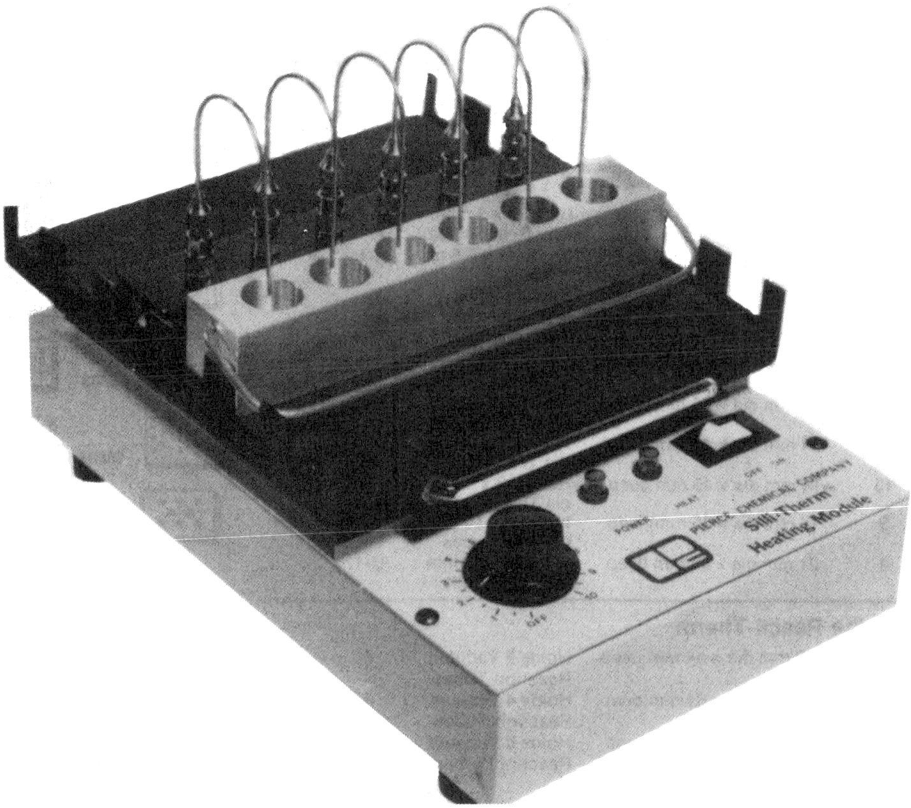


Figure 1.6 Evaporation-heating-stirring device. (Reprinted with permission from Pierce Eurochemie BV, Oud-Beijerland, Netherlands [81].)

ing an evaporator tip above each hole. These evaporator tips can be used to blow nitrogen or helium over the samples for a slow evaporation of the solvent. As a result of the evaporation, the concentration of the analyte and the reagents in the vials will increase and the reaction rate will be enhanced too. The same evaporator tips can be used for evaporation of the contents of the vials to dryness. Nitrogen or helium should be used instead of air to prevent degradation by oxidation.

Simple homemade systems can be used if, for instance, only evaporation of the samples at ambient temperature is applied.

Thermostatted chambers are even described in which thermostatic control within 1 K is possible in the range of 273-313 K. In these chambers the complete chromatographic system, including autosampler, postchromatographic derivatization unit, and detection system, is placed which may be used if temperature control is necessary (IP chromatography, amperometric detection, analyte stability, derivatization speed).

### Sample Manipulation (Addition of Reagents)

The addition of solvents and solutions can be performed with micro- and/or macropipettes, volumes of 5 μl to 5 ml with disposable nonwetting polypropylene tips. The mean accuracy of this type of pipette is 1% in volume of the sample size. Although these disposable pipettes have some distinguished advantages over the classical glass pipettes (e.g., ease of operation, elimination of cross-contamination), their accuracy is unfavorably influenced when solvents with relatively high vapor pressure are processed. In case relatively large sample series are to be analyzed, the use of disposable pipettes with multiheads, five to 10 tips on one pipette, may be advantageous.

### Mixing and Stirring

The mixing and dissolution of samples can be performed by means of vigorous mixing devices (e.g., vortex mixing) applying a mixing speed of 50-1500 rpm for 30 sec to a few minutes. Instead of these vigorous mixing devices, dissolution of samples can also be obtained by the application of microwave radiation for a few seconds up to 1 min or by ultrasonically vibrating the samples for 5-15 min. However, the disadvantage of ultrasonically vibrating the samples is the possible degradation of the samples by the developed heat. An advantage of ultrasonically vibrating the sample is the simultaneous deaeration of the sample. Mixing of relatively large amounts of fluid can be performed with a magnetic stirrer. However, during the stirring of the samples some heat may be developed which can be the reason for degradation of the samples.

Homogenization of samples can be obtained by application of a rotating mixing device using a speed of 1-60 rpm. This simple rotating mixing device is also very suitable for derivatization mixtures which should react for some time at ambient temperature.

During all of these procedures some precautions should be taken to avoid photodegradation or oxidation of the samples by wrapping the reaction or reagent vials/tubes in aluminum foil.

## AUTOMATION OF DERIVATIZATION PROCEDURES

The decision as to whether a system should be automated by using precolumn technology or column-switching techniques depends on:

The number of samples to be analyzed
The number of interfering compounds
The complexity of the matrix
The level(s) of drug and/or degradation products or metabolites
The required sensitivity and selectivity of the detection

Furthermore, the involved chemistry, the stability of the analytes, the available adsorbents, and the possibilities of reusing the precolumns must be taken into consideration.

### Automated Prechromatographic Derivatization

Simple and relatively fast prechromatographic derivatization procedures with reaction times of about 1 min can be automated by using sample processors. These sample processors are able to perform a limited number of sample manipulations such as addition of solvents, and by the presence of one or two six-port switching valves a prechromatographic derivatization procedure can be executed.

Except for these three-dimensional sample processors (combination of autoinjector and switching valve), the cartesian samplers robotic systems (anthropomorphic arms) can be used. The advantages of the cartesian samplers are that they need only limited motion speed and a relatively short cycle time. The main disadvantages are that they can only handle liquid samples, that the number of different sample manipulations (e.g., addition of solvents, heating, mixing) is limited, and that only reactions constants can be obtained. The main advantages of the anthropomorphic systems are that gaseous liquid and solid samples can be processed, that a very wide variety of sample manipulations is possible, and that extremely long reaction times can be dealt with. The disadvantages of these systems are the large motion space that is needed, the long cycle times, and the fact that they are expensive.

Although automated postchromatographic derivatization (see Chap. 4) systems are more frequently described than automated prechromatographic derivatization systems, some data can be found in the literature. One of the main advantages of automated derivatization, except for the much higher throughput of samples, is the degree of standardization of the reaction conditions resulting in a high degree of analytical reproducibility.

One of the possibilities of semiautomation of the derivatization procedure is the application of programmable autosamplers, the sample processors (see Chap. 3). A few examples of (semi)automated prechromatographic derivatization procedures using autosamplers, robotics, or other devices are described below.

The precolumn on-line derivatization of amino acids with *o*-phthaldialdehyde (OPA) using a programmable autosampler involves the withdrawal of the reagent (OPA) from the reagent vial, followed by the uptake of the analyte from the sample vial. The derivatization of the analyte takes place while the sample is transported from the needle tip to the sample loop (reaction time about 1 min). During the transport of the reagent and the sample from the injector to the analytical column, the derivatization is completed. A mixer (e.g., short-coil, packed-bed) can be installed between the column and the injector to prolong the reaction time [82].

A fully automated prechromatographic derivatization can be performed with the "Autoderivat" [83], a system which is equipped with a robotic arm. The robotic arm is able to automatically perform the following actions: remove the caps of the sample container, add the required derivatization reagents and solvents, heat the sample solution to achieve a quantitative derivatization reaction (temperatures up to 450 K), and subsequently cool the sample again to room temperature. In the last step of the procedure the sample container is tightly closed with a cap and transferred to an autosampler.

Another convenient device for in situ prechromatographic derivatization is described by Honda and Kuwada [84], allowing the direct application of samples and introduction of the whole derivatization mixture onto columns (Fig. 1.7). This device is applied for the derivatization of primary amines with fluorescamine. During the first step of the procedure a reagent solution is led into the cuvette from one of the flow lines by using a small injector. Excess solution is drained from the other line. Subsequently an eluent is pumped to the column via the two other flow lines and thereafter the reagent stream is stopped, and a sample is loaded into the cuvette via the Teflon tube by using a microsyringe. The eluent is allowed to continue flowing, and in the next step the Diflon block is turned to the right by 45°. The cuvette is closed tightly and the eluent stream stopped. By allowing the mixture to stand in the cuvette for an appropriate period in this state, derivatization is completed. Finally the block is turned to the right by another 45°, and the eluent flows into the cuvette and washes out the whole derivatization mixture into

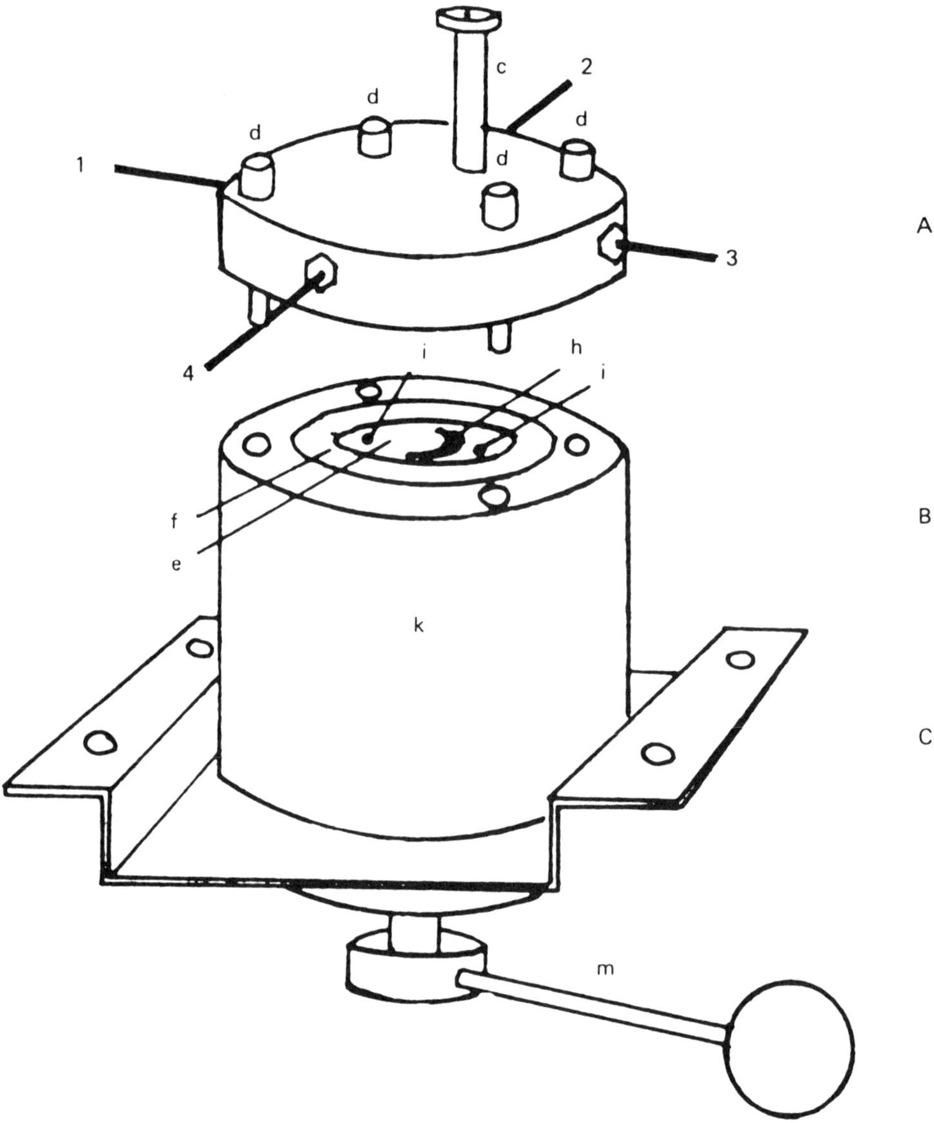


Figure 1.7 Overall view of the device for in situ prechromatographic derivatization. A, top cover. B, rotable body. C, fixing plate. c, syringe guide. d, screw bolt. e, Diflon block. f, block holder. h, derivatization cuvette. i, inlet and outlet of a flow line. k, outer block. m, handle. (Reprinted with permission from Honda, S., and H. Kuwada, *Anal. Chem.*, 55:2466 (1983) [84].)

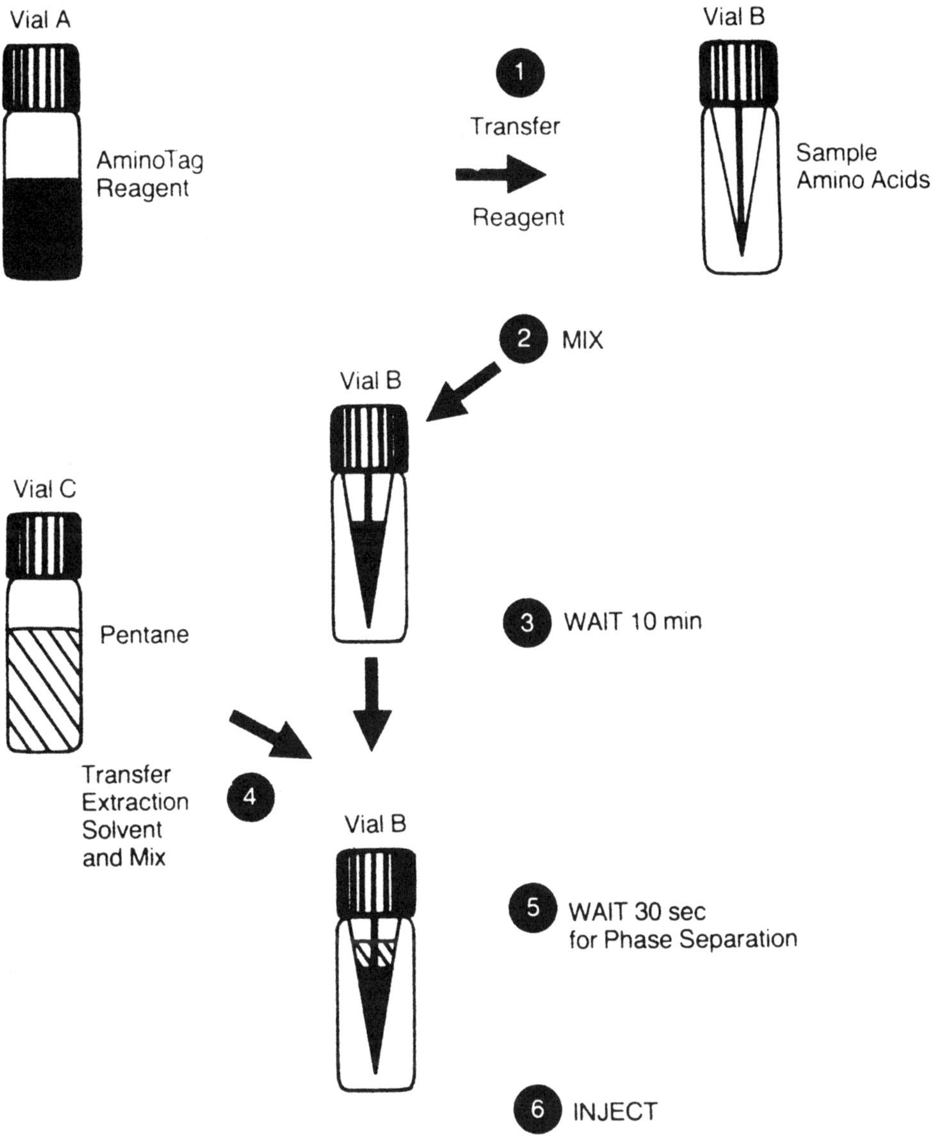


Figure 1.8 Automated 9-fluorenylmethylchloroformate prechromatographic derivatization scheme using automix procedures. (Reprinted with permission from Bell, J. P., R. A. Simpson, and R. L. Cunico, *Int. Lab.*, 16:62 (1986) [86].)

the column. The volume of the derivatization cuvette is 7.5 μl and sample volumes up to 1 μl can be applied.

A special automated prechromatographic derivatization system is developed for the derivatization of amino acids with phenylisothiocyanate or OPA. With this system a sequence-determining analysis of about 100 ng of a protein can be performed within 15 min (PICO-TAG) [85].

A somewhat similar system uses the combination of an autosampler with automix operations (AminoTag), which can be used for the prechromatographic derivatization of amino acids with a fluorescence probe (9-fluorenylmethylchloroformate) [86]. The autosampler is used for autopipetting and automixing, the automix procedure prior to automated sampling and injection into the LC system (Fig. 1.8).

### On-Chromatographic Derivatization

Whereas prechromatographic as well as postchromatographic derivatization procedures have their own characteristic disadvantages, the automated on-chromatographic derivatization may have some advantages in some cases, e.g., in the analysis of the OPA derivatives of amino acids [87]. This derivatization procedure is very rapid (less than 1 min), but the stability of the resulting derivatives is limited. Therefore, the amino acid solution and the reagent are automatically transferred from two different vials, by an injection system, onto the column. The mixing is executed with a flow of 0.02 ml/min and after a reaction time of 1 min the flow is increased to 0.35 ml/min and the derivatives are separated by a gradient program on a RP $C_{18}$ column. Over 100 samples can be analyzed unattendedly with this system.

### Postchromatographic Derivatization

Considerable efforts by several research groups have been put into the development of on-line postchromatographic derivatization methods. Postchromatographic reactions take place in a part of the LC system, called the reactor, between the column and the detector.

Postchromatographic derivatization techniques have some important advantages over the prechromatographic methods. The formation of side products (artifacts) is not a real problem. The reaction need not go to completion, nor should the reaction product be well defined, as long as the procedure is reproducible. The underivatized compounds are separated and eluted from the column and can therefore be detected with other, nondestructive detection methods before being derivatized.

Obvious disadvantages are the restricted freedom in choosing the reaction conditions, band broadening in the reactor with a resulting loss of chromatographic resolution, and possible interference by excess reagent or reagent degradation products with the detection of the derivatized compounds. Another disadvantage is the need for instrument modifications to suit the postchromatographic reaction.

Because of the importance of postchromatographic derivatization procedures, the needed hardware is discussed in Chap. 4 and in the other chapters a considerable number of applications of postchromatographic derivatizations are given.

## OPTIMIZATION DERIVATIZATION PROCEDURES

For the development of a suitable derivatization procedure for a certain analyte or group of analytes, first a number of choices should be made:

Which functional group(s) should be derivatized?
Should the derivatization be performed in an aqueous or nonaqueous medium?
Should the derivatization be performed in the pre- or postchromatographic mode?
Which detection principle should be applied?

Most of the answers to these questions were already given in the beginning of this chapter. However, regarding the selectivity of the derivatization procedure, it should be noted that derivatization of a thiol function is far more selective than derivatization of an amine or carboxylic acid but derivatization of the thiol function is inherently much more difficult. The choice of the derivatization medium strongly depends on the features of the matrix. Furthermore, it will be clear that the choice of detection system mainly depends on the required sensitivity and selectivity of the procedure and whereas the choice between pre- or postchromatographic derivatization is governed by the number of samples that should be analyzed.

One of the important questions to be answered in the process of choosing the proper derivatization procedure is whether a direct or indirect derivatization procedure should be applied.

A possibility of manipulating the selectivity of the derivatization procedure is the application of indirect derivatization methods. With these procedures either one of the functional groups of the analyte or one of the functional groups of the label is activated before the actual derivatization takes place. The advantage of activating the label is that functional groups not reacting spontaneously under these circumstances will react more easily [88].

An example of activating the label is the reaction of aniline with diethylchlorophosphite to form diethylanilinophosphite [89], which reacts in a single step with the analyte, a carboxylic acid, after refluxing in toluene or butyronitrile for 2 hr.

Activation of the carboxylic acid function following by reaction with the label offers the possibility of using different labeling reagents. By choosing the activator, only the reactive functions of the label and the solute are committed; the structure of the label is not. For the derivatization of the carboxylic acid function a few methods have been described. The general reaction of a carboxylic acid function with an alcohol or an amine only yields a considerable amount of the resulting ester or amide after activation of either the carboxylic acid function or the hydroxyl/amine function prior to coupling.

For instance, *N*-ethyl-*N'*-(3-dimethylaminopropyl)carbodiimide hydrochloride, a water-soluble representative of the carbodiimides, is suitable to activate the carboxylic acid function [90]. This activation is relatively slow but can be accelerated by the addition of 1-hydroxybenzotriazole as a catalyst [91]. The activated carboxylic acid reacts with primary alcohols and primary and secondary amines to the corresponding esters and amides, and high reaction yields are usually obtained. Another activation reaction for the carboxylic acid function is the reaction of the acid with 2-bromo-1-methylpyridinium iodide (BMP). After activation of the carboxylic acid function it reacts rapidly with alcohols [92] or amines [93].

## Strategies

Development and optimization of a derivatization procedure can be performed best in the μg/ml scale in the off-line mode (with the LC). During these preliminary experiments the influence of the following important parameters should be investigated carefully and systematically:

- Reaction temperature(s)
- Reaction time(s)
- Reaction solvent(s)
- Concentration of the derivatizing agent(s)
- Influence of activator(s)
- Influence of catalyst(s)
- Sequence of addition of reagent(s)

Investigation of these parameters should result in a clear view on:

- Derivatization yield(s)
- Occurrence of side reaction(s)
- Selectivity of the procedure

This procedure can be performed either in single variate or multivariate approach. The latter offers the best optimal conditions with a limited working load, while the former offers more fundamental information about the individual parameter influences.

In the last stage of the development and optimization of a derivatization procedure, the following features should be considered:

Purification of the derivative(s)
Identity of the derivative(s)
Development of a suitable chromatographic procedure

The experiments needed for the systematic study of the listed parameters can be followed by means of thin-layer chromatography (TLC) or by LC. A general TLC procedure that may be applied includes the use of precoated 20 cm × 20 cm (or 5 cm × 10 cm) silica plates, with a layer thickness of 0.25 mm, which are developed in the ascending mode with eluents consisting of hexane, hexane-chloroform mixtures, chloroform, chloroform-methanol mixtures, or methanol. After evaporation of the eluent the spots can be visualized under ultraviolet radiation of 254 or 365 nm. Nonabsorbing primary and secondary amines can be visualized by spraying the dried TLC plate with 10% (w/v) solution of ninhydrin in ethanol and subsequent heating at 353 K for 15 min. A number of carboxylic acids and phenols can be visualized by spraying with a solution of iron(III) chloride.

During the first experiments it is important that aliquots of the following solutions, originally containing about 10 μg of the analyte, be spotted onto the plate together with solutions of the analyte and the reagents:

1. The derivatization mixture
2. The reagent mixture without the analyte
3. The reagent mixture with the analyte, but without the activator
4. The reagent mixture with the analyte, but without the catalyst
5. The reagent mixture with the analyte, but without the derivatization probe (the label)
6. The derivatization mixture, with an excess of the analyte instead of an excess of derivatization probe

A new spot can be characterized as being the "derivative" if this spot appears only in the chromatograms of the samples 1 and 6 or in the chromatograms of the samples 1, 6, and 4.

Because the main objective of these experiments is to obtain, in a relatively short period of time, some semiquantitative data about the derivatization yield (comparison of samples 1 and 6), the occurrence of side products (samples 1-6), and the polarity and identity of the derivative (samples 1, 5, and 6). TLC will be advantageous compared with LC. However, it will be clear that LC will be used during the further optimization of the derivatization procedure.

Structure identification of the derivative can be performed by means of infrared or nuclear magnetic resonance spectroscopy or by mass spectrometry after isolation of the derivative by means of preparative TLC or preparative LC.

Purification of samples containing 1-5 mg of the derivative under investigation can be performed on precoated silica gel preparative thin-layer plates (20 × 20 cm) with a layer thickness of 1 mm. The chromatographic band containing the derivative is isolated by scraping the silica off the TLC plate and subsequent shaking with methanol. After filtration and evaporation of the methanol under a stream of dry nitrogen, the residue is recrystalized twice from ethanol or toluene and dried and stored over phosphorus pentoxide.

In addition to these experiments, the stability of the derivative should be investigated in several experiments. First of all, the derivative should be isolated, purified, dissolved in a suitable eluent, stored in the dark at temperatures below 273 K, and from time to time injected onto a suitable LC system. A second stability-indicating experiment is the reinjection of chromatographically pure derivatives.

After the optimization of all of the parameters, the initially optimized parameters should be checked as to whether they are still performed under the most ideal conditions.

Another possibility for optimization of the derivatization procedures is the application of computer-based techniques. However, these procedures are only useful when complicated derivatization kinetics are involved.

## Kinetics in Optimization of Derivatization Procedures

During the optimization of the derivatization procedure, a number of features should be taken into consideration. An extensive overview of these parameters is given in Chap. 2.

Important factors in optimizing the derivatization procedure are the stoichiometry of the reaction and the concentration of the compounds in the derivatization mixture. For the most widely used irreversible second-order reaction:

A (analyte) + R (reagent) → D (derivative)

the concentration of the derivative d at a given time t can be obtained from Eq. 1:

$$\frac{2.3}{a - r} \cdot \frac{r(a - d)}{a(r - d)} = kt \qquad (1)$$

in which k represents the reaction rate constant and a and r the initial concentrations of the analyte and the reagent, respectively [94]. Equation 1 shows the time needed for, for instance, a 99% conversion of A into D decreases proportionally with the increase of k. This means that proper manipulation of the parameters (e.g., temperature, solvent) determining k makes it possible to control the speed of the derivatization reaction.

The time required for a 99% conversion of A to D is determined by the nonlinear relationship between a and r of Eq. 1. The time needed for the reaction decreases markedly on increasing the concentration ratio of r/a, up to a ratio 10. Therefore, a larger excess of the reagent is undesirable because it may cause resolution problems during the chromatographic analysis.

Apart from the concentration ratio, the absolute initial concentrations of A and R are very important. For instance, derivatizations in the μg/ml scale will be three orders of magnitude slower than those performed on a mg/ml scale. Therefore, dilute samples must, whenever possible, be concentrated prior to derivatization.

For many reactions a two- to threefold increase of the reaction rate is observed with a temperature rise of 10 K [95,96]. However, the advantage of using higher temperatures to shorten the reaction time may introduce the disadvantage of the formation of thermal degradation products during the derivatization, or when the activation energy for side product formation is larger than that of the desired derivative, possible side reactions will become more important.

The influence of the solvent is rather complex. Because solvent molecules take part in the microscopic reaction steps, a change of solvent will result in a different macroscopic reaction rate and/or reaction yield [97]. According to modern reaction kinetics [98,99], the initial compounds (A and R in Eq. 1) as well as the activated transition intermediate, formed before the final product (D in Eq. 1) is synthesized, can be stabilized or destabilized by means of solvation. If the solvent stabilizes the transition state of the reagent more effectively than the ground state, the reaction rate will be increased. If, for instance, in nucleophilic substitution reactions ions are formed in the transition state or existing charges are concentrated on a smaller surface, the reaction will be faster in protic (e.g., water, methanol) than in (polar) aprotic solvents (e.g., acetonitrile, *N,N'*-dimethylformamide). On the other hand, reactions in which, in the transition state, ionic charges are neutralized or are divided over a larger surface proceed faster in aprotic solvents [100,101]. Thus, alkylation of ionized carboxylic acids with alkyl halides (e.g., phenacylbromide) proceed faster in (polar) aprotic solvents due to a better ground state solvation of the anionic form of the carboxylic acid function in protic solvents [102,103]. Nucleophilic substitution reactions are involved in many derivatization reactions: analytes with an amine, a hydroxyl, or a carboxylic acid function can be derivatized with this type of reaction [97,99,104]. If the analyte is a carboxylic acid, the reaction conditions are usually modified to form a carboxylate anion which is a far better nucleophile than the free acid. Thus, for fast and quantitative reactions the reagent must be susceptible to nucleophilic attack. Reagents with a negative charge are more powerful than uncharged nucleophiles. The reactivity of uncharged nucleophiles decreases in the order:

$$N > O > S$$

Alkylation reactions of dissociated carboxylic acids with alkyl halides are often performed with alkyl iodides because the iodide ion is a good leaving group. For the halides the following decreasing order of leaving group activity exists [105]:

$$I^- > Br^- > Cl^- >> F^-$$

The influence of different substituents in the analyte or in the reagent upon the reaction rate can frequently be predicted by the Hammett-type Eq. 2:

$$\log \frac{k}{k'} = \sigma\rho \tag{2}$$

when k and k' are the rate constants of the substituted and the nonsubstituted compound, respectively, and $\sigma$ is the substituent constant and $\rho$ the Hammett reaction constant. This means that the reactivity of the reagent can often be manipulated in a predictable manner by altering its structure. It is obvious that in case of steric hindrance effects or solubility problems the simple Hammett Eq. 2 cannot be used to predict the reactivity of newly synthesized reagents.

Furthermore, many of the existing derivatization reagents can be modified in a predictable way to optimize the structure of the reagent in order to achieve a more rapid derivatization reaction under suitable reaction conditions, a higher derivatization yield, better chromatographic properties, or enhanced detectability.

## DERIVATIZATION IN LIQUID CHROMATOGRAPHY/ MASS SPECTROMETRY

One of the relatively new detection based derivatization techniques that will not be discussed extensively in this monograph is the coupling of LC and mass spectrometry (MS) [106-107]. Various new interfaces (direct liquid introduction [108], moving belt [109], thermospray [110]) are developed in the last decade and LC/MS is not in its infancy any more. Still it cannot be considered as a routine method but impressive results have been shown in a broad field of analytical chemistry [111].

Derivatization can be applied in LC/MS to improve the chromatographic, the interfacing or the mass spectrometric behavior of the compounds of interest. The ideal cases all three aspects are positively influenced by (a) chemical reaction(s). A chemical step can be introduced in the beginning (prechromatographic) of an analysis procedure or between the chromatographic separation and the interface (postchromatographic).

Postchromatographic derivatization is directly related to the improvement of either the detection properties of the mass spectrometer, or the combination of the interface and the mass spectrometer. Therefore, the well-known approaches from gas chromatography (GC)/MS, aimed at the improvement of selectivity and sensitivity, can be applied such as the increase of the solute volatility, the introduction of an electrophoric group for negative chemical ionization, the increase of molecular weight for obtaining more selectivity, or the generation of more stable molecular weight related ions.

Examples of postchromatographic derivatization systems are the on-line extraction device for RP chromatography [112,113], the use of alkylsulfonate and sulfate counter-ions in both electron impact and chemical ionization modes [114-115], the alkylation of analytes with trimethylanilinium hydroxide [111], and the derivatization of primary amines with carbondisulfide in order to form isothiocyanates [116].

As prechromatographic derivatization influences both the chromatographic, the interface and the detection properties of the LC/MS system careful consideration of all effects is necessary.

Interesting examples are the use of silylation reactions, as often applied in GC/MS, for the LC/MS analysis of fatty acids [117], the use of anthrylmethyl esters of fatty acids with LC/MS using a non-aqueous RP system [118], the analysis of N-acetyl-N,O,S-permethylated derivatives of oligopeptides [119], and the derivatization of carbonyl compounds with 2,4-dinitrophenylhydrazine [120].

Considering the developments in LC/MS interfaces LC/MS itself, and consequently derivatization in LC/MS, will find its own place in the field of hyphenated chromatographic/detection techniques.

## ACKNOWLEDGMENTS

The authors are grateful to Prof. Dr. J. van der Greef and Dr. W. M. A. Niessen (Center for Bio-Pharmaceutical Sciences, Leiden, University, The Netherlands) for their contribution for the section on derivatization in liquid chromatography/mass spectrometry of this chapter.

## REFERENCES


1. Silva, J. A. F. de, *J. Chromatogr.*, *273*:19 (1983).
2. Lingeman, H., W. J. M. Underberg, A. Takadate, and A. Hulshoff, *J. Liq. Chromatogr.*, *8*:789 (1985).
3. Lawrence, J. F., and R. W. Frei (eds.), *Chemical Derivatization in Liquid Chromatography*, J. Chromatogr. Libr. Vol. 7, Elsevier Amsterdam, 1976.
4. Frei, R. W., and J. F. Lawrence (eds.), *Chemical Derivatization in Analytical Chemistry*, Vol. 2, *Separation and Continuous Flow Techniques*, Plenum Press, London, 1982.
5. Blau, K., and G. S. King (eds.), *Handbook of Derivatives for Chromatography*, Heyden & Son, London, 1978.
6. Knapp, D. R. (ed.), *Handbook of Analytical Derivatization Reactions*, Wiley-Interscience, New York, 1979.
7. Morozowich, W., and M. J. Cho, *GLC and HPLC Determination of Therapeutic Agents, Part I* (W. Morozowich and M. J. Cho, eds.), Chromatogr. Sci. Series, Vol. 9, Marcel Dekker, New York, 1978.
8. Nicholson, J. D., *Analyst*, *103*:1,193 (1978).
9. Drozd, J. (ed.), *Chemical Derivatization in Gas Chromatography*, J. Chromatogr. Libr. Vol. 19, Elsevier, Amsterdam, 1981.
10. Little, C. J., J. A. Whatley, and A. D. Dale, *J. Chromatogr.*, *171*:63 (1979).
11. Meyer, S. D., and D. E. Tallman, *Anal. Chim. Acta*, *146*:227 (1983).
12. Lingeman, H., U. R. Tjaden, C. M. B. van den Beld, and J. van der Greef, *J. Pharm. Biomed. Anal.*, *6*:687 (1988).
13. Hulshoff, A., and H. Lingeman, in *Molecular Luminescence Spectroscopy: Methods and Applications, Part I* (S. G. Schulman, ed.), Wiley-Interscience, New York, 1985.
14. Wal. Sj. van der, *J. Liq. Chromatogr.*, *61*:37 (1984).
15. Frei, R. W., *Chromatographia*, *15*:161 (1982).
16. Toren, E. R., Jr., and D. N. Vacik, *Anal. Chim. Acta*, *152*:1 (1983).
17. Schwedt, G., and E. Reh, *Chromatographia*, *14*:317 (1981).
18. Hulshoff, A., and H. Lingeman, *J. Pharm. Biomed. Anal.*, *2*:337 (1984).
19. Rosenfeld, M., M. Mareika-Russell, and A. Phatak, *J. Chromatogr.*, *183*:127 (1984).
20. Lawrence, J. F., and R. W. Frei, *J. Chromatogr.*, *98*:253 (1974).
21. Roberts, J. D., and M. C. Caserio (eds.), *Modern Organic Chemistry*, W. A. Benjamin, New York, 1967, p. 361.
22. Huber, J. F. K., J. A. P. J. Hulsman, and C. A. M. Meyers, *J. Chromatogr.*, *62*:79 (1971).

23. Karger, B. L., M. Martin, and G. Guiochon, *Anal. Chem.*, *46*:389 (1974).
24. Hamilton, R. J., and P. A. Sewell (eds.), *Introduction to High Performance Liquid Chromatography*, Chapman & Hall, London, 1977.
25. Kirchner, J. G., in *Techniques of Chemistry*, Vol. 14, 2nd ed. (A. Weissberger, ed.), Wiley-Interscience, New York, 1978.
26. Snyder, L. R., and J. J. Kirkland (eds.), *Introduction to Modern Liquid Chromatography*, 2nd ed., Wiley-Interscience, New York, 1980.
27. Hearn, M. T. (ed.), *Ion-pair Chromatography, Theory and Biological and Pharmaceutical Applications*, Chromatogr. Sci. Series Vol. 31, Marcel Dekker, New York, 1985.
28. Rogers, D., *Int. Lab.*, *15*:12 (1985).
29. Kissinger, P. T., L. J. Felix, D. J. Miner, C. R. Preddy, and R. E. Shoup, in *Contemporary Topics in Analytical and Clinical Chemistry*, Vol. 2 (D. M. Hercules, G. M. Hieftje, L. R. Snyder, and M. A. Evenson, eds.), Plenum Press, New York, 1978.
30. Kucera, P., and H. Umagat, *J. Chromatogr.*, *255*:563 (1983).
31. Kucera, P. (ed.), *Microcolumn High-Performance Liquid Chromatography*, J. Chromatogr. Libr. Vol. 28, Elsevier, Amsterdam, 1984.
32. Little, J. N., and G. J. Fallick, *J. Chromatogr.*, *112*:389 (1975).
33. Verzele, M., J. van Dijk, P. Mussche, and C. DeWaele, *J. Liq. Chromatogr.*, *5*:1431 (1982).
34. Schwarzenbach, R., *J. Chromatogr.*, *334*:35 (1985).
35. Buchanan, D. N., and J. G. Thoene, *Anal. Biochem.*, *124*:108 (1982).
36. Lingeman, H., H. A. van Munster, J. H. Beynen, W. J. M. Underberg, and A. Hulshoff, *J. Chromatogr.*, *352*:261 (1986).
37. Baty, J. D., S. Pazouki, and J. Dolphin, *Chromatogr. Int.*, *20*:28 (1986).
38. Lingeman, H., A. Hulshoff, W. J. M. Underberg, and F. B. J. M. Offermann, *J. Chromatogr.*, *290*:215 (1984).
39. Scott, R. P. W., and P. Kucera, *J. Chromatogr. Sci.*, *12*:473 (1974).
40. Snyder, L. R., *J. Chromatogr.*, *92*:233 (1974).
41. Snyder, L. R., *J. Chromatogr. Sci.*, *16*:223 (1978).
42. Horvath, C., W. Melander, and I. Molnar, *Anal. Chem.*, *49*:142 (1977).
43. Karger, B. L., L. R. Snyder, and C. Eon, *Anal. Chem.*, *50*:2126 (1978).
44. Dolan, J. W., L. R. Snyder, and M. A. Quarry, *Int. Lab.*, *17*:66 (1987).

45. Berridge, J. C. (ed.), *Techniques for the Automated Optimization of HPLC Separations*, John Wiley and Sons, New York, 1985.
46. Wouters, J., S. Hendrickx, J. Hoogmartens, and H. Vanderhaeghe, *J. Chromatogr.*, *291*:59 (1984).
47. Schoenmakers, P. J. (ed.), *Optimization of Chromatographic Selectivity*, J. Chromatogr. Libr. Vol. 35, Elsevier, Amsterdam, 1986.
48. Goldberg, A. P., E. Nowakowska, P. E. Antle, and L. R. Snyder, *J. Chromatogr.*, *316*:241 (1984).
49. Glajch, J. L., J. J. Kirkland, K. M. Squire, and J. M. Minor, *J. Chromatogr.*, *199*:59 (1980).
50. Nickel, J. H., and S. N. Deming, *LC*, *1*:414 (1983).
51. Meyer, V., *Swiss Chem.*, *7*:7 (1985).
52. Frei, R. W., *Int. J. Environ. Anal. Chem.*, *5*:143 (1978).
53. Erni, F., R. W. Frei, and W. Lindner, *J. Chromatogr.*, *125*:265 (1976).
54. Frei, R. W., L. Michel, and W. Santi, *J. Chromatogr.*, *142*:261 (1977).
55. Tjaden, U. R., J. de Jong, and C. F. M. van Valkenburg, *J. Liq. Chromatogr.*, *6*:2225 (1983).
56. Karasek, F. W., *Res./Develop.*, *28*:38 (1977).
57. Karger, B. L., J. R. Gant, A. Hartkopf, and P. Weiner, *J. Chromatogr. Sci.*, *128*:65 (1976).
58. Kraak, J. C., and P. Bijster, *J. Chromatogr.*, *143*:499 (1977).
59. Majors, R. E., *J. Cromatogr. Sci.*, *18*:571 (1980).
60. Kirkland, J. I., *Analyst*, *99*:859 (1974).
61. Lankelma, J., and H. Poppe, *J. Chromatogr.*, *159*:587 (1978).
62. Frei, R. W., and U. A. Th. Brinkman, *Trends Anal. Chem.*, *1*:45 (1981).
63. Gavis, G., *Anal. Chem.*, *51*:1960 (1979).
64. Hageman, R., *J. Chromatogr.*, *274*:239 (1983).
65. Nielen, M. W. F., A. J. Valk, R. W. Frei, U. A. Th. Brinkman, Ph. Mussche, R. de Nijs, B. Ooms, and W. Smink, *J. Chromatogr.*, *393*:69 (1987).
66. Nielen, M. W. F., U. A. Th. Brinkman, and R. W. Frei, *Anal. Chem.*, *57*:806 (1985).
67. Jong, J. de, M. W. F. Nielen, R. W. Frei, and U. A. Th. Brinkman, *J. Chromatogr.*, *381*:431 (1986).
68. Programmable On-line Solid Phase Extraction Technique (PROSPEKT), Spark Holland, Emmen, The Netherlands.
69. Major, R., *LC*, *2*:16 (1984).
70. Benson, J. R., and D. J. Woo, *J. Chromatogr. Sci.*, *22*:386 (1984).
71. Atwood, J. G., G. L. Schmidt, and W. Slavin, *J. Chromatogr.*, *171*:109 (1979).
72. Rabel, F. M., *J. Chromatogr. Sci.*, *18*:394 (1980).

73. Dong, M. W., and J. R. Gant, *LC, 2*:294 (1984).
74. Dolan, J. W., and V. V. Berry, *LC, 1*:542 (1983).
75. Major, R., *LC, 3*:10 (1985).
76. Reed, G. D., and L. R. Loscomb, *Chromatographia, 15*:15 (1982).
77. Ueland, P. M., and E. S. Olheim, *J. Chromatogr., 276*:157 (1982).
78. Nielen, M. W. F., R. C. A. Koordes, R. W. Frei, and U. A. Th. Brinkman, *J. Chromatogr., 330*:113 (1985).
79. Heaton, J., *Int. Labmate, 11*:23 (1986).
80. Tjaden, U. R., R. A. M. van der Hoeven, J. A. C. Bierman, H. J. E. M. Reeuwijk, H. Lingeman, and J. van der Greef, *Chromatographia, 24*:597 (1987).
81. Reacti-Vials, Pierce Eurochemie BV, Oud-Beijerland, The Netherlands.
82. Liang, Y., *Chromatogr. Rev., 13*:10 (1986).
83. Autoderivat AD 100, Siemens AG, Karlsruhe, FRG.
84. Honda, S., and H. Kuwada, *Anal. Chem., 55*:2466 (1983).
85. PICO-TAG Amino Acid Analysis System, Waters Associates, Milford, MA, USA.
86. Bell, J. P., R. A. Simpson, and R. L. Cunico, *Int. Lab., 16*:62 (1986).
87. Schuster, R., *Chromatogr. Int., 20*:9 (1986).
88. Goya, S., A. Takadate, and H. Fujino, *Yakugaku Zasshi, 102*:63 (1982).
89. Anderson, G. W., J. Blodinger, R. W. Young, and A. D. Welcher, *J. Am. Chem. Soc., 74*:5304 (1952).
90. Goto, J., N. Goto, A. Hikichi, T. Nishimaki, and T. Nambara, *Anal. Chim. Acta, 120*:187 (1980).
91. Goto, J., N. Goto, and T. Nambara, *J. Chromatogr., 239*:559 (1982).
92. Saigo, K., M. Usui, K. Kikuchi, E. Shimada, and T. Mukaiyama, *Bull. Chem. Soc. Jpn., 50*:1863 (1977).
93. Bald, E., K. Saigo, and T. Mukaiyama, *Chem. Lett. Jpn.*, 1163 (1975).
94. Frost, A. A., and R. G. Pearson (eds.), *Kinetics and Mechanism*, 2nd ed., John Wiley and Sons, New York, 1961.
95. Martin, A. N., J. Swarbrick, and A. Cammarata (eds.), *Physical Pharmacy*, 2nd ed., Lea & Febiger, Philadelphia, 1969.
96. Eriksen, S. P., and H. Stelmach, *J. Pharm. Sci., 54*:1029 (1965).
97. Reichardt, *Lösungsmitteleffekten in der Organischen Chemie*, Verlag Chemie, Weinheim, 1969.
98. Kaufmann, E. D. (ed.), *Advanced Concepts in Physical Chemistry*, McGraw-Hill, New York, 1966.
99. Connors, K. A. (ed.), *Reaction Mechanisms in Organic Analytical Chemistry*, Wiley-Interscience, New York, 1973.

100. Pocker, Y., in *Progress in Reaction Kinetics*, Vol. 1 (G. Porter and B. Stevens, eds.), Pergamon Press, New York, 1961.
101. Ingold, C. K. (ed.), *Structure and Mechanism in Organic Chemistry*, Cornell University Press, Ithaca, 1953.
102. Durst, H., M. Milano, D. J. Kikta, S. A. Connely, and S. Grushka, *Anal. Chem.*, *47*:1799 (1975).
103. Parker, A. J., *Chem. Rev.*, *69*:1 (1969).
104. Hendrickson, J. B., D. J. Cram, and G. S. Hammond (eds.), *Organic Chemistry*, 3rd ed., McGraw-Hill, New York, 1970.
105. Hulshoff, A., and A. D. Förch, *J. Chromatogr.*, *272*:275 (1981).
106. Anderegg, R. T., *Mass Spectrom. Rev.*, *7*:395 (1988).
107. Greef, J. van der, W. M. A. Niessen, and U. R. Tjaden, *J. Chromatogr.*, in press.
108. Niessen, W. M. A., *Chromatographia*, *21*:277, 342 (1986).
109. Alcock, N. J., C. Eckers, D. E. Games, M. P. L. Games, M. S. Lant, M. A. McDowall, M. Rossister, R. W. Smith, S. A. Westwood, and H.-Y. Wong, *J. Chromatogr.*, *251*:165 (1982).
110. Blakley, C. R., and M. L. Vestal, *Anal. Chem.*, *55*:750 (1983).
111. Vouros, P., E. P. Lankmayr, M. J. Hayes, B. L. Karger, and J. M. McGuire, *J. Chromatogr.*, *251*:175 (1982).
112. Karger, B. L., D. P. Kirby, P. Vouros, R. L. Foltz, and B. Hidy, *Anal. Chem.*, *51*:2324 (1979).
113. Karger, B. L., D. P. Kirby, and P. Vouros, *J. Chromatogr. Sci.*, *18*:111 (1980).
114. Kirby, D. P., P. Vouros, and B. L. Karger, *Science*, *209*:495 (1980).
115. Kirby, D. P., P. Vouros, B. L. Karger, B. Hidy, and B. Petersen, *J. Chromatogr.*, *203*:139 (1981).
116. Vouros, P., A. Sahil, C.-P. Tsai, and B. L. Karger, Proceedings 33th ASMS Meeting, San Diego, May 26-31 (1985).
117. Quilliam, M. A., and J. M. Yaraskavitch, *J. Liq. Chromatogr.*, *8*:449 (1985).
118. Baty, J. D., R. G. Willis, and R. Tavendale, *Biomed. Mass Spectrom.*, *12*:565 (1985).
119. Yu, T. J., H. A. Schwartz, S. A. Cohen, P. Vouros, and B. L. Karger, *J. Chromatogr.*, *301*:425 (1984).

# 2

# Derivatization Reactions and Kinetics in Liquid Chromatography

ADRIANUS N. FRUIJTIER and WILLY J. M. UNDERBERG
*State University of Utrecht, Utrecht, The Netherlands*

HENK LINGEMAN* *Leiden University, Leiden, The Netherlands*

JOS H. BEIJNEN *Slotervaart Hospital, The Netherlands Cancer Institute, Amsterdam, The Netherlands*

## INTRODUCTION

When developing a derivatization procedure for a certain compound, the influence of all reaction conditions (e.g., solvent, temperature, reaction time, concentration of reactants and catalyst(s), and substituent effects) must be thoroughly investigated in order to obtain the optimal reaction conditions.

The purpose of this chapter is to aid in performing derivatization reactions, as the necessary adjustments for optimization of a certain procedure can be derived from the theoretical backgrounds.

In this chapter no fundamental organic chemistry will be discussed because it does not fit within the structure of this handbook and it is discussed in a number of excellent books on mechanisms and structures in organic chemistry [1-5].

## KINETICS

In the past 20 years a great number of derivatization procedures in combination with chromatographic separations have been described.

**Present affiliation*: Free University, Amsterdam, The Netherlands

Most of these methods are based on nonspecific alkylation and acylation reactions. The advantage of using these nonspecific derivatization reactions is that all compounds with a reactive hydrogen (e.g., hydroxyl, phenol, amine, sulfhydryl) can be derivatized with the same reagent. The disadvantage, however, is that in complex matrices a great number of interfering compounds will be derivatized as well. Therefore, in complex matrices (e.g., pharmaceutical formulations, biological materials, environmental samples) it is advisable to use more selective derivatization procedures such as cyclization and condensation reactions.

In addition to the reaction mechanism, the physicochemical properties of the solvent (e.g., polarity, dielectric constant, optical density) are very important in choosing the right derivatization procedure for a particular analyte. For instance, for the analysis of a solute present in a biological sample a derivatization technique which can be performed in a polar protic solvent (e.g., water) may be advantageous in comparison with methods that require nonpolar or aprotic solvents.

## Alkylation Reactions

All alkylation methods have in common the replacement of an active hydrogen by an alkyl group. In principle, compounds with a primary, secondary, or tertiary hydroxyl function, a phenol, a carboxyl, a sulfhydryl, a primary or secondary amine, a carboxamide, a sulfonamide, an imido group, and, under certain conditions, a tertiary amino function can be alkylated. One of the features of alkylation reactions is that lesser polar derivatives are formed, often resulting in better chromatographic properties of the derivatives than the nonderivatized solutes [4].

### *Diazoalkane Alkylation Reactions*

From the diazoalkane reagents, for the alkylation of a great variety of drugs diazomethane is the most important representative. Diazomethane has been applied to the methylation of bile acids [5] and prostaglandins [6] and the corresponding derivatives are separated by LC.

The methylation reaction of a compound with an active hydrogen, Y-H, is represented by Eq. 1:

$$\text{Y–H} + (\text{H}_2\text{C=N}^+\text{=N}^- \leftrightarrow \text{H}_2\text{C}^-\text{-N}^+\!\equiv\text{N}) \rightarrow \text{Y–CH}_3 + \text{N}_2 \qquad (1)$$

The transfer of the proton Y–H to the diazo compound causes the immediate elimination of nitrogen ($N_2$), which is an excellent leaving group, and the immediate reaction between the resulting carbonium ion ($CH_3^+$) and the nucleophile ($Y^-$).

The procedure is relatively simple; the reagent, a gas, is bubbled through a solution of the analyte until a persistent yellow color indicates the completeness of the reaction. Excess of the reagent can be removed easily under a stream of dry nitrogen. The solvent usually is diethyl ether; however, much higher reaction rates are obtained after addition of about 10% methanol, water, or ethanol to the solution [7]. The role of methanol seems to be purely catalytic [7]. Dioxane (with methanol) and tetrahydrofuran have also been used as reaction solvents. The alkylation rate of the higher, more hydrophobic diazoalkanes can be considerably enhanced by the addition of boron trifluoride to the reaction mixture. Sometimes, e.g., with phenobarbital, side reactions occur. Usually, the procedure is smooth and rapid [4,8], but the diazoalkane reagents are toxic and explosive and should be handled with care.

### *Alkylation by Acid-Catalyzed Reactions (Esterification)*

The esterification reaction of carboxylic acids with low molecular primary alcohols proceeds by a bimolecular alkyl-oxygen cleavage mechanism (Eq. 2). The rate of this reaction decreases sharply with increasing chain branching of the alcohol. This is probably caused by a combination of steric and ponderal effects [1]. The esterification reaction is catalyzed by mineral acids since after protonation of the carbonyl moiety the carboxylic acid function is more susceptible to nucleophilic attacks, whereas it also promotes the loss of the leaving group, since water is lost more easily than the hydroxyl ion. Catalysis with sulfuric acid results in slow and often

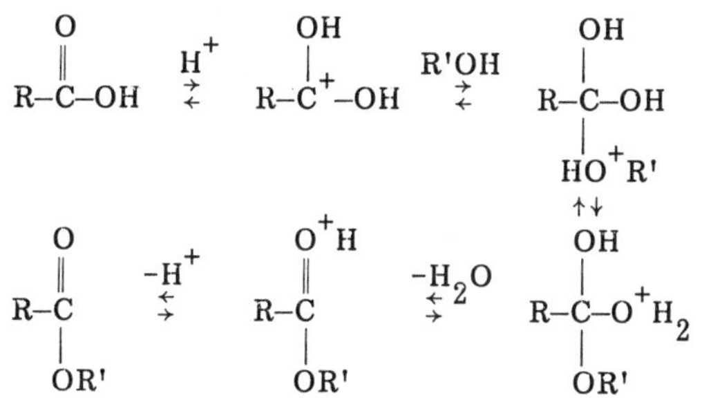

(2)

incomplete reactions. Esterification reactions with dry hydrogen chloride dissolved in an appropriate alcohol gave much better results. The reagent can be prepared either by bubbling hydrogen chloride through the alcohol or by adding acetyl chloride or thionyl chloride to the alcohol. Methylation is usually the derivatization procedure of first choice with this method, but in principle a wide range of different esters can be prepared. The reaction is sometimes completed within a few minutes, but usually the reaction mixture must be heated at 333-373 K for 5 min to 2 hr. Isolation of the ester before chromatography is often required.

The advantages of this technique are as follows:

- Several functional groups other than the carboxylic acid function can be derivatized.
- The derivatives have strong electron-capturing properties enabling sensitive detection [7].

### Alkylation with Alkyl Halides

Alkylation reactions with alkyl halides are nucleophilic substitution ($S_N$) reactions. A nucleophile (Nu or $Nu^-$) displaces a leaving group (L or $L^-$), which is normally a halide ion, from the alkylating agent (Eq. 3):

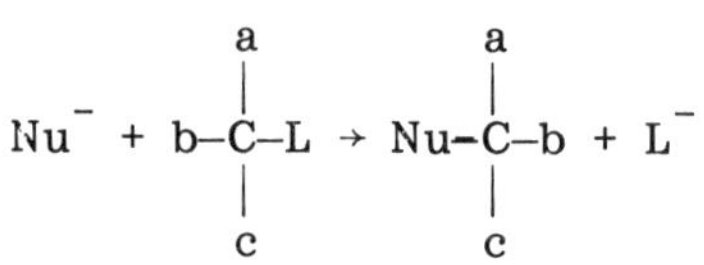

(3)

The nucleophile and the leaving group possess principally the same general features: they are anionic or neutral bases with unshared electron pairs. Two mechanistic routes have been clearly identified for $S_N$ reactions: $S_N1$ and $S_N2$ substitution reactions.

#### *$S_N1$ Nucleophilic Substitution Reactions*

The rate of this reaction can be expressed by the following equation:

$$\text{Rate} = k_1[R - L] \tag{4}$$

Because the rate is independent of the nucleophile concentration, the rate-determining step is the ionization of R-L (Eq. 3) followed by a rapid, almost immediate, attack of the nucleophile as can be seen in Eq. 5 [1,2]:

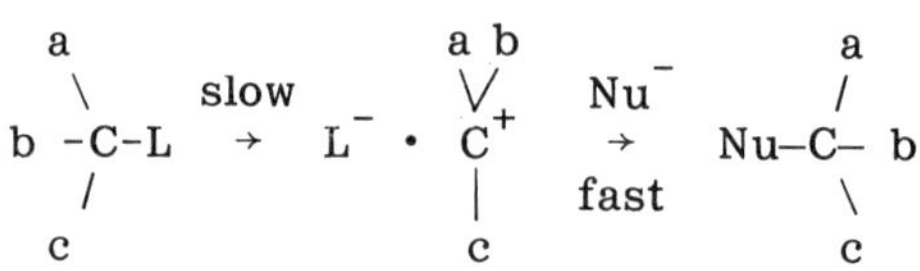

(5)

Since the carbonium ion formed during the rate-determining step is planar in structure, the subsequent attack of the nucleophilic reagent will take place with equal chance from either side of the planar intermediate. In the derivatization of a chiral center this may lead to a product with a 50/50 mixture of species having the same and

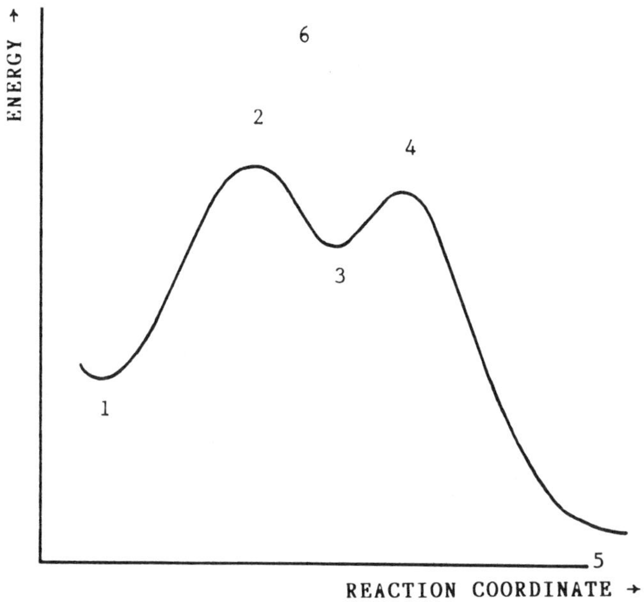


Figure 2.1 Energy versus reaction coordinate plot for $S_N1$ reaction. 1, $Nu^- + R - L$; 2, $Nu^- + R^{\delta+} \ldots L^{\delta-}$; 3, $Nu^- + R^+ + L^-$; 4, $Nu^{\delta-} \ldots R^{\delta+} + L^-$; 5, $Nu - R + L^-$.

the opposite stereochemical configuration compared to the starting material. Therefore, complete racemization is expected, although this is not always the case. It may occur that the leaving anion ($L^-$) initially is only a few molecular diameters away, which makes an attack on that side more difficult. In this case partial inversion is observed [2,3,8].

The reaction rate can be decreased by addition of the leaving anion $L^-$ because in that case competition with $Nu^-$ takes place, reversing the first step of the procedure (Eq. 5). This effect is called the mass law effect and it becomes more important as the reaction proceeds as a result of the increasing concentration of the leaving group [1].

The energy profile for the $S_N1$ reaction is presented in Fig. 2.1.

### *$S_N2$ Nucleophilic Substitution Reactions*

The reaction rate of an $S_N2$ nucleophilic substitution reaction is expressed as follows:

$$\text{Rate} = k_2[\text{R} - \text{L}][\text{Nu}^-] \tag{6}$$

indicating that both the halide and the nucleophile participate in the rate-determining step, which is represented by Eq. 7:

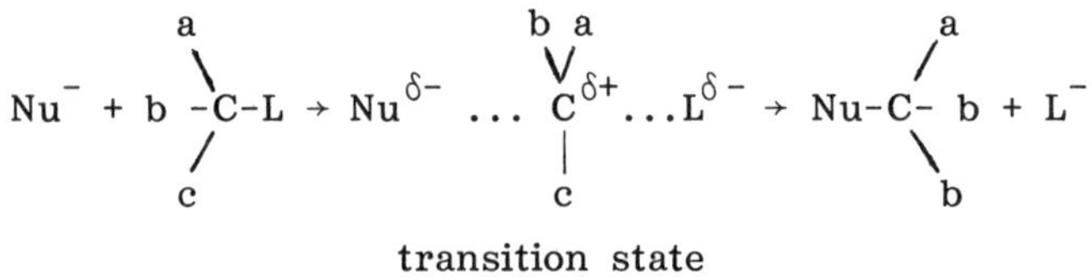


(7)

Because the nucleophile approaches the molecule from the rear to cause halide expulsion from the front, in this case complete inversion is observed. The attack from the front side is prevented by the halide.

In some cases no second-order kinetics are observed. This might be the case when the solvent can act as a nucleophilic reagent as well, e.g., water under $S_N2$ conditions:

$$\text{Rate} = k_2[\text{R} - \text{L}][\text{H}_2\text{O}] \tag{8}$$

Because the concentration of water is constant, Eq. 8 simplifies to

$$\text{Rate} = k_3[\text{R} - \text{L}] \qquad \text{with } k_3 = k_2\,[\text{H}_2\text{O}] \tag{9}$$

These are called pseudo-first-order kinetics and are often observed with derivatization reactions owing to the large excess of alkylating agents.

The energy profile for the $S_N2$ nucleophilic substitution reaction is presented in Fig. 2.2.

The reaction rate is relatively low when the alkyl halide possesses bulky groups at the carbon atom under attack. In this case an $S_N1$ reaction becomes more likely. However, $S_N1$ reaction conditions are generally unsuitable for alkylation with alkyl halides. Many side reactions occur and this frequently results in low derivatization yields.

### *Structure of the Alkyl Halides*

For the alkylation reaction with alkyl halides, primary alkyl halides are predominantly used and the reaction is therefore usually of the $S_N2$ type. Inductive effects are the reason that the reactivity of primary *N*-alkyl halides is lower than that of methyl halides, i.e., because of the increased electron density at the attacked carbon atom.

Resonance effects have a relatively small effect on reaction rate; the benzyl halides show about a fourfold higher reaction rate than the corresponding methyl halides.

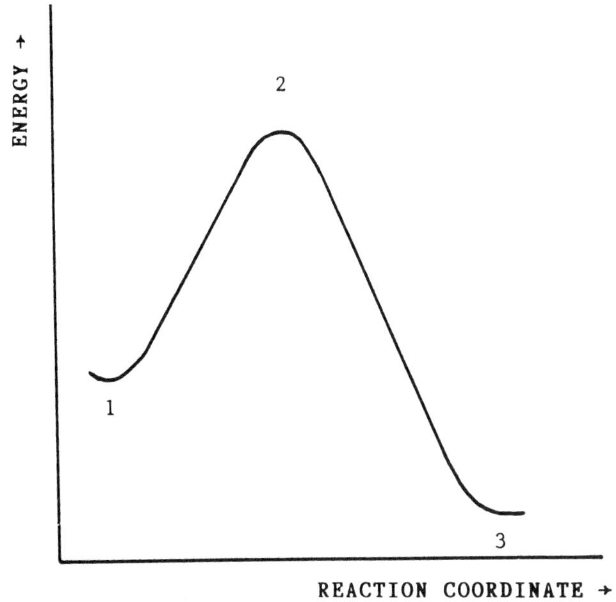


Figure 2.2 Energy versus reaction coordinate plot for $S_N2$ reaction. 1, $Nu^- + R - L$; 2, $Nu^{\delta-} \ldots R^{\delta+} \ldots L^{\delta-}$; 3, $Nu - R + L^-$.

The group of alkyl halides with a carbonyl function at the α-position to the reactive group are the most reactive substrates in $S_N2$ reactions, apparently because the attacking nucleophile transfers its charge to the carbonyl function as well as to the adjacent reaction site [4,8].

*Relative Reactivity of Anions*

The conjugate bases of acids can also act as nucleophiles. Within a series of structurally related compounds, the nucleophilicity of the anions parallels their basic strength. The following order of decreasing nucleophilicity exists for anions with oxygen as the reactive atom:

$$C_2H_5O^- > OH^- > C_6H_5O^- > CH_3COO^-$$

No alkylation of alcohols will take place under the normally applied mild reaction conditions because the alcohols are not deprotonated to any significant extent. With nucleophiles containing more than one atom with active electron pairs under $S_N2$ conditions, the more polarizable and less electronegative atom is preferentially alkylated [4,8].

*Reactive Leaving Group Ability*

The more polarizable atoms yield the better leaving groups. For halides the following order of decreasing leaving group activity prevails:

$$I^- > Br^- > Cl^- >> F^-$$

Derivatizations with alkyl halides are therefore performed almost exclusively with iodides or bromides.

## Base-Catalyzed Alkylation Reactions

In principle, alkylation reactions can only be performed if the analyte under investigation, the acid, is deprotonated first by an organic or inorganic base. The choice between organic or inorganic base strongly depends on the reaction solvent. If an organic base is preferable, normally in combination with organic solvents, tertiary amines must be applied because primary and secondary amines may also be alkylated under certain conditions. The inorganic bases, usually alkali carbonates, can be combined with derivatization procedures in water or in aqueous solutes containing some ethanol, methanol, or acetone.

*Carbonate-Catalyzed Alkylation Reactions*

A well-known procedure is the reaction of the analyte in a polar aprotic solvent with an alkyl halide in the presence of an alkali carbonate. This reaction has been used for the formation of highly fluorescent derivatives of monocarboxylic acids. The alkyl halide is 4-bromomethyl-7-methoxycoumarin and the derivatives can be analyzed by liquid chromatography (LC) [9].

The reaction proceeds as follows:

$$\begin{aligned} &2\text{R-COOH} + K_2CO_3 \longrightarrow 2\text{R-COO}^-K^+ + H_2O + CO_2 \\ &\text{R-COO}^-K^+ + R_1\text{-L} \longrightarrow \text{R-COO-}R_1 + K^+\text{-}L^- \end{aligned} \qquad (10)$$

Concentration of the sample is easily achieved by evaporating the solvent under a stream of dry nitrogen or by partial refluxing [10].

Apart from potassium carbonate, other alkali carbonates may be used. The reaction rate in, for instance, the methylation of a carboxylic acid is found to be highest with cesium carbonate [11], which can be explained by the differences in the radii of different solvated alkali atoms.

It is also possible to add a crown ether to the reaction mixture. The cation of the cation-anion pair is complexed by the crown ether (Eq. 11), where the ion is bound inside the cavity of the ring. These complexing agents have been used for the derivatization of carboxylic acids. As a result of the complexation of the cation,

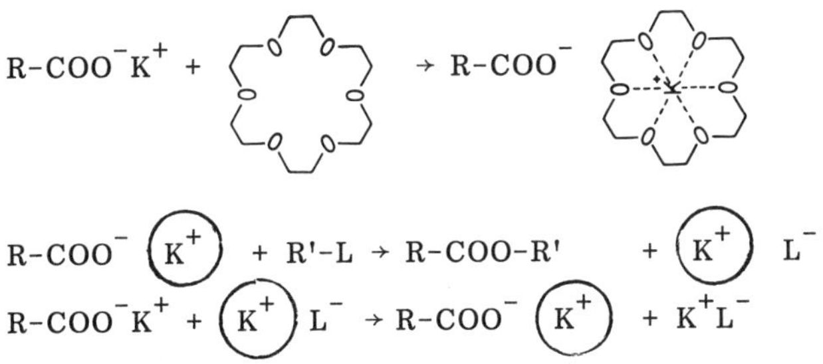


(11)

the resulting "naked" anion is very reactive. This results in a rapid derivatization with a high yield [12]. Because both the solvent and the reagent are volatile, the reactions should be performed in an airtight closed container. For complete derivatization the reaction mixture must be heated. The reaction solvents *N,N'*-dimethylacetamide [13,14] and acetone [15] have been applied in combination with methanolic solutions of tetramethylammonium hydroxide or another quaternary ammonium hydroxide as catalyst.

The carbonate-catalyzed alkylation reaction is very attractive because it is a rapid method and furthermore a wide range of derivatives can be prepared.

A number of other bases have been used in combination with various polar solvents in derivatization reactions prior to and after LC separations. Among these bases are triethylamine and *N,N'*-diisopropylethylamine. A comparison between triethylamine and the crown ether 18-crown-6 complexing a potassium cation in the derivatization of some penicillins in acetonitrile shows that both catalysts gave the same reaction rates and derivatization yields [16].

### Phase Transfer-Catalyzed Alkylations Reactions

In bioanalysis as well as in environmental analysis it may be advantageous to combine sample clean-up, sample concentration, and derivatization into one single procedure. The application of extractive alkylation, well known in GC analysis, is an elegant way of combining these sample pretreatment procedures [4].

An example of extractive alkylation is the one-step derivatization procedure for disulfiram in plasma samples [17]. In this procedure the analyte, containing a sulfhydryl function, is converted to its alkyl derivative and subsequently separated from its metabolites and analyzed with a reversed phase (RP) LC system. The procedure is simple. To a plasma sample, buffered to pH 8-9 and stabilized with ethylenediaminetetraacetic acid, an excess of the alkyl iodide is added and after mixing, heating, and cooling, the mixture is extracted with diethyl ether. After evaporation of the organic phase to dryness the residue is dissolved in the eluent and injected into the LC system.

*Extractive Alkylation Reactions*

The anionic compound under investigation is extracted as an ion pair (IP) with a tetraalkylammonium cation from an aqueous solution, at a suitable pH, into an aprotic solvent, usually dichloromethane. Because of the poorly solvating properties of the solvent, the anions are very reactive and an $S_N2$ reaction with the alkyl halide takes place.

The extractive alkylation of a compound with an active hydrogen (Y-OH) is given by the following reaction sequences:

$$\begin{array}{ll} \text{Aqueous phase} & \text{Y-OH} + R_4N^+X^- \rightleftarrows YO^-R_4N^+ + X^- + H^+ \\ & \qquad\qquad\quad \uparrow\downarrow \qquad\qquad\quad \uparrow\downarrow \\ \hline \text{Organic phase} & \text{Y-OR}' + R_4N^+X^- \rightleftarrows YO^-R_4N^+ + R'X \end{array} \tag{12}$$

The extraction equilibrium of an anion ($X^-$) with a tetraalkylammonium ion ($R_4N^+$) can be written as follows:

$$X^-_{aq} + R_4N^+_{aq} \rightleftarrows R_4N^+X^-_{org} \tag{13}$$

The equilibrium constant $K_{ex}$ is defined as

$$K_{ex} = \frac{(R_4N^+X^-)_{org}}{(X^-)_{aq}(R_4N^+)_{aq}} \tag{14}$$

If no side reactions occur, such as protonation of $X^-$, the distribution ratio $D_x$ of $X^-$ is expressed by

$$D_x = \frac{(R_4N^+X^-)_{org}}{(X^-)_{aq}} = K_{ex}(R_4N^-) \tag{15}$$

The distribution ratio and therefore the extraction efficiency are proportional to $K_{ex}$ and the $R_4N^+$ concentration. Because $K_{ex}$ increases as the lipophilicity of the IP increases, tetraalkylammonium ions with alkyl chains shorter than *n*-butyl are not used, since the extraction recoveries of the IPs are not sufficient. Tetrabutylammonium ($TBA^+$), tetrapentylammonium ($TPA^+$), and tetrahexylammonium

($THxA^+$) are the most frequently applied symmetrical quaternary ammonium cations. With larger counterions (e.g., cetrimide) even hydrophilic anions can be extracted.

The pH of the aqueous phase should be at least two units above the $pK_a$ value of the acid when the partition coefficient is unity, whereas a high partition coefficient means that the pH can be kept lower.

Usually the optimal conditions for extractive alkylation do not coincide with the conditions in obtaining the highest extraction efficiency and the optimum reaction rate. A major drawback may be the coextraction of hydroxyl ions as IP, with the quaternary ammonium counterions because the hydroxyl ions may hydrolyze the reagent. Therefore, it is not advisable to extract at high pH values and with $R_4N^+$ ions of increasing hydrophobicity because this enhances the coextraction of the hydroxyl ions [4,8].

It has been found that without the addition of quaternary ammonium counterions phenolic compounds can be alkylated in a biphasic system, but so far no explanation for the mechanism of this procedure has been given [18,19]. Furthermore, the extractive alkylation reaction is inhibited by low levels of chloride ions. The chloride concentration can be reduced by mercury complexation or by dilution of the mixture [20].

Optimization of the extractive alkylation procedures requires a great number of experiments. To limit this a multivariate optimization (simplex optimization) was developed [21] which allows a quick approach to the optimal reaction conditions. Moreover, better insight into the derivatization mechanism can be obtained.

Advantages of the extractive alkylation are the mild reaction conditions, the relatively harmless reagents, the short reaction times, and the combination of extraction and alkylation in one step [7].

### *Alkylation by Solid Phase Transfer Catalysis Reactions*

This type of alkylation reaction may be considered a special form of extractive alkylation. The method has been developed for compounds which are extremely sensitive to hydrolysis [22-24] or for solutes which are hydrophilic in nature.

The compound to be alkylated (H-Y) is dissolved in a suitable solvent, e.g., dichloromethane, containing $TBA^+$ hydrogen sulfate ($TBA^+$ $HSO_4^-$) and the alkyl halide (R-L). After addition of solid sodium hydrogencarbonate, shaking, and centrifugation, the organic phase is analyzed. The occurring reactions may be denoted as in Eq. 16.

An example of a solid phase transfer catalysis reaction is the selective derivatization of carboxylic acids with 2-bromo-1-methylpyridinium iodide (BMP). The carboxylic acid, present in an aqueous biological matrix, is extracted with dichloromethane after addition

$$\begin{array}{lllll} NaHCO_{3\,solid} & + TBA^+HSO_4^- & \rightarrow TBA^+HCO_{3\,org}^- & + NaHSO_{4\,solid} & \\ TBA^+HCO_{3\,org}^- & + HA & \rightarrow TBA^+A^-_{org} & + H_2CO_3 & (16) \\ TBA^+A^-_{org} & + RX & \rightarrow RA & + TBA^+X^-_{org} & \end{array}$$

of $TBA^+HSO_4^-$ and sodium hydrogen carbonate. The derivatization is performed in the dichloromethane phase, in which the derivatization reagent (amine) and the activator (BMP) are present. After a reaction time of about 5 min the organic phase is evaporated to dryness and the residue is dissolved in the LC eluent [25].

### Miscellaneous Alkylation Reactions

Although the application of tetrafluoroborate-catalyzed derivatization reactions in GC has been well investigated, it is a less commonly used alkylation procedure in LC. The use of "Meerwins" reagent (triethyloxonium tetrafluoroborate) in dichloromethane is one of the exceptions [26]. The advantage of this procedure is that the reaction is very fast (less than 5 sec) due to the excellent leaving group properties of the tetrafluoroborate anion. The disadvantage, however, is that the reagent is unstable in solution. Therefore, the reagent should be prepared freshly every day.

The procedure is applied for the analysis of diethylcarbamate in plasma samples and is performed by mixing the plasma and the reagent solution in dichloromethane. After extraction the derivative is analyzed by RP-LC.

### Acylation Reactions

Acylation reactions are normally nucleophilic addition elimination reactions in which a tetrahedral intermediate occurs (Eq. 17). The result is an overall substitution of the nucleophile $Y^-$ for the leaving group X [1].

$$R\text{-}\overset{\overset{\displaystyle O}{\|}}{C}\text{-}X \;\underset{}{\overset{Y^-}{\rightleftarrows}}\; R\text{-}\underset{\underset{\displaystyle Y}{|}}{\overset{\overset{\displaystyle O^-}{|}}{C}}\text{-}X \;\overset{-X^-}{\rightleftarrows}\; R\text{-}\underset{\underset{\displaystyle Y}{|}}{\overset{\overset{\displaystyle O}{\|}}{C}} + X^- \qquad (17)$$

Compounds with functional groups bearing a reactive hydrogen can act as nucleophiles, e.g., alcohols, enols, and phenols, thiols, amines, and amides. Another possibility less common in chromatographic derivatization is the addition of an acyl group, R-CO, across a double bond or an aromatic ring. The relative reactivities of the acylating agents, R-CO-X, depend on

The electron-donating or withdrawing properties of the leaving group toward the carbonyl group
The ability of X as a leaving group
The electronic and steric effects of R [8]

Acylation is a simple derivatization method. The reactions normally result in a quantitative formation of the derivatives with a better chromatographic behavior than the original compounds [7].

### *Acyl Chlorides and Acid Anhydrides*

Acyl chlorides are extremely reactive because chloride is a good leaving group. They are therefore suitable for compounds which are not easily acylated. The mechanism of this derivatization reaction, which must be performed in polar solvents with good ion-solvating properties, might be an $S_N1$ reaction. The rate-limiting step is the loss of the chloride ion, followed by a rapid reaction of the acyl carbonium ion, $R\text{-}C^{+}\text{=}O$, with the nucleophile. Acyl halides undergo hydrolysis readily, which makes them difficult to handle in these procedures.

Acid anhydrides, R-CO-O-CO-R, can also react with comparatively weak nucleophiles, e.g., water and aliphatic alcohols, though less avidly than acyl chlorides.

Reactions with acid chlorides or anhydrides can be carried out by treatment of the drug with the pure reagent or with the reagent in a suitable solvent, such as pyridine, tetrahydrofuran, dichloromethane, or toluene. The derivatization yield often depends dramatically on the solvent. A proton acceptor is usually needed with acyl chlorides in order to neutralize the hydrochloric acid produced during the reaction and thus to force the reaction to completion. For this purpose *N*-methylimidazole [27] and 4-dimethylaminopyridine [28] have been used. Pyridine has a much weaker catalytic power. However, quantitative reactions with anhydrides without a base catalyst are also possible. Removal of the excess reagent can be effected by evaporation under a stream of dry nitrogen. Because many acylated compounds are very susceptible to hydrolysis, in extreme cases even nitrogen purging may not prevent the degradation of the derivative. Therefore, storage of the derivatives is best performed in organic solvents with an excess of the reagent. It is absolutely essential that the solvents in which the derivative is stored be kept anhydrous.

Acylation with acyl chlorides and acid anydrides is an important derivatization procedure, as can be seen in the chapters on ultraviolet/visible and fluorescence derivatization in this book. However, as mentioned before, the stability of many acyl derivatives is limited.

Table 2.1 Derivatization Procedures in Liquid Chromatography

| Procedure | Reacting groups | Reaction mechanism | Stability reagents[a] | Stability derivatives[a] |
|---|---|---|---|---|
| Diazoalkane alkylation | Active hydrogen | $S_N2$ | - | + |
| Esterification | Active hydrogen | $A_{AC}2$ | + | +/- (volatile) |
| Alkylation with alkyl halides | Active hydrogen | $S_N2$ | - | + |
| Carbonate-catalyzed alkylation | Active hydrogen | $S_N2$ | + (volatile) | + |
| Miscellaneous base-catalyzed alkylation | Active hydrogen | $S_N2$ | + (volatile) | + |
| Extractive alkylation | Active hydrogen | $S_N2$ | - | + |
| Alkylation by solid phase transfer catalysis | Active hydrogen | | | |
| Acylation with carbonyl chlorides or anhydrides | Active hydrogen | $S_N1$ | -- | -- |
| Acylation with sulfonyl chlorides | Phenol. amine (prim., sec.), hydroxyl | $S_N1$ | +/- | + |
| Cyclic boronate formation | *o*-Dihydroxyl, *o*-hydroxylamine | $A_{AC}2$ | - | + |
| Hydrazone and oxime formation | Ketone, aldehyde | $A_N$ | + | - |
| Isothiocyanate derivatizations | Amine (prim.) amino acid, hydroxamic acid | | | + (volatile) |
| Nitrophenyl derivatizations | Amine (prim., sec.), amino acid, hydroxyl, sulfhydryl, imidazoline | $S_N1$ $S_N2$ | - (hydrolysis) | + |
| Multiple derivatizations | Amino acid | | | |
| a. Alkylation | Carboxyl | $S_N1$ | + | + |
| b. Acylation | Amine (prim.) | $S_N2$ | + | + |

[a]Stability: --, less than 1 hr; -, between 1 hr and 24 hr; +, over 24 hr.

[b]Reaction speed: ++, less than 0.5 hr; +, between 0.5 and 1 hr;

[c]Toxicity: ++, very toxic; +, toxic; +/-, slightly toxic; -, nontoxic.

| Reaction medium | Reaction speed[b] | Reaction temperature (K) | Reaction time (hr) | Toxicity[c] | Ref. |
|---|---|---|---|---|---|
| Aqueous or diethyl ether | + | 293 | 0-1 | ++ | 7 |
| Aqueous | - | 298-373 | 0-16 | +/- | 4,7,8 |
| Aprotic solvent | +/- | 293-333 | 0-3 | - | 4,8,31, 32 |
| Aprotic solvent | + | 323-393 | 0-1 | - | 4,7,8 |
| Water or aprotic solvent | | | 0-1 | - | 4,7,8,33 |
| Water and organic solvent | ++ | 293-323 | 0.1 | | 4,7,8 |
| Dichloromethane or ketone | | | | | 4,8 |
| Aqueous (basic) or organic solvent | + | 293-373 | 0.1-1 | - | 7,8,31, 32 |
| Aqueous (basic) or organic solvent | - | 293-343 | 0-24 | +/- | 31,32,33 |
| Aqueous or organic solvent | ++ | 293-373 | 0.1 | - | 7,31,33 |
| Organic solvent | -- | 293-343 | 0-100 | - | 7,31,32, 33 |
| Ethyl acetate | ++ | 293 | 0.5 | +/- | 7,33 |
| Water or acetonitrile | + | 333-373 | 0.5-0.75 | + | 32,33,34 |
| Aprotic solvent | - | 293-343 | 0-7 | +/- | 4,8,31, 33 |
| Aprotic solvent | - | 293-343 | | | |

24 hr; +/-, depends on reagent.

+/-, between 1 hr and 3 hr; -, between 3 hr and 24 hr; --, over

### Miscellaneous Derivatization Methods

Along with the discussed alkylation and acylation procedures, a number of other derivatization procedures have been described. Examples of most methods, including alkylation and acylation reactions, can be found in the other chapters of this book.

One specific example is the complexation with diethylammonium-diethyldithiocarbamate (DDTC). DDTC is a complexing (chelating) agent that can be used for the derivatization of platinum(II) compounds. The derivatization reaction can be performed directly in alkalized urine samples by heating the solution for 10 min at 365 K. The resulting derivative is subsequently extracted with chloroform and analyzed with RP-LC [29]. DDTC reacts with a great number of metal ions, but the complexes with, for instance, calcium(II), zinc(II), iron(II), and manganese(II) are very unstable.

The advantages and disadvantages of the nonspecific alkylation and acylation reactions over specific procedures, e.g., cyclization and amidification, are discussed in Table 2.1.

## SOLVENT EFFECTS

The solvent can exert a distinguished effect on the rate and mechanism of a certain derivatization reaction. The influence of the solvent is rather complex. Solvent molecules take part in the microscopic reaction steps and, furthermore, a change of the solvent will result in a different macroscopic reaction rate and/or a change in reaction yield [35]. According to modern kinetics [36-39], the initial compounds a,b,c-C-L and $Nu^-$ in Eq. 3 as well as the activated transition intermediate, formed before the final product (Nu-C-c,b,a in Eq. 3) is formed, can be stabilized or destabilized by means of solvation or association processes with the solvent.

If the solvent stabilizes the transition state of the reagent more effectively than the ground state, the rate of reaction increases. If, for instance, in nucleophilic substitution reactions ions are formed in the transition state or existing charges are concentrated on a smaller surface, the reaction will be faster in polar and protic (e.g., water, methanol) than in polar and aprotic solvents (e.g., acetonitrile, *N,N'*-dimethylformamide). On the other hand, reactions during which in the transition state ionic charges are neutralized or divided over a larger surface proceed faster in aprotic solvents due to a better ground state solvation of the reactants in protic solvents [40,41].

In terms of the extent of the solvation of charged initial compounds in protic solvents, the following assumptions can be made:

The solvation increases with increasing charge.

Table 2.2 Influence of Solvent on Nucleophilic Reactions

| Reaction time | Initial compound | Activated complex | Change in ionic charge | a |
|---|---|---|---|---|
| $S_N1$ | R-X | $R^{\delta+} \ldots X^{\delta-}$ | Separation of unlike charges | Increase |
| $S_N2$ | $R\text{-}X^+$ | $R^{\delta+} \ldots X^{\delta+}$ | Slight dispersal of positive charge | Decrease |
| $S_N2$ | Y + R-X | $Y^{\delta+} \ldots R \ldots X^{\delta-}$ | Separation of unlike charges | Increase |
| $S_N2$ | $Y^- + R\text{-}X$ | $Y^{\delta-} \ldots R\text{-}X^{\delta-}$ | Slight dispersal of negative charge | Decrease |
| $S_N2$ | $Y + R\text{-}X^+$ | $Y^{\delta+} \ldots R\text{-}X^{\delta+}$ | Slight dispersal of positive charge | Decrease |
| $S_N2$ | $Y^- + R\text{-}X^+$ | $Y^{\delta-} \ldots R\text{-}X^{\delta+}$ | Neutralization of charge | Decrease |

[a]Influence of increasing solvent polarity at reaction rate.

The solvation decreases with increasing division of the charge over a larger surface.

The decrease in solvation resulting from the division of charge over a larger surface is less than the decrease in solvation resulting from the neutralization of ionic charge.

In reactions in which the formation of the activated complex requires creation or destruction of charges, the effects of solvent polarity are very striking. For example, the specific rate constant for substitution may change by three to six orders of magnitude when the reaction is transferred from water to absolute ethanol. On the other hand, the influence of the solvent polarity on nucleophilic substitution reactions for which activation only requires a dispersal of charge is much smaller. In this case the specific rate constant for substitution is 3-10 times as rapid in absolute ethanol than in water [2].

The solvent may also influence the stereochemistry of the reaction. However, such an influence has only been observed with electrophilic substitutions [35], and will not be discussed further in this chapter.

In bimolecular substitution reactions hydrophobic interactions may enhance the reaction rate due to the fact that in water-type solvents the attraction between hydrophobic molecules is much larger. The reaction will be facilitated in this way [35], which may be important if, for instance, a number of fatty acids containing different hydrophobic alkyl chains are derivatized. Therefore, it is important to know which type of reaction applies in order to evaluate the influence of the solvent. These influences on the various types of nucleophilic substitutions are listed in Table 2.2. The solubility parameters and the solvent polarity values of various solvents serving as measures for the internal pressure or polarity are listed in Table 2.3.

### Influence of Anion Solvation on Rate of Nucleophilic Substitution Reactions

$S_N1$ nucleophilic substitution reactions are the most sensitive to solvent changes because the transition state is highly ionic and requires a solvent with a high ionizing power. Two parameters are important for the ionizing power of the solvent. A high dielectric constant facilitates the separation of ions, since the force between charged particles depends inversely on the dielectric constant. Therefore, a higher dielectric constant leads to an increasing ionizing power.

The second parameter is the ability of the solvent to solvate the separate ions. Cations are most effectively solvated by compounds of elements in the first row of the periodic system which have unshared electron pairs, e.g., ammonia, alcohols, water, carboxylic acids, sulfur dioxide, and diethylsulfoxide. Anions are most effectively solvated by solvents with strongly polarized bonds between hydrogen and a strongly negative element (protic solvents); they tend to form hydrogen bonds. Because the possibility of a protic solvent accepting hydrogen bonds increases with increasing charge density (the quotient of charge and volume), the solvation with these solvents decreases in the following order:

$$F^- > Cl^- > C_6H_5O^- > Br^- > OH^- > CH_3O^- > N_3^- > SCN^- > I^- > CN^- > C_6H_5S^-$$

In this order the nucleophilicity increases, since the solvent molecules surrounding the anion must be removed. The stronger the solvent molecules are bound to the anion, the greater the activation energy and the slower the reaction rate will be. The lower the degree of solvation, the greater the nucleophilicity and reactivity. Because dipolar aprotic solvents cannot form hydrogen bonds, they do not solvate anions effectively.

The best solvation is obtained with large and polarizable anions. The listed order of increasing nucleophilicity is therefore reversed

Table 2.3 Solvent Polarity (P') and Solubility Parameters (δ)

| Compound | δ | P' | Compound | δ | P' |
|---|---|---|---|---|---|
| *n*-Butane | 7.1 | | Anthracene | 9.9 | |
| *n*-Hexane | 7.3 | 0 | Carbon disulfide | 10.0 | |
| *n*-Heptane | 7.5 | | Dioxane | 10.0 | 4.8 |
| *n*-Hexadecane | 8.0 | | Nitrobenzene | 10.0 | |
| Cyclohexane | 8.2 | 0 | *t*-Butyl alcohol | 10.6 | |
| Carbon tetrachloride | 8.6 | 1.7 | Benzoic acid | 11.3 | |
| Toluene | 8.9 | 2.3 | *n*-Propyl alcohol | 11.9 | 4.3 |
| Ethyl acetate | 9.1 | 4.3 | Ethyl alcohol | 13.0 | 5.2 |
| Benzene | 9.2 | | Acetic acid | 13.2 | |
| Chloroform | 9.3 | 4.4 | Iodine | 14.1 | |
| Aniline | 9.8 | | Methyl alcohol | 14.5 | 6.6 |
| Phenanthrene | 9.8 | | Glycerin | 16.5 | |
| Acetaldehyde | 9.9 | | Water | 23.4 | |
| Naphthalene | 9.9 | | Dichloromethane | | 3.4 |
| Acetone | | 5.4 | Acetonitrile | | 0.2 |
| | | | Tetrahydrofuran | | 4.2 |

in dipolar aprotic solvents. A small quantity of water can restore the given order [35].

During the solvation process a considerable amount of energy is liberated for the initial ionization, which is thus promoted by the solvation process [1].

Water seems to be the best solvent in fulfilling the requirements regarding the dielectric constant and solvating ability. It can solvate both anions and cations.

In $S_N2$ nucleophilic substitution reactions protic solvents with good anion-solvating properties will decrease the reaction rate [35]. However, a poorly solvating solvent will raise the energy of the reactants more than it raises that of the transition state. The solvation decreases with increasing division of charge over a larger surface resulting in an increase of the reaction rate. It appears that polar aprotic solvents which have a high dielectric constant but which lack hydrogen-bonding ability to solvate anions well have a large effect on the reaction rate [42,43]. Since the anions in polar aprotic solvents are much less effectively solvated, there is a dramatic increase in the reaction rate of $S_N2$ reactions going from a protic solvent to a polar aprotic solvent. For example, in Eq. 18 the reaction rate is six orders of magnitude higher in dimethylacetamide than in methanol.

$$Cl^- + CH_3I \xrightarrow[298\ K]{k_2} Cl\text{-}CH_3 + I^- \qquad (18)$$

In Eq. 19 the reaction rate is about five orders of magnitude higher in dimethylacetamide than in methanol.

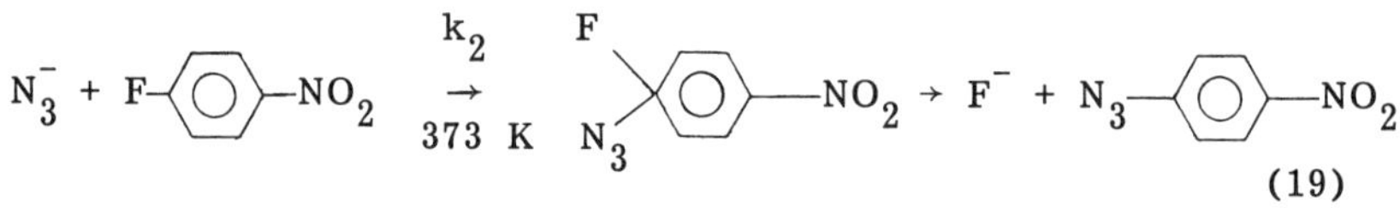

(19)

Usually dimethylsulfoxide and hexamethylphosphoramide (HMP) are used as solvents in order to accelerate $S_N2$ nucleophilic substitution reactions [3].

## Influence of Cation Solvation on Rate of Nucleophilic Substitution Reactions

The nucleophilic reactivity of anions depends not only on the solvation but also on the degree of association with the complementary cation, since a free anion is much more nucleophilic in nature than an anion bound in an IP. The smaller the ions in such an IP, the more stable the IP. The reactivity of anions in a poorly dissociating solvent depends on the nature of the added salt and, thus, on the cation [35].

It has been found that in the reaction of *N*-butyl-*p*-bromobenzolsulfonate with lithium or tetra-*N*-butylammonium halide in the poorly dissociating solvent acetone, the nucleophilic reactivity of the lithium salts decreases in the same order as with protic solvents:

$$I^- > Br^- > Cl^-$$

On the other hand, the decrease in reactivity with quaternary ammonium salts is of the same order as with polar aprotic solvents. The explanation for this phenomenon might be the fact that the lithium cation forms an IP whereas the $R_4N^+$ cation does not, due to the small attraction between the $R_4N^+$ cation and the anions [44].

Solvents with electron-donating properties will solvate cations effectively and therefore increase the degree of dissociation of an IP, since the electrostatic attraction between solvated ions is smaller than that between nonsolvated ions. Dissociation of the IP enhances the nucleophilicity of the anion and as a consequence the reaction rate of the nucleophilic substitution will increase. Indeed, it has been found that the addition of a small amount of a polar solvent to a nonpolar solvent like benzene increases the reaction rate [45].

### Influence of Ambivalent Anions on Reaction Rate

A special group of reagents is that with two active centers. When the active centers of such an anion react differently on solvent effects, it is possible to guide the reaction to the final product by choosing the right solvent. For example, with the alkylation of metal salts of phenols with alkyl halides in alcohols, ethers, or dimethylformamide, only O alkylation takes place. In a protic solvent the oxygen of the phenolate ion is solvated because it is the atom with the highest charge density. The nucleophilicity of the oxygen is strongly decreased and therefore C alkylation takes place [46].

### Influence of Solvation on Mechanism of Chemical Reaction

As mentioned before, the solvent can exert a profound effect on the mechanism by which a reaction takes place. Considering the solvolysis of a methyl halide, the reaction is more likely to proceed via the $S_N1$ rather than the $S_N2$ route with increasing polarity of the solvent. This is partially caused by the high dielectric constant of the solvent, promoting ionization and solvation of the separate ions. On the other hand, a number of reactions of *t*-butyl halides proceeding by an $S_N1$ mechanism in aqueous alcoholic solutions will show an $S_N2$ mechanism in acetone [2]. In choosing a solvent it is possible to guide a reaction via the $S_N1$ or $S_N2$ route [1].

## TEMPERATURE EFFECTS

The effect of temperature on reaction rate is given by the Arrhenius relation:

$$k = Ae^{-E_a/RT} \tag{20}$$

where k is the specific reaction rate, A is a constant called the frequency factor, $E_a$ is the activation energy, R is the gas constant, and T is the absolute temperature.

In the transition state theory [47] the constant A of Arrhenius is related to the entropy of activation of the transition state:

$$A = \nu e^{\Delta S^*/R} \tag{21}$$

where $\nu$ is the frequency with which the reaction (activated complex) goes to completion. In the collision state theory [47] $\nu$ is a good approximation of RT/Nh (N number of Avogadro, h constant of Planck). $\Delta S^*$ is the difference between the standard free entropy in the transition state and in the normal reactant state. For most practical purposes Eq. 22 is valid:

$$E_a = \Delta E^* \tag{22}$$

The influence of the temperature on the reaction rates in the transition state theory is given by Eq. 23:

$$k = (\nu e^{\Delta S^*/R})e^{-\Delta E^*/RT} \tag{23}$$

For many reactions a two- to threefold increase of the reaction rate is observed [47,48] at a temperature increase of 10 K. The rate may thus be increased by a factor of 30-240 in raising the temperature from 298 K to 348 K. However, the advantage of using higher temperatures may be limited by the introduction of complications. The derivatized compound should be stable at elevated temperatures to avoid the formation of thermal degradation products during the derivatization. This is particularly important in drug stability studies, especially in cases where the thermal degradation products formed in the derivatization and products formed during degradation of the parent compound are the same.

When the activation energy $E_a$ for the side product formation is considerably larger than that of the desired derivative, the side reactions will become more important at temperature increase. For example, when the values of the frequency factors A are $10^7$ and $10^{10}$ $sec^{-1}$ $M^{-1}$ and the activation energies are 41.8 and 62.7 kJ/mol for the desired and the side reactions, respectively, the extent of side product formation increases from 18 to 42% in going from 298 to 348 K [35]. Therefore, in practice the formation of thermal degradation products and the side product formation should be determined when working at elevated temperatures. This can be done by using an internal standard in LC or by a simple TLC monitoring of the derivatization reaction.

## CONCENTRATION AND STOICHIOMETRY EFFECTS

Important factors in optimizing the derivatization procedure are the stoichiometry of the reaction and the concentration of the compounds in the derivatization mixture. For the most widely applied irreversible second-order reaction:

$$\text{A (analyte)} + \text{R (reagent)} \longrightarrow \text{D (derivative)} \tag{24}$$

the concentration of the derivative d at a given time t is expressed by Eq. 25:

$$\frac{2303}{a - r} \log \frac{r(a - d)}{a(r - d)} = kt \tag{25}$$

or after rearrangement by Eq. 26:

$$k = \frac{2303}{t(a - r)} \log \frac{r(a - d)}{a(r - d)} \tag{26}$$

in which k represents the reaction rate constant and a and r the initial concentrations of the analyte and the reagent, respectively [47,49].

The time required for conversion of A to D to any percentage (e.g., 99%) can be calculated by solving Eq. 25 with respect to t. Equation 25 shows that the time needed for a certain percentage conversion of A to D decreases proportionally to the increase of k. This means that proper manipulation of the parameters (e.g., temperature, solvent) determining k gives a possibility to control the speed of the derivatization reaction.

The time required for the conversion of A to D is determined by the nonlinear relationship between a and r of Eq. 25. The time

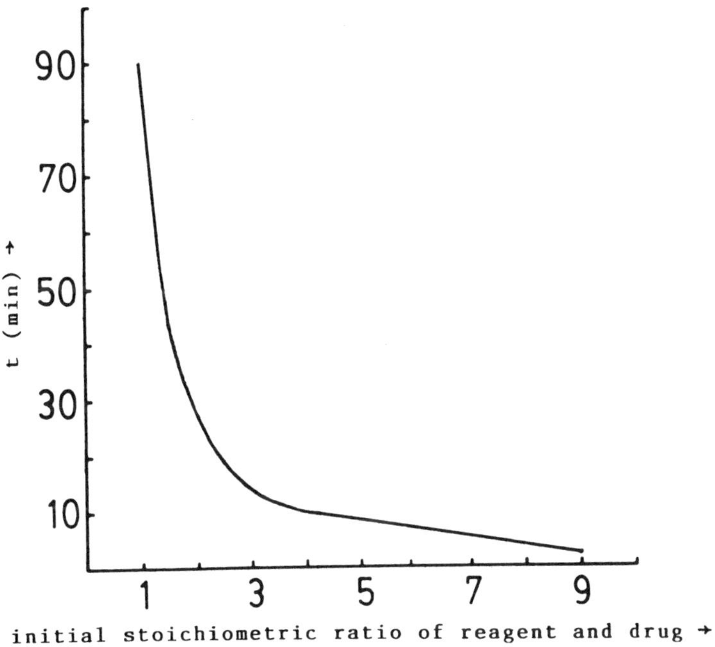


Figure 2.3 Reaction rate versus initial stoichiometric ratio of reagent and drug plot.

required for the reaction to be completed for 99% decreases markedly as the stoichiometric ratio is increased only from 1:1 to 1:3. The reaction takes 90 min at 1:1 versus 20 min at 1:3 which is shown in Fig. 2.3. However, this figure shows that a ratio higher than 10 does not yield much advantage in terms of the time needed for a 99% conversion. When a large excess of the reagent is used, the reaction will be pseudo-first-order with respect to the reagent. Therefore, as the ratio r/a becomes larger, $t_{99}$ decreases only linearly. It can be concluded that for practical purposes a ratio higher than 10 is not necessary. It may even be undesirable because of possible analytical problems such as resolution during the chromatographic analysis.

The absolute initial concentration of a drug and a reagent are important as well. As can be seen from Eq. 26, $t_{99}$ decreases as a and r decrease. According to the law of mass action, the rate of a reaction is proportional to the products of the molar concentration of the reactants each raised to a power equal to the number of molecules undergoing the reaction. For the reaction in Eq. 25 the reaction rate can be expressed as:

$$\text{Reaction rate} = k[a][r] \tag{27}$$

For instance, derivatizations in the μg/ml scale will be three orders of magnitude slower than those conducted on the mg/ml scale. Therefore, it is advisable to concentrate the samples prior to the derivatization reaction whenever possible.

Apart from the slower reaction rate in scaling down to, for instance, the nanogram range, problems may arise when an IP extraction must be performed before the derivatization. Knowledge of the various parameters involved in extraction reactions is therefore necessary. Contaminants also can create problems [50].

## SUBSTITUENT EFFECTS

Knowledge of substituent effects can be important in adapting reaction conditions for compounds related to a compound of which the reaction conditions are known. Substituent effects are also important in manipulating the reactivity of a reagent by changing its structure.

As was shown before, many derivatization reactions involve nucleophilic substitution reactions. With this type of reaction, analytes with an amine, hydroxyl, or carboxyl function can be derivatized [39,42,47]. If the drug is a carboxylic acid, then the reaction conditions are usually modified to form a carboxylate anion which is a far better nucleophile than the free acid. The reagent must be very susceptible to nucleophilic attack to achieve fast and quantitative reactions. Reagents with an unsaturation at the carbon atom in β position to a leaving group are very susceptible to nucleophilic attack. This is caused by the interaction between the attacking nucleophile and the $\pi$-electron system present in the reagent which lowers the activation energy leading to the transition state. For this reason, phenacyl, benzyl, and allyl compounds are frequently used in derivatization procedures. Usually carbonyl compounds are derivatized by an initial attack to a nucleophilic reagent followed by dehydration of the intermediate adduct. Nucleophilic attack will control the reaction rate in the presence of acid because the dehydration step is fast.

The nucleophilicity can be related to the basicity because the carbonyl atom bears a partial positive charge and the addition of the nucleophile resembles protonation. Reagents with a negative charge are always powerful nucleophiles. The reactivity of uncharged nucleophiles decreases in the order N > O > S [35].

The Hammett-type equation can frequently predict the influence of different substituents in the analyte or in the reagent:

$$\log \frac{k}{k'} = \sigma\rho \tag{28}$$

where k and k' are the rate constants of the substituted and nonsubstituted compound, respectively, $\sigma$ is the substituent constant, which depends only on the nature and position of the substituent, and $\rho$ is the Hammett constant, which depends on the reaction, the reaction conditions, and the nature of the side chain. The absolute value of $\rho$ shows the degree of influence of substituents on the system.

For benzoic acid $\rho$ is defined as 1. A value of $\rho$ over 1 indicates that the reaction rate is increased by substituents which withdraw negative charge from the reaction center. The Hammett equation still is empirical, although many attempts have been made to provide a good theoretical foundation. Log k/k' is proportional to the difference in energies of activation of substituted and unsubstituted compounds. Therefore, the Hammett equation is referred to as a "linear free-energy relationship."

Hammett suggested that reaction constants have the following form:

$$\rho = \left(\frac{\beta_1}{D} + \beta_2\right) RTd^2 \tag{29}$$

in which R is the gas constant, T is the temperature, D is the dielectric constant of the solvent, and d is the distance from the substituent to the reaction center. $\beta_1$ was assumed to depend on purely electrostatic interaction between the reacting benzene derivative and the medium. $\beta_2$ was assumed to measure the susceptibility of

Table 2.4 Reaction Constants Based on Hammett Constants

| Functional group | Solvent | T | $\rho$ | s | r | n | $-\log k^o$ | Ref. |
|---|---|---|---|---|---|---|---|---|
| *Esterification* | | | | | | | | |
| $ArCOOH + CH_3OH + H^+ \rightarrow ArCOOCH_3$ | Methanol | 298 | -0.229 | 0.141 | 0.457 | 16 | 3.841 | 52,53 |
| | | 313 | -0.216 | 0.138 | 0.444 | 16 | 3.352 | |
| | | 323 | -0.209 | 0.150 | 0.404 | 16 | 3.029 | |
| | | 333 | -0.213 | 0.148 | 0.429 | 15 | 2.738 | |
| $ArCOOH + C_6H_{11}OH + H^+ \rightarrow ArCOOC_6H_{11}$ | Cyclohexanol | 328 | 0.555 | 0.089 | 0.875 | 14 | 4.826 | 54 |
| | | 338 | 0.518 | 0.091 | 0.855 | 14 | 4.432 | |
| | | 348 | 0.498 | 0.109 | 0.798 | 14 | 4.100 | |
| | | 358 | 0.479 | 0.089 | 0.842 | 14 | 3.772 | |
| *Alkylation with alkyl halides* | | | | | | | | |
| $Ar^1Ar^2CHCl + CH_3OH \rightarrow Ar^1Ar^2CHOCH_3$ | Methanol | 298 | -4.072 | 0.507 | 0.932 | 5 | 2.792 | 55 |
| $Ar(C_6H_5)CHCl + C_2H_5OH \rightarrow Ar(C_6H_5)CHOC_2H_5$ | Ethanol | 298 | -5.090 | 0.330 | 0.957 | 10 | 2.062 | 56-60 |
| $Ar^1Ar^2CHCl + (CH_3)_2CHOH \rightarrow Ar^1Ar^2CHOCH(CH_3)_2$ | Isopropanol | 298 | -2.755 | 0.570 | 0.651 | 6 | 4.621 | 55 |

| Reaction | Solvent | T | ρ | s | r | n | $k^o$ | Ref. |
|---|---|---|---|---|---|---|---|---|
| $Ar(C_6H_5)_2CCl + C_2H_5OH \longrightarrow Ar(C_6H_5)_2COC_2H_5$ | Methanol/ether (6:4) | 273 | -3.923 | 0.256 | 0.960 | 8 | 3.608 | 56-60 |
| *Acylation with carbonyl halides* | | | | | | | | |
| $ArCOCl + CH_3OH \longrightarrow ArCOCOOCH_3$ | Methanol | 273 | 1.469 | 0.097 | 0.982 | 10 | 3.362 | 61 |
| $ArCOCl + C_2H_5OH \longrightarrow ArCOOC_2H_5$ | Ethanol | 273 | 1.542 | 0.145 | 0.960 | 13 | 4.074 | 61,62 |
| | Ethanol/ether (6:4) | 273 | 1.922 | 0.070 | 0.994 | 8 | 3.997 | 63 |
| | | 289 | 1.499 | 0.086 | 0.987 | 8 | 3.969 | |
| $ArCOCl + (CH_3)_2CHOH \longrightarrow ArCOOCH(CH_3)_2$ | Isopropanol | 298 | 1.280 | 0.061 | 0.991 | 6 | 4.271 | 64 |
| $ArCOCl + C_6H_5NH_2 \longrightarrow ArCONHC_6H_5$ | Benzene | 298 | 1.219 | 0.051 | 0.997 | 4 | 2.980 | 65 |
| *Acylation with sulfonyl chlorides* | | | | | | | | |
| $ArSO_2Cl + C_6H_6 + AlCl_3 \longrightarrow ArSO_2C_6H_5$ | Benzene | 303 | -1.802 | 0.119 | 0.981 | 6 | 4.314 | 66 |

Temperature (K); ρ, reaction constant; s, standard error; r, correlation coefficient; n, number of compounds; $k^o$, velocity constant (observed); Ar, aromatic nucleus.

the reaction to changes in the charge density at the reaction site.

Hammett suggested further that $\rho$ depends on the electrical effects of the substituents [51].

## Reaction Constants

As mentioned before, the Hammett constant ($\rho$) shows the influence of the substituents on the system. Three factors are important for this influence:

The transmission of electrical effects to the reaction site
The influence of changes of electron density at the reaction site
The effect of reaction conditions [51]

Existing data on reaction constants are listed in Table 2.4. These reaction constants are based on rate measurements.

Since $\rho$ is an expression of the electrical effect of substituents, it must depend on the effectiveness with which the side chain can transmit the electrical effect of the substituent to the reaction site. Therefore, Hammett included the factor $1/d^2$ in Eq. 29 in order to relate the magnitude of the reaction constant with the distance between the substituent and the reaction site. Hammett found that the $1/d^2$ factor is only applicable when the aromatic nucleus and side chain are conjugated [51]. It has been found that the polarizability of M has a decisive influence on the magnitude of the reaction constant [67].

The electron density at the reaction site has an effect on the reaction rate. A positive $\rho$ indicates that the reaction is facilitated by low electron density at the reaction site: a negative $\rho$ indicates that a high electron density at the reaction site increases the reaction rate [51].

It appears that Eq. 29 usually expresses the relation between $\rho$ and T quite well. However, in some cases the reaction constants increase with temperature. As shown in Eq. 29, the reaction constants increase with decreasing dielectric constant. However, again there are exceptions. By changing the solvents, large variations in behavior are observed. No general rules can be given for the relation between the increasing solvent polarities and reaction constants [51].

## Substituent Constants

The substituent constants used to be defined by Eq. 28. Assuming $\rho$ to be 1 for benzoic acid, this equation can be transformed to:

$$\log \frac{k_b}{k_b'} = \sigma \qquad (30)$$

with $k_b$ and $k_b'$ as the rate constants for benzoic acid and its monosubstituted derivative, respectively. This definition has been abandoned. Usually the value of $\sigma$ is defined as that which best fits the entire body of experimental data [51].

*Deviations from the Hammett Equation*

Some deviations from the Hammett equation can be explained by polarizability effects. In this case the Hammett equation can be modified in the following way:

$$\log \frac{k}{k'} = (\sigma + \sigma' f(\rho))\rho \qquad (31)$$

where $\sigma'$ expresses the polarizability effects and vanishes when these effects are not important. According to Hammett there is no theoretical reason why substituent constants should depend on the temperature [68] and such a dependence has not been observed. However, sometimes substituent constants depend on the solvent, e.g., with increasing alcohol concentration in ethanol-water mixtures the substituent constants of benzoic acid decrease [51].

Except for deviations related to experimental errors and reaction conditions, deviations might arise from the approximation inherent in the Hammett equation. The assumptions of the independence of substituent constants of the nature of the side chain, the reaction, and the solvent are only approximations. An underlying assumption of the Hammett equation is that all members of a given series undergo a reaction by the same mechanism. A change in mechanism can be detected by a change in the kinetic order of a reaction. No comparison can be made between rate constants from different orders, as they have different dimensions. However, reaction series are usually studied under reaction conditions, which lead to pseudo-first-order kinetics, in this case changes in mechanism need not affect the kinetic order.

Unfortunately, the reactions of the alkyl and acyl halides as well as the esterifications do not appear to fit the Hammett equation very well. The reactions involve bond-breaking and bond-forming steps, and probably neither step alone is rate determining [51].

It has been shown that the effect of multiple substituents on rate constants is the sum of the effects of the individual substituents. Deviation from additivity can be caused by steric inhibition of resonancy but is rare. The effect of multiple substitution can be expressed in the following way [51]:

$$\log \frac{k}{k'} = \rho\Sigma\sigma \qquad (32)$$

### Taft Equation

In many cases, e.g., the formation of hydrazones, oximes, semicarbazones, *o*-substituted compounds, and aliphatic compounds, the Hammett equation is not valid. For aliphatic and *o*-substituted aromatic compounds the Taft equation can be used:

$$\log \frac{k}{k'} = \rho^* \sigma^* \tag{33}$$

This equation is only valid for a restricted class of reaction series. The fluorine, amino group, and hydrogen must be excluded from the ortho substituents.

The definition of $\sigma^*$ (polar substituent constant) differs markedly from the definition of $\sigma$. The values of $\sigma^*$ are based on the acid and based-catalyzed hydrolysis of ethyl ethers of acetic acid (of the type $R\text{-}CH_2\text{-}COOC_2H_5$) in the following way:

$$\sigma^* = \frac{1}{2.5}\left[\log\left(\frac{k}{k'}\right)_b - \log\left(\frac{k}{k'}\right)_a\right] \tag{34}$$

where k and k' are the hydrolysis rate constants of the substituted and nonsubstituted esters, respectively. a and b indicate the acid- and base-catalyzed reaction. The factor 1/2.5 is introduced to correlate the values of $\sigma$ and $\sigma^*$ with each other.

The Hammett and Taft equations show that the reactivity of the reagent can be manipulated in a predictable way by altering its structure. This might lead to a more rapid derivatization reaction under suitable reaction conditions, a higher derivatization yield, better chromatographic properties, or an enhanced detectability.

## CONCLUSIONS

In this chapter the important derivatization reactions in LC as well as the factors that may affect the reaction rate are discussed. Derivatization can be performed by different alkylation methods. All alkylation methods replace active hydrogens. In nonspecific reactions such as alkylation and acylation, all compounds with a reactive hydrogen can be derivatized. In specific reactions, e.g., cyclization and condensation, only a specific functional group will react.

Diazoalkane alkylation is a smooth and rapid procedure, but diazoalkanes are highly toxic. The derivatization does not always require heat, different functional groups can be alkylated at the same time, and derivatives with strong electron-capturing properties are produced. However, relatively long reaction times are required.

Alkylation with alkyl halides is often performed. The reaction takes place in the presence of carbonate. It can be accelerated and the yield increased by the addition of a crown ether. The carbonate-catalyzed alkylation is attractive because it is rapid and a wide range of derivatives can be prepared. A number of other bases in combination with various polar solvents can also be used.

Extractive alkylation has the advantage of combining extraction and derivatization in one step. The reaction conditions are mild, the reagents are not dangerous, and the procedure is usually more rapid than comparable base-catalyzed alkylation procedures.

For compounds which are extremely sensitive to hydrolysis, alkylation can be performed by solid phase transfer catalysis.

Acylation is a smooth and rapid procedure with readily available reagents. However, acylating reagents and acyl derivatives are often unstable.

The solvent has a large influence on the rate and mechanism of a reaction. If in nucleophilic substitution reactions ions are formed in the transition state or existing charges are concentrated on a smaller surface, the reaction rate will be higher in protic than aprotic solvents. If charges are neutralized or divided over larger surfaces, the reaction will proceed faster in polar aprotic solvents than in protic solvents. In choosing a solvent, it is also possible to guide a reaction via the $S_N1$ or $S_N2$ route.

The temperature is another important factor in derivatization reactions. At a 10° increase in temperature, most reactions show a two- to threefold increase in reaction rate. However, the derivative may be unstable at elevated temperatures and side product formation may be increased.

The stoichiometric ratio of reagent to drug should not be higher than 10. Higher ratios do not give much advantage in terms of the time needed for a complete conversion, whereas a large excess of reagent can cause analytical problems. Concentrating prior to derivatization is advantageous due to the increase in reaction rate in concentrated solutions of analyte and reagents.

Reaction rates can be increased by choosing the right substituent. The increase in reaction rate can be calculated by the equations of Hammett and Taft. However, not all derivatizations fit the equations well.

## REFERENCES

1. Sykes, P. (ed.), *A Guidebook to Mechanism in Organic Chemistry*, 5th ed., Longmans, London, 1963.
2. Gould, E. S. (ed.), *Mechanism and Structure in Organic Chemistry*, Holt, Rinehart & Winston, New York, 1959.
3. Roberts, J. D., R. Stewart, and M. C. Casserio (eds.), *Organic Chemistry: Methane to Macromolecules*, 7th ed., W. A. Benjamin, Menco Park, 1971.

4. Hulshoff, A., and A. D. Förch, *J. Chromatogr.*, *220*:275 (1981).
5. Jefferson, W. E., and F. C. Chang, *Anal. Lett.*, *9*:429 (1976).
6. Fitzpatrick, F. A., M. A. Wynalda, and D. G. Kaiser, *Anal. Chem.*, *49*:1032 (1977).
7. Nicholson, J. D., *Analyst*, *102*:1, 193 (1978).
8. Hulshoff, A., and H. Lingeman, *J. Pharm. Biomed. Anal.*, *2*:337 (1984).
9. Dünges, W., and N. Seiler, *J. Chromatogr.*, *145*:483 (1978).
10. Dünges, W. (ed.), *Prae-Chromatographische Mikromethoden*, Hüthig Verlag, Heidelberg, 1979.
11. Thio, A. P., M. J. Kornet, H. S. I. Tan, and D. H. Tompkins, *Anal. Lett.*, *12*:1009 (1979).
12. Knapp, D. R. (ed.), *Handbook of Analytical Derivatization Reactions*, Wiley-Interscience, New York, 1979.
13. Greeley, R. H., *Clin. Chem.*, *20*:192 (1974).
14. Roseboom, H., and A. Hulshoff, *J. Chromatogr.*, *173*:65 (1979).
15. Kapetanovic, I. M., J. S. Dutcher, and J. M. Strong, *Anal. Chem.*, *49*:1843 (1977).
16. Lam, S., and E. Grushka, *J. Liq. Chromatogr.*, *1*:33 (1978).
17. Masso, P. D., and P. A. Kramer, *J. Chromatogr.*, *224*:457 (1981).
18. Adinolfe, N. A., and M. K. L. Bicking, *J. Chromatogr. Sci.*, *23*:407 (1985).
19. Rosenfeld, J. M., and J. L. Crocco, *Anal. Chem.*, *50*:701 (1978).
20. Rosenfeld, J. M., J. L. Crocco, and T. L. Ling, *Anal. Lett.*, *13*:283 (1980).
21. Bicking, M. K. L., and N. A. Adinolfe, *J. Chromatogr. Sci.*, *23*:348 (1985).
22. Arbin, A., *J. Chromatogr.*, *144*:85 (1977).
23. Arbin, A., H. Brink, and J. Vessman, *J. Chromatogr.*, *170*:25 (1979).
24. Arbin, A., H. Brink, and J. Vessman, *J. Chromatogr.*, *196*:255 (1980).
25. Lingeman, H., *Pharm. Weekbl., Sci. Ed.*, *9*:183 (1987).
26. Lieder, P. H., and R. F. Borch, *Anal. Lett.*, *18*:57 (1985).
27. Wachowiak, R., and A. K. Connors, *Anal. Chem.*, *51*:27 (1978).
28. Rowe, E. L., and S. M. Machkovic, *J. Pharm. Sci.*, *66*:273 (1977).
29. Borch, R. F., J. H. Markovitz, and M. E. Pleasants, *Anal. Lett.*, *12*:917 (1979).
30. Nicholson, J. D., *Analyst*, *102*:193, 1 (1978).

31. Morozowich, W., and M. J. Cho, in *GLC and HPLC Determination of Therapeutic Agents*, Part I (K. Tsuji and W. Morozowich, eds.), Marcel Dekker, New York, 1978.
32. Lingeman, H., W. J. M. Underberg, A. Takadate, and A. Hulshoff, *J. Liq. Chromatogr.*, *5*:789 (1985).
33. Lawrence, J. F., in *Chemical Derivatization in Analytical Chemistry*, Vol. 2 (R. W. Frei and J. F. Lawrence, eds.), Plenum Press, New York, 1982.
34. Barends, D. M., J. S. F. Sandt, and A. Hulshoff, *J. Chromatogr.*, *182*:201 (1980).
35. Reichardt, *Lösungsmitteleffekten in der Organischen Chemie*, Verlag Chemie, Weinheim, 1969.
36. Kaufmann, E. D. (ed.), *Advanced Concepts in Physical Chemistry*, McGraw-Hill, New York, 1966.
37. Amis, E. S. (ed.), *Solvent Effects on Reaction Rates and Mechanisms*, Academic Press, New York, 1966.
38. Connors, K. A. (ed.), *Reaction Mechanism in Organic Analytical Chemistry*, Wiley-Interscience, New York, 1973.
39. Hendrickson, J. B., D. J. Cram, and G. S. Hammond (eds.), *Organic Chemistry*, 3rd ed., McGraw-Hill, New York, 1970.
40. Pocker, Y., in *Progress in Reaction Kinetics*, Vol. 1 (G. Porter and B. Stevens, eds.), Pergamon Press, Ithaca, 1953.
41. Ingold, C. K. (ed.), *Structure and Mechanism in Organic Chemistry*, Cornell University Press, Ithaca, 1953.
42. Durst, H., M. Milano, D. J. Kikta, S. A. Conney, and S. Grushka, *Anal. Chem.*, *47*:1797 (1975).
43. Parker, A. J., *Chem. Rev.*, *69*:1 (1969).
44. Winstein, S., L. G. Sauedoff, S. Smith, I. D. R. Stevens, and J. S. Gall, *Tetrahedron Lett.*, *9*:24 (1960).
45. H. E. Zaugg, B. W. Horron, and S. Borgwardt, *J. Am. Chem. Soc.*, *82*:2895 (1960).
46. Kornblum, N., P. J. Berrigan, and W. J. Le Noble, *J. Am. Chem. Soc.*, *85*:1141 (1963).
47. Martin, A. N., J. Swarbrick, and A. Cammarata (eds.), *Physical Pharmacy*, 2nd ed., Lea & Febiger, Philadelphia, 1969.
48. Ericsen, S. P., and H. Stemach, *J. Pharm. Sci.*, *54*:1029 (1961).
49. Frost, A. A., and R. G. Pearson (eds.), *Kinetics and Mechanism*, 2nd ed., John Wiley and Sons, New York, 1961.
50. Greving, J. E., J. H. G. Jonkman, and R. A. de Zeeuw, *J. Chromatogr.*, *148*:389 (1978).
51. Jaffe, H. H., *Chem. Rev.*, *53*:191 (1953).
52. Hartman, R. J., and A. J. M. Borders, *J. Am. Chem. Soc.*, *59*:2107 (1937).
53. Hartman, R. J., and A. G. Gassmann, *J. Am. Chem. Soc.*, *62*:1599 (1940).

54. Hartman, R. J., H. M. Hoogsteen, and J. A. Noede, *J. Am. Chem. Soc.*, *66*:1714 (1944).
55. Altscher, S., R. Baltzly, and S. W. Blackman, *J. Am. Chem. Soc.*, *74*:3649 (1952).
56. Fariacci, N. T., and L. P. Hammett, *J. Am. Chem. Soc.*, *59*:2542 (1937).
57. Hughes, E. D., C. K. Ingold, and N. A. Taher, *J. Chem. Soc.*, *62*:1949 (1940).
58. Norris, J. F., and C. Banta, *J. Am. Chem. Soc.*, *50*:4804 (1928).
59. Norris, J. F., and J. T. Blake, *J. Am. Chem. Soc.*, *50*:1808 (1928).
60. Ward, A. M., *J. Chem. Soc.*, *49*:2285 (1927).
61. Norris, J. F., and H. H. Young, Jr., *J. Am. Chem. Soc.*, *57*:1420 (1935).
62. Norris, J. F., and W. H. Strain, *J. Am. Chem. Soc.*, *57*:187 (1935).
63. Branch, G. E. K., and A. C. Nixon, *J. Am. Chem. Soc.*, *58*:2499 (1936).
64. Norris, J. F., and D. V. Gregory, *J. Am. Chem. Soc.*, *50*: 1813 (1928).
65. Williams, E. G., and C. N. Hinshelwood, *J. Chem. Soc.*, *56*: 1079 (1934).
66. Olivier, S. C. J., *Rec. Trav. Chim.*, *33*:244 (1914).
67. Jaffe, H. H., L. D. Freedman, and G. D. Noak, *J. Am. Chem. Soc.*, *75*:2209 (1953).
68. Hammett, L. P., *J. Am. Chem. Soc.*, *59*:916 (1937).

# 3

# Sample Pretreatment Procedures

HENK LINGEMAN* and UBBO R. TJADEN *Leiden University, Leiden, The Netherlands*

## INTRODUCTION

In order to achieve accurate and reliable liquid chromatographic (LC) data, relatively "pure" samples must be analyzed. Therefore, sample pretreatment is an essential part of every chromatographic procedure.

The main reasons for the implementation of a sample pretreatment procedure in an analytical method are:

- To improve accuracy
- To increase the column's lifetime (e.g., by removal of solid contaminants
- To increase reproducibility
- To improve detectability (e.g., by concentration of the analyte)
- To improve precision
- To improve selectivity (e.g., by removal of interfering matrix components)
- To increase stability

Fitting the derivatization step into a complete (bio)analytical procedure in a harmonious way is a major analytical challenge. Aqueous biological matrices such as plasma or serum, urine, tissue homogenates, and saliva contain a multitude of endogenous compounds, e.g., proteins in plasma samples and fatty acids in urine samples, which are often present in much higher concentrations than the analytes or

**Present affiliation*: Free University, Amsterdam, The Netherlands

their metabolites [1]. Many of these endogenous compounds have reactive functional groups (e.g., the carboxylic acid function of amino acids and fatty acids) which can participate in the derivatization reactions and can interfere with the analysis of the compounds of interest, if not sufficiently separated by the ensuing chromatographic procedure(s). Furthermore, most derivatization reactions are best performed in nonaqueous solvents and the presence of the anhydrous medium may even be obligatory. The injection of nonaqueous samples into certain LC systems, e.g., reversed phase (RP) chromatography, is in some cases not compatible with the applied system and in other systems the injection of (semi)aqueous samples, e.g., normal phase (NP) chromatography, may be troublesome. These factors necessitate some form of sample clean-up prior to or in conjunction with the derivatization reaction and the LC system.

For different types of samples different strategies must be used. In bioanalysis, with sample volumes of 0.1-1 ml, the majority of the procedures applied nowadays are executed in an off-line mode with the LC analysis, using, for instance, disposable cartridges. For the pretreatment of environmental samples, with sample volumes of 10-1000 ml, usually on-line procedures are preferred. The sample size strongly depends on the amount of available sample (e.g., limited in bioanalysis) as well as the combination of the analyte concentration in the sample and the limits of determination (e.g., rare earth metals in wastewater). Although for the analysis of large samples on-line procedures are to be preferred, in terms of reproducibility off-line systems can also be used.

For the pretreatment of pharmaceutical formulations, the strategy will be quite different from the strategy for biological samples. In pharmaceutical preparations (e.g., creams, syrups, ointments, tablets), in principle one single analyte in relatively high concentration should be analyzed in a matrix consisting of a limited number of components [2,3]. Contrarily to the biological samples, the majority of the pretreatment procedures for pharmaceutical formulations are performed with simple techniques in an off-line mode with the LC analysis. However, the basic principles and techniques for sample clean-up and sample concentration for most of the matrices will be the same.

In developing a suitable sample pretreatment procedure a number of features should be considered, such as:

The physicochemical properties of the analyte
The chemical composition of the sample
The degree of drug-protein binding
The type of interferences from the matrix
The stability of the analyte during sample pretreatment
The presence of contaminants from the vials used during collection, storage, or pretreatment

The recovery, which should be (nearly) quantitative
The solvent applied in the final step, which should be compatible with the chosen LC system
The simplicity of the sample pretreatment procedure
The reproducibility, accuracy, and precision of the sample pretreatment procedure, which should be as high as possible
The enrichment (concentration) of the analyte in the final step

Of course, it is almost never possible to obtain the ideal conditions for the sample pretreatment. The first five features listed above are physicochemical parameters that cannot be changed during the analysis, and the rest of the features are the variables in the sample pretreatment procedure that should be optimized in order to obtain the most suitable conditions for the sample pretreatment and the subsequent derivatization and chromatography.

However, until a few years ago no sensational progress had been made in the development of sample pretreatment procedures. Most of the applied techniques are still laborious and time consuming, lacking accuracy and precision. The new developments of recent years in sample clean-up methods are closely related to the development of new stationary phases in LC, which allows the direct injection of, for instance, plasma samples, the introduction of automated sample processors, and the application of robotic systems.

One of the future trends in LC will be the miniaturization of the systems, but miniaturized LC in bioanalysis will not be the first way to go. Therefore, automated precolumn switching systems which can be used for the preconcentration and clean-up of plasma, serum, or uring samples in combination with microbore or narrow-bore technology will not be discussed in this book [4-6].

Because the necessary pretreatment for analysis of organic compounds in a biological matrix, being dirty and complex, is one of the most troublesome to perform, the emphasis in this section will be on the sample clean-up and sample enrichment procedures for analytes in biological matrices [7]. The degree of sample preparation strongly depends on the concentration of the analyte, the composition of the matrix, the number of samples, and the chosen chromatographic/detection system.

Because of the fact that the degree of sample clean-up depends on the matrix under investigation, the choice of the matrix can be very important, e.g., in bioanalysis where sometimes a choice can be made between the analysis of plasma or urine samples. The advantages of analyzing urine samples instead of plasma are the following:

Availability of (almost) unlimited amounts of uring (contrarily to plasma samples)
Higher urine concentrations for a great number of analytes

Lower patient load
Relatively low protein content of urine samples

However, for drug monitoring and pharmacokinetic studies analysis of plasma samples is often preferred over analysis of urine samples. The most important disadvantage of urinalysis is the great number of components (over 600), with molecular weights ranging from about 17 up to 70,000, which may interfere with the selective and sensitive analysis of the solute.

Sometimes analysis of saliva can replace the analysis of plasma samples, which is especially advantageous when studies in children should be performed. For a number of drugs it has been shown that a constant ratio between plasma and saliva concentrations exists. The amount of analyte excreted in saliva depends on the degree of protonation and the extent of protein binding, meaning that neutral non-protein-binding analytes are found in relatively high concentrations in the saliva.

## STRATEGIES FOR SAMPLE PRETREATMENT

Before starting a sample pretreatment procedure, the necessary data about the physicochemical properties of the analyte, the stability of the analyte in the matrix, the storage of the samples, the procedures of sampling, the used vials, the presence of interfering components, the chosen LC and detection technique, and so on, must be gathered and studied (see Introduction of this chapter).

To start a sample pretreatment procedure in order to analyze the compound of interest (analyte, solute) in a protein-rich matrix (e.g., plasma, feces, saliva), the degree of drug-protein binding should also be checked [8], since it can never be assumed that the applied sample pretreatment procedure (e.g., extraction, precipitation) denaturates the proteins quantitatively.

The easiest way to destroy the binding of the solutes with proteins is to dilute the sample with a physiological salt solution or with a modifier containing LC eluent. For instance, for isoxazolyl penicillins, with a binding percentage of over 95%, a dilution with only 9 volumes of 0.9% (w/v) sodium chloride solution is sufficient [9].

Protein denaturation is very important. The presence of proteins on the analytical LC column causes a rapid deterioration of the column. In pharmacological and pharmacokinetic experiments sometimes the total (bound and free) concentration and sometimes the free fraction of the analyte should be measured [8].

Determination of the degree of protein binding can be performed with dialysis or ultrafiltration techniques [10,11].

### *Denaturation*

Denaturation of the samples is important to avoid artifact formation during analysis because of unwanted metabolism of the analytes. The procedures usually applied for stopping enzyme activity in enzyme-containing matrices are rapid cooling in liquid nitrogen, microwave denaturation, homogenization/precipitation procedures, the addition of enzyme inhibitors, and boiling of the samples [12,13].

### *Freezing*

Freezing of the samples to block the enzyme activity or to store the samples should be executed carefully. During the freezing process of some analytes (e.g., mitomycin C), sample losses can occur due to crystallization of the analyte as large particles, which do not redissolve after thawing (microwave irradiation) the samples. The explanation is that during the freezing process a part of the water freezes out and the analyte concentration of the remaining solution rises until crystallization occurs. This means that samples should be frozen very rapidly at much lower temperatures than 250 K, for instance at 200 K [14].

### *Homogenization*

A homogenization procedure may be necessary for samples containing insoluble structures (e.g., tissues, feces) containing, for example, insoluble proteins. Treatment of the sample with 1 mol/liter hydrochloric acid or methanol in a blade homogenizer is normally sufficient to solubilize the sample. In the case of gelatinous samples (e.g., seminal fluids), the required solubilization can be obtained by sonification of the sample.

### *Solubilization*

Solubilization of tissue samples can also be achieved by the addition of an enzyme (e.g., subtilisin A) [15]. However, a limitation of this technique may be the introduction of a relatively large amount of enzyme and the high optimal pH (>8), which may interfere with the chromatographic analysis or the derivatization.

Because of the great diversity of interfering compounds, e.g., urea, electrolytes, proteins, hormones, lipids, drugs, in biological samples a considerable number of sample pretreatment procedures and protein denaturation procedures exist:

Chemical modification techniques
Direct injections on special sorbents
Microwave denaturations
Lyophilization techniques

Hydrolysis reactions
Centrifugation (precipitation) procedures
Filtration techniques
Electrophoresis
Head space techniques
Liquid-liquid extractions
Solid phase isolations
Derivatization procedures

Unfortunately, the analysis of samples without a pretreatment procedure is normally only possible when a radioimmunoassay is applied and not whenever a LC procedure is used. Most of the listed techniques can be executed in an off-line or on-line mode with the chromatographic separation/derivatization and can be performed in an automated or nonautomated procedure.

The most important methods will be discussed in the following sections.

### Direct Analysis of Protein-Containing Samples

Normally protein-containing samples cannot be injected directly onto a LC system because accumulation of the proteins deteriorates the LC column and/or the frits by clogging the pores [16]. This means that applying a postchromatographic derivatization procedure before the LC separation, a sample clean-up should be executed. Using a prechromatographic derivatization technique, it is usually also advisable to remove the interfering proteins before the derivatization takes place. Therefore, for the analysis of protein-containing samples a number of procedures are applied such as precipitation of the proteins, isolation of the analyte with a (disposable) solid phase cartridge, and the application of a precolumn in a column-switching device [17-19].

One of the newer techniques is the application of internal surface reversed phase materials, the so-called Pinkerton phase (Fig. 1.3) [20]. It is a silica-based material with small pores (5.3 nm) which is subsequently chemically modified with a glycerylpropyl phase and a polypeptide. Subsequently, this phase is treated with a carboxypeptidase. The result is a stationary phase material with a glycerylpropyl structure outside the pores (hydrophilic exterior surface) and a peptide structure within the pores (hydrophobic interior surface). The injected serum proteins are not retained by the glycerylpropyl phase and are unable to enter the pores. The analytes, however, can enter the pores and are able to interact with the peptide phase, resulting in a selective retaining of the solute molecules.

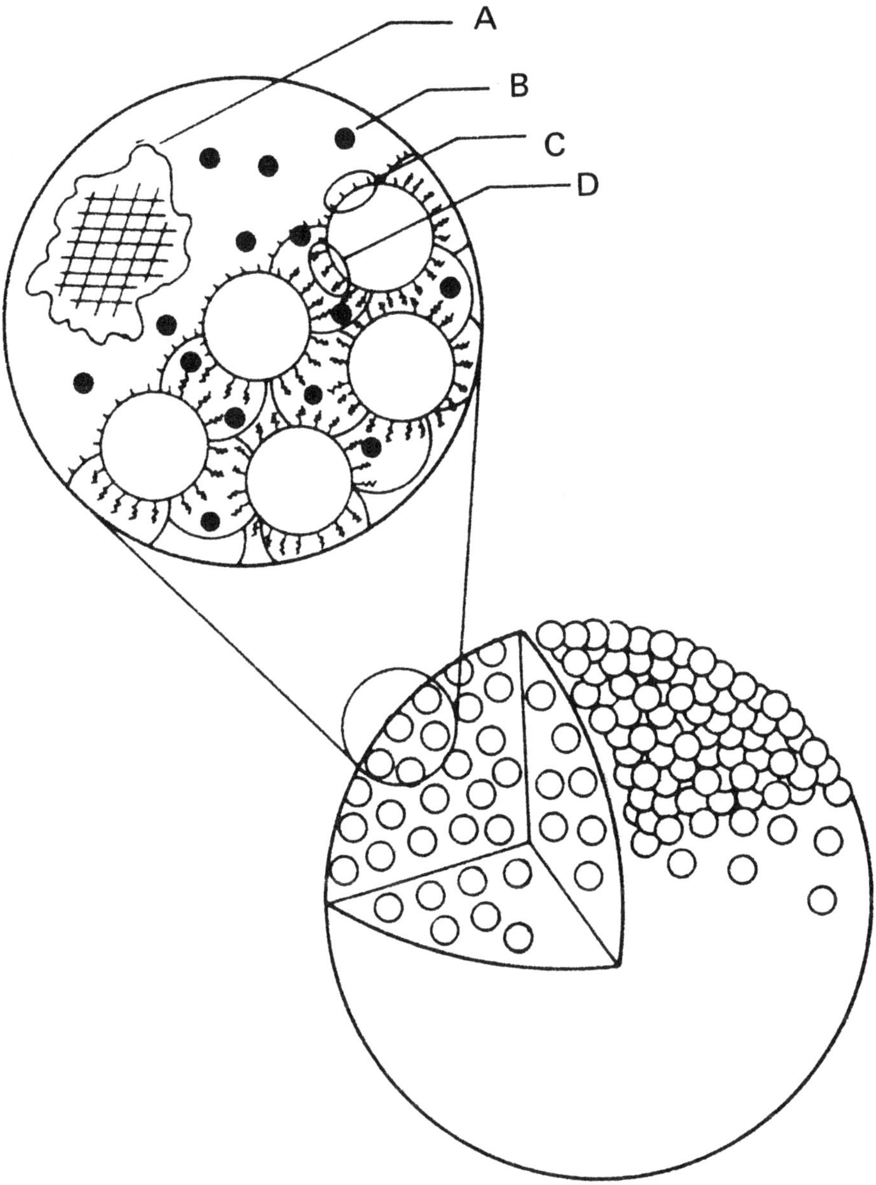


Figure 3.1 Structure of Internal Surface Reversed-Phase Silica Support ("Pinkerton-Phase"). (A) protein, (B) analyte, (C) hydrophilic glycerylpropyl bonded external phase, and (D) hydrophobic internal partitioning phase. (From ref. 20.)

This stationary phase material can be used for acidic as well as basic compounds. The sorbent is used for the LC analysis of phenytoin-containing plasma samples by the direct injection of the samples onto a precolumn. The actual analysis is executed with a $C_{18}$ analytical column [20]. However, normally column-switching is applied (see subsection on "(Semi)-Automated Sample Pretreatment Procedures" in this chapter). When the phase is used in a guard column (see Chapter 1) of 10 × 3.0 mm I.D. packed with 5-μm spherical particles, about 1000 serum injections can be made before the column should be replaced.

One of the possibilities for direct analysis of analytes in plasma omitting a denaturation procedure can be executed by treating a $C_{18}$ column with human plasma. After this treatment the column no longer adsorbs plasma proteins from an aqueous solution, but it still shows hydrophobic characteristics for smaller analytes [21].

### Off-Line Sample Pretreatment Procedures

The simplest approach when some form of sample clean-up is necessary is protein removal from the biological matrix by microwave denaturation, lyophilization, (ultra)filtration, hydrolysis, or by precipitation reactions [22-25]. If afterward the analyte has to be transferred to a nonaqueous medium, i.e., because the derivatization can only be performed in an organic solvent, freeze-drying is a possibility (Fig. 3.2). However, when the analyte must be determined in a more complex matrix, more advanced sample pretreatment procedures such as liquid-liquid extractions, solid phase isolations, or combinations of these techniques should be applied.

(Vacuum) distillation techniques [27], dehydration methods [2], and Soxhlet extraction procedures [28] are not very popular in bioanalysis and are therefore not discussed in this chapter.

### Chemical Modification Techniques

Chemical modification techniques are an interesting group of sample pretreatment techniques [29]. Although, in general the strategy of sample preparation is to remove interfering compounds to enhance the sensitivity and/or the selectivity of an analytical procedure, this selectivity can also be enhanced by a number of physical or chemical manipulations such as pH changes, oxidation-reduction reactions, hydrolytic reactions, irradiations, and chemical derivatizations.

In a mixture of carboxylic acids, for instance, one group of acids is selectively oxidized and subsequently selectively extracted. A second possibility is that a fluorescent analyte is chromatographed and after postchromatographic pH adjustment (fluorescence quenching), the resulting mixture is chromatographed again. The difference between the two signals is a measure for the concentration.

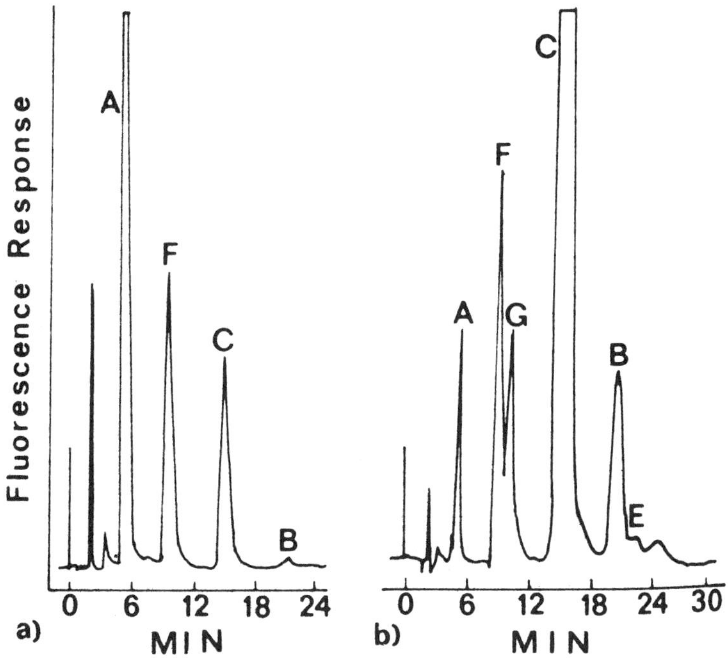


Figure 3.2 Chromatograms obtained after normal-phase LC of warfarin metabolites in plasma and urine samples after extraction and freeze-drying. (A) warfarin, (B) 6-OH-warfarin, (C) 7-OH-warfarin, (D) 8-OH-warfarin, (E) 4'-OH-warfarin, (F) (R,S)- or (S,R)-warfarin alcohol, and (G) (R,R)- or (S,S)-warfarin alcohol: a) plasma, and b) urine. (From ref. 26.)

A third procedure is the selective covalent labeling of, say, primary amines followed by a selective isolation of the derivatives. The covalent labelings procedure (chemical derivatization reaction) can also be useful when unstable solutes must be isolated [30].

*Microwave Denaturations*

Microwave irradiation as a solubilization method is a technique in which energy (heat) can be used for the extraction of a number of compounds from soil, seeds, foods, and feeds. The samples are mixed with an appropriate solvent (e.g., methanol, hexane) and irradiated for 30 sec. After cooling and repeated irradiation, the samples are centrifuged and an aliquot of the resulting supernatant is injected into the LC system.

In bioanalysis microwave irradiation is not only used as a solubilization technique but also to denaturate the quaternary and tertiary structures of proteins. Because enzyme denatures within 1 sec, it is a relatively fast and clean procedure for the digestion of proteins [12].

Microwave irradiation is furthermore an excellent way of thawing samples frozen at 253 K. For instance, doxorubicin and epirubicin are stable when stored at 253 K and by thawing with microwave irradiation no significant degradation is observed [31].

### *Lyophilization Techniques*

Lyophilization, or freeze-drying, is a satisfactory way of preparing biological samples for storage or for an analytical procedure. Relatively large samples, containing water-soluble analytes, are frozen in a Dry Ice-acetone bath or in liquid nitrogen. Subsequently, the frozen samples are placed in the freeze dryer where water is removed by vacuum sublimation [32].

After freeze-drying the residues can be dissolved in a suitable organic solvent and in this way freeze-drying is a suitable way of cleaning and concentration of aqueous samples that cannot be pretreated by, for instance, liquid-liquid extractions.

Freeze-drying is furthermore a helpful technique when the necessary derivatization procedure should be applied in a nonpolar medium or when chemically or thermally unstable compounds must be analyzed. However, sometimes derivatization of the analyte before freeze-drying is necessary to avoid sample loss during the last stages of the lyophilization process.

### *Hydrolysis Reactions*

The application of hydrolysis techniques, especially in bioanalysis, can be divided in two different categories: hydrolysis of the serum proteins and hydrolysis of the drug conjugates.

Subtilisin A as well as proteinase K are enzymes catalyzing the hydrolysis of proteins. These nonspecific proteolytic enzymes hydrolyze peptide bonds [33-37]. This approach has been applied to the determination of the antineoplastic agents etoposide and teniposide in blood [38]. The treatment has proved to be sufficient to prevent clogging of the LC system, meaning that it can be used as a single clean-up step in combination with a LC separation.

During the metabolism of a great number of endogenic and exogenic organic compounds, conjugates with, for example, glucuronic acid and sulfate are formed. However, the isolation and quantitation of the glucuronides itself is usually laborious and time consuming because they are hydrophilic in nature and are ionized at physiological pH values.

The most widely applied approaches for the analysis of these conjugates, because they are not amenable to liquid-liquid extractions, are enzymatic hydrolysis (glucuronidase or sulfatase enzymes) or chemical hydrolysis (heating in hydrochloric acid) procedures.

However, techniques based on the hydrolysis of the conjugates, followed by the analysis of the parent compounds, may fail to yield important information about metabolic pathways or lack sufficient selectivity or sensitivity [39,40]. Consequently, specific quantitation methods based on the analysis of the intact conjugates are preferable [41,42].

### *Precipitation Procedures*

Precipitation of proteins from biological samples can be performed with a great number of reagents [43], e.g., acetonitrile [44,45], acetone [46], ethanol, methanol [47], trichloroacetic acid [48], perchloric acid [49,50], tungstic acid [51,52], ammonium sulfate [53], ammonium chloride, aluminum chloride [54], sulfosalicylic acid [55], basic lead acetate or hydrochloric acid [56].

Applying inorganic salts (ammonium sulfate or ammonium chloride), a reversible protein precipitation is achieved, meaning that after dilution of the samples the biological activity of the proteins is still present.

However, by using acids or organic solvents for protein precipitation the biological activity of the proteins is destroyed irreversibly. The advantage of using methanol for protein precipitation is that a clear supernatant is obtained and a flocculent precipitate formed. Acetonitrile, to the contrary, gives a hazy supernatant with a fine precipitate. Methanol and acetonitrile are frequently used as protein denaturants because of their compatibility with the applied LC eluents.

In order to increase the stability of the analytes during the denaturation process or to enhance recovery, the denaturation can be performed in an ice bath with an ice-cold (273 K) denaturation solvent (e.g., ethanol, acetonitrile) using a refrigerated centrifuge (Fig. 3.3) [57].

The major disadvantage of protein precipitation reactions in combination with a chromatographic separation technique is that in the majority of analyses a second sample pretreatment procedure should be incorporated in the procedure, in order to increase the column's lifetime, to avoid interfering peaks in the chromatograms, or to concentrate the samples.

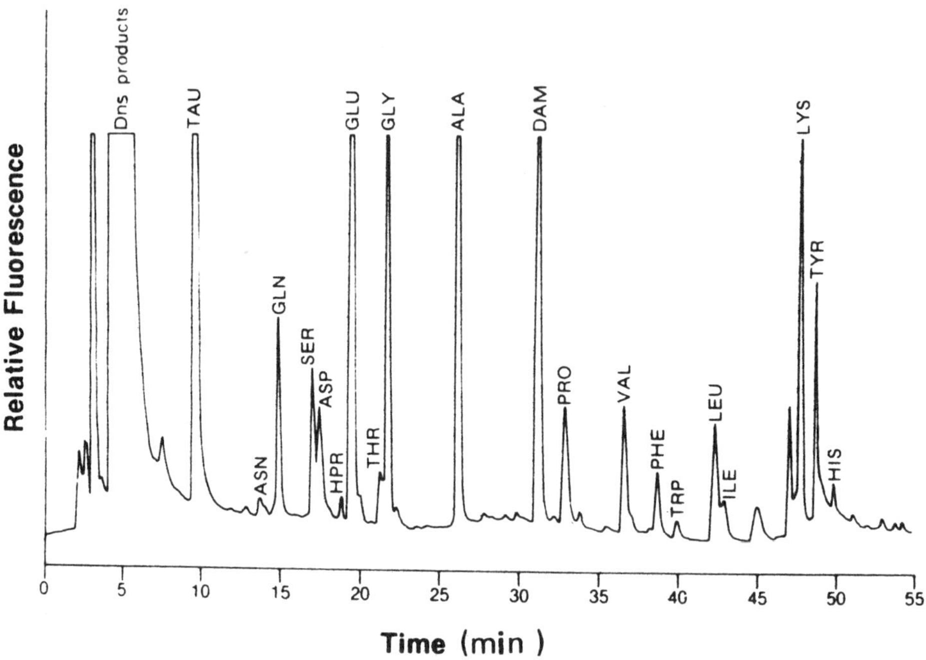


Figure 3.3 Chromatogram obtained with reversed-phase LC of dansyl-amino acids in rat liver homogenates after cooled protein precipitation. The abbreviations are according to the official amino acid nomenclature. (From ref. 57.)

### *Ultrafiltration Techniques*

Ultrafiltration (UF) is a sample pretreatment technique somewhat similar to dialysis, but with some important advantages over the present dialysis methods [9,10]. Dialysis has a number of disadvantages: it is rather time consuming, the transport over the membrane is dependent on the type of analyte, there is a significant temperature dependence, the system must be renewed frequently, and automation is difficult (except for the continuous flow systems). An advantage of dialysis is the independent transport of the analytes from the retained macromolecules. On the contrary, in UF systems the transport of the analyte is dependent of the analytes, the method is relatively fast, the temperature effects are less pronounced, only a minimum of exchange solvent is necessary, and automation is simple. However, at higher macromolecular concentrations the transport rate of the analyte is reduced.

UF can be applied for the concentration of samples by removal of the solvent, for the desalting of samples, and for the fractionation of samples by separating molecular species of different size or shape, e.g., in protein-binding studies [11], and for protein removal [58, 59].

UF is based on the selective retainment of analytes by convective solvent flow through an anisotropic membrane. Compounds with dimensions larger than the specified membrane cutoff are nearly quantitative retained and compounds with smaller dimensions pass the membrane together with the solvent. Nowadays UF membranes are present with cutoff values ranging from a molecular weight of 500 to 300,000. To achieve the desired UF, the equipment must be designed to obtain a high transport flow over the membrane and to diminish the effect of concentration polarization (increase of macromolecular concentration just above the membrane), which is the major problem in UF [60].

The choice of the right membrane for a certain application depends on a number of parameters such as the degree of retention of the analyte, flow rate, some environmental characteristics, and the analyte recovery. Although most membranes are made of inert polymers, undesired adsorption of analytes onto the membrane surface may be observed, especially in dilute solutions and when certain buffer systems are applied.

One of the major advantages of the UF technique is its speed, e.g., applying a hollow fiber UF system a flow rate of 100 liters/hr can be obtained. Furthermore, the combination of several UF membranes allows the isolation of a group of analytes with a very narrow molecular weight range. Protein removal by UF avoids the interferences and low recoveries which are inherent to most of the chemical precipitation methods [61,62]. A limitation of UF may be that sufficient amounts of sample are not always available when highly protein-bound drugs must be analyzed [63].

### *Head Space Techniques*

Another relatively new sample pretreatment procedure in LC is the head space technique [64]. The advantage of this simple sample pretreatment procedure is the possibility of analyzing the volatile and nonvolatile components of a sample in a single chromatographic system. The sample container is connected by a stainless steel capillary with a second container, and by heating the first sample container to 310-420 K and cooling the second container to 270-190 K, the volatile components are gathered in the second container (in a suitable solvent) and can be injected into the LC system.

### *Electrophoresis*

Electrophoresis is one of the alternative techniques for the isolation of ionogenic compounds (e.g., amines) from the complex samples [65].

The technique allows a selective separation between ionogenic and nonionogenic components of the sample as well as a separation between the different ionogenic components in a sample. Although the main separation principle is based on the presence of ionogenic properties of the analytes, the separation is also based on their relative molecular weights. After protein denaturation and lyophilization, the blood samples are subjected to a electrophoretic separation on, for instance, cellulose acetate strips in a methanolic environment. After the electrophoresis, the parts of the strip containing the analytes are cut out and after extraction, the solutes are injected into the LC system [65].

By using zone electrophoresis in capillaries of nonconducting material, this technique can be used on-line with the LC separation [66]. This technique is especially suitable for the isolation of analytes with relatively high molecular weights.

*Liquid-Liquid Extraction*

The most classical approach for sample pretreatment, clean-up, and concentration is some form of solvent extraction [67]. In combination with a derivatization procedure, liquid-liquid extraction (LLE) is together with solid phase isolation the most obvious sample pretreatment procedure. LLE is based on the distribution of a compound over two immiscible solvents. The choice of the organic extraction solvent determines the recovery of the analyte along with the number and amount of the coextracted compounds. An important parameter in LLE is the pH of the aqueous phase because only uncharged (neutral) compounds can be transferred from the aqueous phase to the organic phase. The addition of highly ionized salts (e.g., sodium chloride) can facilitate the extraction of the analyte to the organic phase by means of the salting-out effect.

The most pronounced advantages of LLE are its

- Adjustable selectivity (by the choice of organic solvent)
- Applicability to fairly polar to nonpolar analytes
- Applicability to all types of aqueous matrices
- Applicability of ion-pairing reagents for the extraction of ionic solutes [68]

Apart from aqueous mixtures, extraction procedures can also be applied for nonaqueous or even solid samples. The nonaqueous or almost nonaqueous samples are dried by evaporation, freeze-drying, etc., and the resulting solid residue is dry-extracted with a suitable organic solvent.

The list of disadvantages of LLE is considerably larger than the advantages. The most important disadvantages are as follows:

Formation of emulsions (sample losses by occlusions)
The use of toxic and inflammable solvents
The necessity of evaporation of the organic solvent (possible degradation of the analyte)
Adsorption onto solid (e.g., glass) surfaces
High cost (due to the use of relatively large amounts of organic solvents and equipment)
Its time-consuming and laborious nature

Emulsification can be avoided by using larger volumes of the extracting solvent or by less vigorous mixing of the samples.

Another disadvantage inherent to LLE is the fact that some extraction solvents may be involved in a chemical reaction with the analyte, e.g., when chloroform is used without a preservative (ethanol) [69,70]. A last disadvantage is that it is almost impossible to transfer all of the organic solvent from one reaction vial to another, resulting in sample losses. Freezing and aqueous phase in liquid nitrogen and subsequently decanting the extraction solvent can solve this problem. However, under circumstances that the detectability of the analyte is not the limiting step in the whole analytical procedure, an accurately known aliquot can be taken.

The most direct approach is the combination of an extraction and a derivatization step. Extractive derivatization procedures, particularly extractive alkylation methods, are very popular in combination with gas chromatography (GC) [71], but they can also be applied in combination with LC [72]. Fatty acids can be analyzed by using an extractive alkylation procedure. The acids are extracted as ion pairs with a quaternary ammonium compound from an aqueous phase into dichloromethane, where the alkylation takes place [73].

However, frequently an additional sample pretreatment step other than protein removal by a precipitation reaction should be included in the procedure to avoid interfering peaks in the chromatograms. Examples of these clean-up steps are the preliminary extraction of the solute from the sample with an organic solvent and the extraction of the analyte from the sample followed by back-extraction to an aqueous phase with a suitable pH (Fig. 3.4).

Occasionally the analyte is chemically unstable in the pH range that is necessary for an efficient LLE. To overcome this problem the analyte can be derivatized in the biological sample with an appropriate reagent to form a suitable derivative that can be successfully extracted.

The most frequently applied approach is extraction of the drug followed by a derivatization procedure. Although less direct, this approach has some potential advantages over the extractive derivatization methods. One is the freedom of choice in extraction solvent.

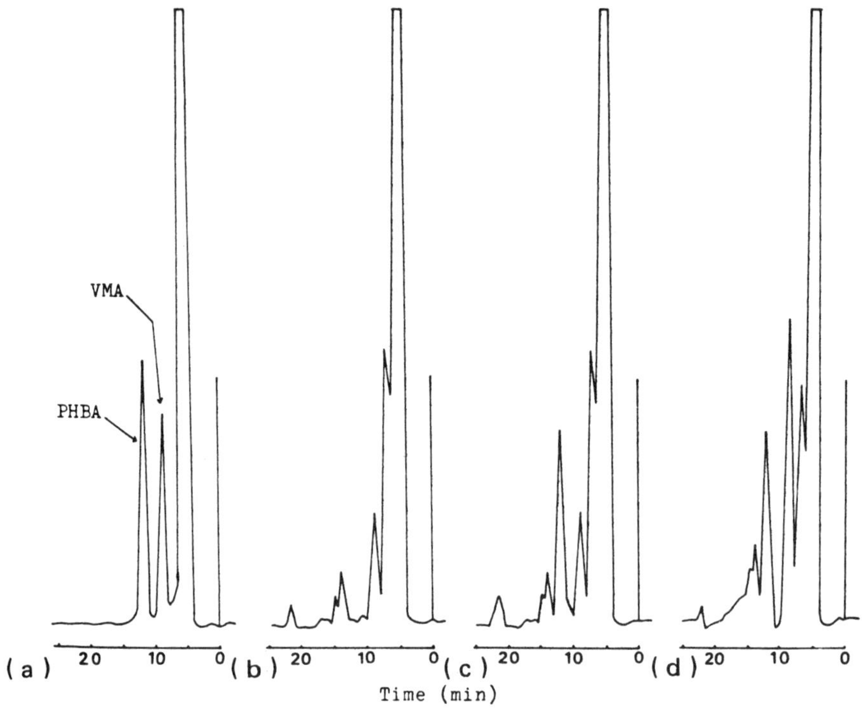


Figure 3.4 Chromatograms obtained with normal phase LC of vanillylmandelic acid (VMA) in urine samples after extraction and back-extraction. (a) academic solution containing VMA and p-hydroxybenzoic acid (PHBA), (b) urine blank, (c) urine containing PHBA, and (d) urine containing PHBA and VMA. (From ref. 74.)

The most nonpolar solvent with which the drugs can be extracted with sufficient efficiency is usually the best one because of the gain in selectivity of the procedure.

Many endogenous compounds, especially metabolites, are very polar due to the presence of ionogenic functions and cannot be extracted with relatively nonpolar organic solvents. Ion-pairing techniques can often be adapted for the extraction of ionic solutes from an aqueous solution when classic LLE fails [75]. In ion-pairing extractions a more lipophilic organic solvent-soluble species is created by the interaction of the charged analyte with a suitable counterion. This combination is usually called the ion pair. The extent of extraction is controlled predominantly by the nature of the organic solvent and the ion-pairing agent, and furthermore by the concentration of the agent. The procedures are easy to develop and to adapt for both hydrophilic and hydrophobic analytes, but they are particularly useful for the extraction of charged aprotic compounds and hydrophilic acids and bases.

The selectivity can be controlled by the choice of counterion and extraction solvent, e.g., for ammonium ions the best selectivity is obtained with a hydrogen-accepting counterion and a hydrogen-donating solvent. An example is the ion pair extraction of some corticosteroid glucuronic acid conjugates with the tetrapentylammonium cation. The resulting ion pairs are fairly lipophilic and readily extractable with chloroform [76].

For difficult separations multiple extractions may be needed, although in many cases the background is preferentially coextracted. The subsequent derivatization step can sometimes be performed directly in the analyte extract, but usually concentration of the sample by means of evaporation of the organic solvent is necessary. The residue is then available for derivatization under optimum conditions. An elegant extraction-concentration procedure which is suitable for the extraction of many acidic drugs prior to an alkylation reaction consists of the extraction of the analyte with toluene followed by back-extraction to a very small volume of a methanolic solution of tetramethylammonium hydroxide. An aliquot of this solution is mixed with *N,N'*-dimethylacetamide and an alkylating agent to perform a prechromatographic alkylation step [77,78].

Special extraction procedures are necessary for the sample pretreatment of biological samples containing strongly lipophilic analytes (e.g., K vitamins). In the digestive tract these analytes are solubilized in mixed micelles consisting of bile salts, monoglycerides, and free fatty acids [79,80]. The extraction procedure should therefore effectuate a complete disruption of the lipoprotein structure, leading to the liberation of the analyte and associated lipids. The extraction from serum samples can be executed after destroying the lipoproteins (Fig. 3.5) with an alcohol (ethanol or isopropanol) and subsequent vigorous mixing with hexane.

A special form of LLE is the use of time- and solvent-saving diatomaceous earth (e.g., Celite, Extrelut) columns [81,82]. The use of the solid material diatomaceous earth is not considered to be a solid phase isolation because the transfer of the analytes from the aqueous phase to the organic phase obeys all the laws of classic LLE. The advantage of the use of diatomaceous earth in comparison with classic LLE is that emulsification during the extraction can be reduced and that higher recoveries can be obtained by the application of these prefabricated or manually packed diatomaceous earth columns [83,84]. The sorbent in the column is kieselguhr with a large pore size [85]. An aqueous sample is applied to the sorbent

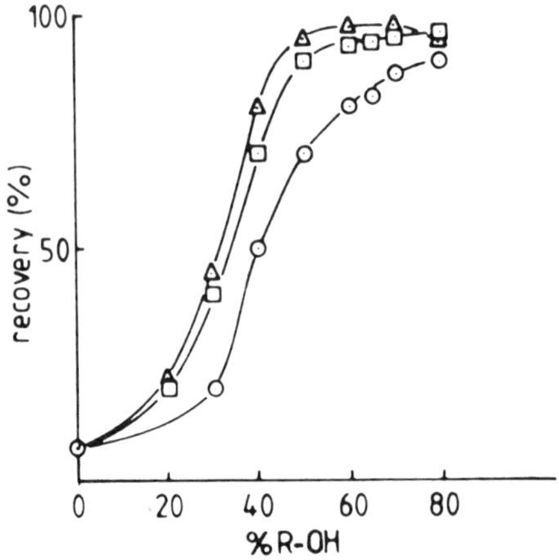


Figure 3.5 Recovery (%) of the extraction of vitamin K with n-hexane from a serum sample after denaturation with various percentages of methanol (o), ethanol (⊡), and isopropanol (△). (From ref. 79.)

and the sample is partitioned over the porous matrix as a stationary phase. The analytes are eluted with a water-immiscible organic solvent. This technique is especially applicable to the isolation of lipophilic compounds from aqueous (e.g., urine) solutions because the material can hold four times its weight of water [82,84]. In principle every classic LLE procedure can be transferred to a diatomaceous earth procedure, just as in classic LLE the pH of the sample is very critical for the recovery of the analyte. For the analysis of blood and plasma samples a dilution step is normally required because of the high viscosity of the sample and because of protein binding of the analytes. As already mentioned, in comparison with LLE cleaner samples, an increased reproducibility, and a total lack of emulsions are the most pronounced features. The main disadvantage is that relatively large volumes of the organic solvent are needed. For the analysis of a number of antiepileptic drugs in serum by LC, 0.5 ml of the serum is eluted with 6 ml of chloroform [86].

A rapid isolation procedure for the analysis of acidic and basic analytes in human tissues using diatomaceous earth columns is described in the literature (Fig. 3.6) [87]. The acidic and basic analytes are analyzed in postmortem material after an enzymatic digestion and multicolumn extraction on a number of diatomaceous earth columns. The obtained samples are better in purity and higher recoveries are obtained than with the use of XAD-2 resins (see following section).

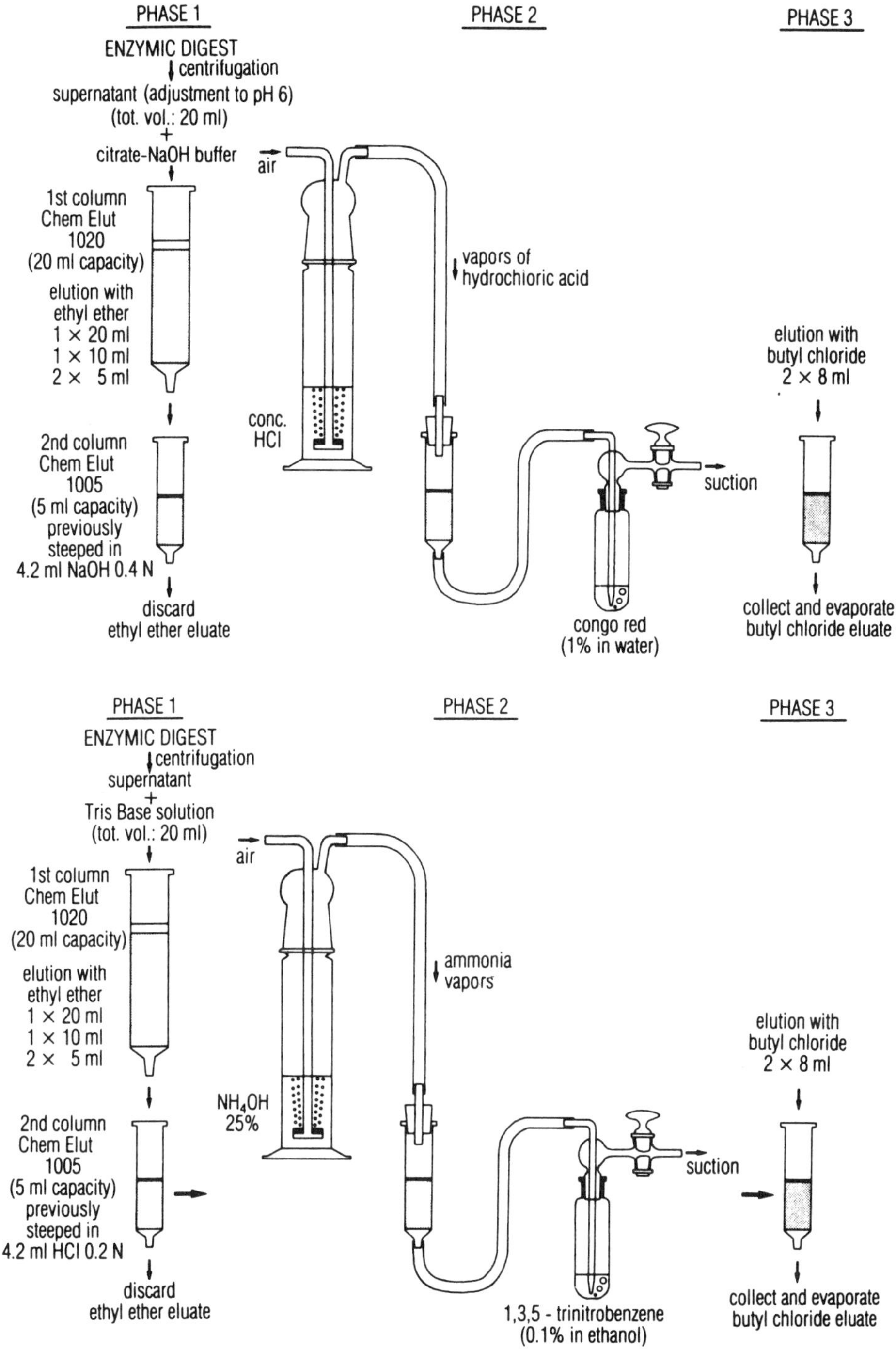


Figure 3.6 Extraction procedures for the extraction of acidic and basic analytes from biological samples by coupled diatomaceous earth columns. (Top) basic analytes; (bottom) acidic analytes. (From ref. 87.)

One of the most promising chromatographic techniques for the near future will be supercritical fluid chromatography (SFC) and in sample pretreatment the application of a supercritical solvent represents a new extraction technology [88]. SFC is one of the newer chromatographic techniques and can be considered a combination of LC and GC, possessing the advantages of both techniques (see Chap. 1). The sample containing the analyte is placed in an extraction cylinder and the supercritical solvent introduced by means of a controlled constant pressure pump. All of the system piping is installed within a temperature-regulated oven to assure precise control of the supercritical solvent conditions. Dissolved analytes are passed through a detection system to check for desorbed analytes before they are transferred to a precision nozzle-orifice where depressurization occurs, resulting in a precipitation of the analyte in a collector trap. The solute is retained in this collection trap while subcritical gaseous solvent is vented or collected separately. In this way supercritical fluid extraction technology provides an on-line quantitative measure of extraction efficiency, while performing the extraction or purification.

### *Solid Phase Isolation*

Because sample preparation procedures based on LLE are tedious to perform, time consuming, susceptible to errors and problems (e.g., phase emulsions, evaporation of large volumes of organic solvents), costly (organic solvents), and often result in a loss of the analyte because of incomplete extraction or formation of artifacts, solid phase isolation (SPI) procedures have become very popular in the last decade. Excellent papers about the features of SPI were recently published [67,89].

SPI techniques can be applied either before or after the derivatization reaction. For instance, before the derivatization procedure the sample is only cleaned from a number of interfering compounds during the derivatization and applying a SPI after the derivatization means that the sample is cleaned from a number of interferences during the chromatography of the samples. On special occasions, i.e., dirty matrix and complex derivatization procedures, both SPI procedures should be applied.

Contrary to LLE the combination of a SPI and a derivatization reaction in one single step is not routinely applicable nowadays. However, the introduction of the immobilized bed reactors will also make this extremely elegant pretreatment-derivatization technique possible within a couple of years.

SPI has some important advantages over the classic LLE:

The SPI systems are relatively fast, e.g., for the extraction of 10 samples of corn meal, containing aflatoxins, only 15 min is required.

High recoveries, normally over 90%, are obtained.
The procedures have high precision.
Cleaner samples are obtained because of the minimal introduction of impurities.
Small biological sample volumes can be contemplated.
The columns are disposable.
Volatile solutes can be analyzed as well.
Emulsification of the samples is eliminated.
A high degree of selectivity is obtained by the wide variety of sorbents.

By applying selective sorbents (e.g., metal-loaded columns) or by combination of several columns packed with different stationary phases, every group of analytes can be isolated. However, the sequence of the different precolumns is very important, and a preconcentration factor of $10^4$ is now possible.

The selectivity of the sample pretreatment procedure can also be increased by the combination of a LLE and a SPI (Fig. 3.7), especially when the LLE is used to remove some strongly hydrophobic components from the matrix which may interfere with LC analysis [41].

The SPI procedures are relatively simple; a disposable manually packed or prefabricated cartridge with a suitable sorbent is solvated (wetted) and conditioned with appropriate solvents and the sample is applied. Subsequently the column is washed, and the analyte or the analytes are eluted selectively from the column with less than 1 ml of the isolation solvent. Normally, the columns are packed with 100-1000 mg of the sorbent (40-μm particles of 60 or 275 Å) and up to about 60 mg of the analyte can be extracted with 1 ml to about 5 ml of the elution solvent. In general it can be stated that for the chemically bonded silica phases about 1-5% of the sorbent weight can be retained, that 20-bed volumes of washing solvent can be applied, and that elution should be executed with less than 5-bed volumes. However, in complicated matrices the number and concentration of the interfering matrix components should also be taken into consideration.

Applying SPI systems a number of functions can be fulfilled such as sample concentration (trace enrichment), sample clean-up, storage of samples, guard column functions, and support material for on-column derivatizations [90-93].

SPI procedures can be performed off-line and on-line with the chromatographic separation. The major drawbacks of the off-line application of SPI systems in comparison with the on-line application are:

An increase of the determination limit
The possibility of sample losses during the evaporation steps

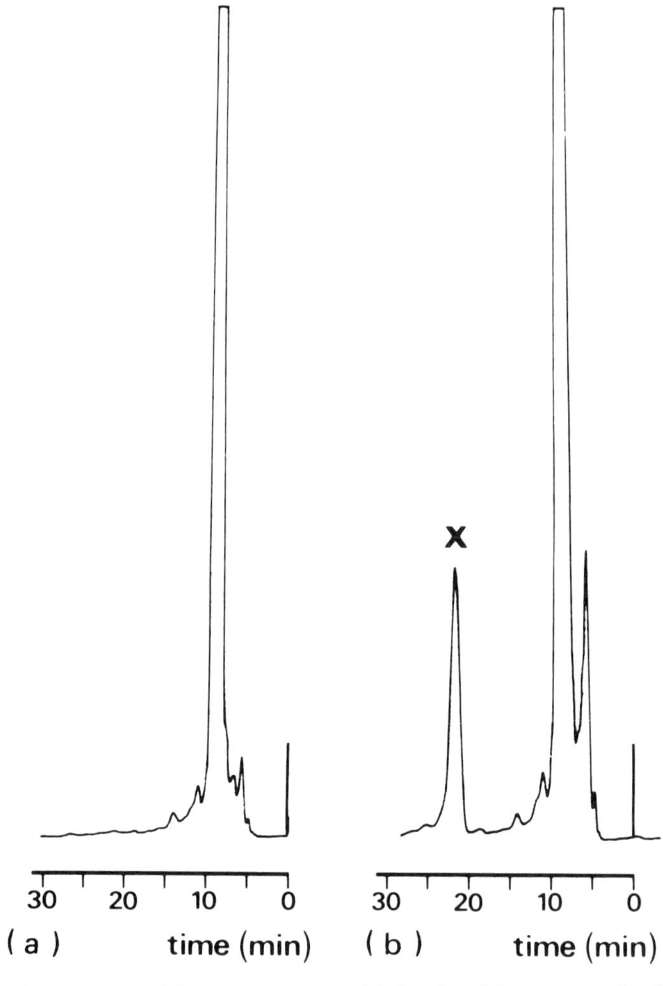


Figure 3.7 Chromatograms obtained with reversed-phase/ion-exchange analysis of β-oestradiol-17-β-D-glucuronic (17-EG) acid in plasma after pre-extraction with dichloromethane and SPI with a $C_{18}$ cartridge and pre-chromatographic fluorescence derivatization. (A) blank chromatogram and (B) 100 μl plasma sample; x, derivative of 17-EG and N-(1-naphthyl)ethylenediamine. (From ref. 41.)

The possibility of contamination errors
Difficulties of automation

The main advantage of the application of on-line SPI is the ease of automation. In general, off-line methods can be easily transformed to on-line procedures, although some problems may arise in the compatibility of the solvents used for the sample pretreatment column and the analytical column.

Other advantages of SPI are the possibility of performing a preconcentration and a clean-up of the sample in one analytical proceeding, the possibility of using several precolumns in series, and the possibility of incorporating the precolumn on-line in the LC system (see section on "(Semi)Automated Sample Pretreatment Procedures" in this chapter). One of the drawbacks of the system is the fact that frozen and thawed plasma sometimes contains solid particles (fibrins) which can block the SPI cartridges, but the application of specially constructed frits can solve this problem.

The sequence of manipulations (Fig. 3.8) in an off-line SPI, and in principle in an on-line system also, starts with the activation (wetting) and conditioning (equilibration) of the sorbent. The wetting, especially of the chemically bonded hydrophobic columns, is very important because it opens the hydrocarbon chains (solvation) and increases the surface area of the sorbent [94]. Subsequently the sample is applied to the column and the interfering compounds are washed off the column (clean-up). Finally, the analyte is eluted from the column and, if necessary, the column is reconditioned. Elution of the analytes can be executed with reduced pressure, pressurized (syringe), or by centrifugation. However, the speed of the elution is very important in obtaining high recoveries and reproducible results. Applying sample blocks or centrifuges for the elution up to 12 samples can be analyzed simultaneously. For the analysis of small amounts of the analytes it may be advantageous to use polypropylene or polyethylene columns to avoid irreversible adsorption of the analytes onto glass surfaces.

Extensive discussion about which solvents, which concentration, and which volumes should be used for the isolation of an analyte by means of SPI is given in the literature [94,95].

The polarity of the sorbent and the choice of the eluting and washing solvents is very critical when SPI procedures are applied. The explanation is that the meaning of SPI is to isolate an analyte or a group of solutes within a narrow polarity range, while all the other compounds are totally unretained or completely retained. A number of physicochemical and other parameters should therefore be taken into consideration [94,95]:

The nature of the analyte (e.g., molecular weight, solubility)

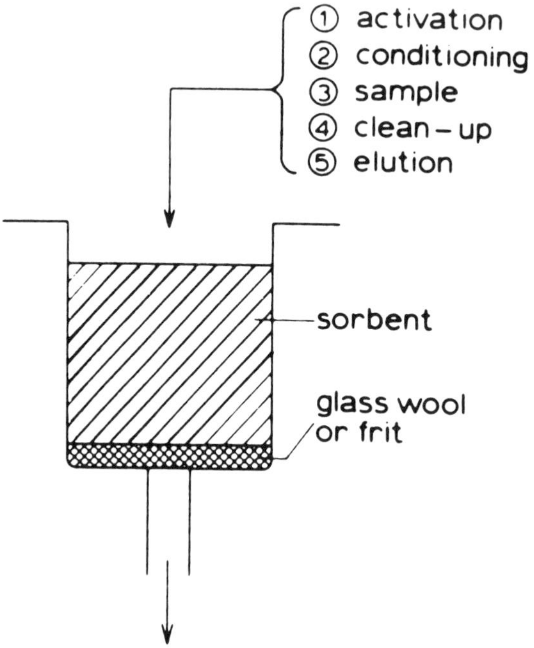


Figure 3.8 Solid-phase isolation column/cartridge) composition and handling sequence.

The nature of the functional group(s) of the analyte (e.g., polarity, acidity)
The nature of the sorbent (e.g., particle size)
The nature of the matrix (e.g., physicochemical properties)
The forces of the different interactions (e.g., dispersion, dipole-induced dipole, dipole-dipole, hydrogen bond, ionic, covalent)
Additional interactions between the analyte and the sorbent (e.g., complex retention mechanism)
Interactions between the analyte and the matrix (the resulting solutions may not be too diluted)
Interactions between matrix components and the sorbent
The properties of the applied LC mobile phase
The recoveries of the analyte from the aqueous (academic) as well as from the biological matrix

Concerning the energies of the different forces it should be noted that the dispersion forces or hydrophobic interactions (5-20 kJ/mol) belong to the weakest and that the covalent forces (650-3350 kJ/mol) are responsible for the strongest interactions. These interaction forces, as listed above, are given in such a way that the weakest force is the first and the strongest force is the last one in the summation.

Because SPI is a combination of (ad)sorption and desorption it cannot be stated that covalent interactions are to be preferred. Although the analytes will be quantitatively (ad)sorbed to the solid phase material, the desorption of the analytes is nearly impossible.

The correlation between solvent strength and solvent polarity depends on the polarity of the applied sorbent. For polar sorbents solvent strength is increased by increasing the solvent polarity. However, for nonpolar solvents, the solvent strength is increased by decreasing the solvent polarity [94,95]. Some general procedures for using the SPI systems for sample pretreatment are given in Table 3.1.

Two very important parameters which should be established during the development of every SPI procedure are the breakthrough volume and the sample capacity of the columns. The capacity of most of the commercially available columns will almost never limit the applicability of these systems. For instance, for the chemically bonded silica materials, normally up to 1 mg analyte can be pretreated with 20 mg of sorbent [95].

The breakthrough (retention) volume is a more critical parameter [39], controlling the amount of sample, the solute in a defined solvent, and the solvent volume that can be applied to the column before the analyte is washed off. The breakthrough volume of a sample is determined by the capacity ratio (k') of the analyte on the sorbent, and quantitative sample retention can only be achieved with k' values of 10 and higher [97].

For every class of analytes a special sorbent should be used [94,95]. Nowadays a great number of sorbents allowing manual packing of SPI columns as well as prefabricated cartridges are available. The isolation of the analytes on these materials is based on the same principles as normal LC and an overview of possible sorbents, solvents, and applications is listed (see Chap. 1). The best known principles are adsorption, NP partitioning, RP partitioning, ion exchange, or size exclusion chromatography. Moreover, some selective interacting materials (e.g., metal-loaded phases, alumina, diphenylborate [98]) will be discussed.

During the first years of SPI a limited number of adsorbents have been applied (charcoal, silica, alumina, or the earlier described diatomaceous earth), but the real breakthrough of SPI started with the introduction of the chemically bonded phase materials [99]. In

Table 3.1 General Procedures for Using Solid Phase Isolation Systems for Sample Pretreatment

| Handling sequence | Sorbent | | | | | |
|---|---|---|---|---|---|---|
| | Reversed phase | Adsorption | Ion exchange | | Hydrophobic interaction | |
| | | | Cationic | Anionic | | |
| Packing polarity | Low | High | High | | Low | |
| Analyte polarity | Nonpolar | Polar | Acidic | Basic | Acidic | Basic |
| Column conditioning<br>Solvent polarity | High [methanol] | Low [chloroform] | Water | Water | Water | Water |
| Sample application<br>Solvent polarity | High [water] | Low [hexane] | Basic | Acidic | Acidic | Basic |
| Analyte elution<br>Eluting solvent | Decreased solvent strength [methanol] | Increased solvent strength [methanol] | | | | |
| Elution order | Most polar first | Least polar first | Most ionized first | | Least ionized first | |

the following paragraphs a number of sorbents and groups of sorbents will be discussed in more detail.

Charcoal. Charcoal (graphite, carbon) can be used for the clean-up of plasma and urine samples containing relatively hydrophobic analytes [100-102].

An example of the application of charcoal in the pretreatment of plasma samples is described for nitrendipine [103]. A plasma sample of 1 ml is mixed with a phosphate buffer and a slurry of the charcoal (active charcoal, Norit A) in the same phosphate buffer. After vortex-mixing for 30 sec, the sample is centrifuged and subsequently the charcoal is extracted with a suitable organic solvent.

Hydrophobic Sorbents. Hydrophobic sorbents (e.g., copolymers of styrene and divinylbenzene, octadecyl, diphenyl, octyl, cyclohexyl, phenyl, butyl, ethyl, methyl) can be applied for retaining nonpolar to medium polar analytes. The styrene divinylbenzene copolymer phases (e.g., PRP-1, XAD-2) are the most nonpolar phases and the methyl phase is the last nonpolar phase on this list. The analytes are adsorbed onto the hydrophobic materials by means of van der Waals interactions and to some extent by hydrophobic bonding or dipole-dipole interactions.

Amberlite XAD-2, a neutral hydrophobic form of XAD, has the interesting characteristic that it hardly swells or shrinks upon hydration of its structure, and because of its macroreticular structure it can be applied for the clean-up of urine and plasma samples. The analytes are adsorbed onto the column by means of the same interactions described for the chemically bonded phases [104] and it can be used for the simultaneous isolation of acidic, basic, and neutral compounds in a sample. An interesting application of the XAD-2 resin is the isolation of sulfate and glucuronic acid conjugates from biological samples, which is troublesome with the classic LLE methods [105]. One of the disadvantages of XAD-2 over the chemically bonded stationary phases is that the pretreatment, cleaning, wetting, and equilibration of the column itself is rather time consuming. However, in comparison with the chemically bonded phases, XAD-2 and the copolymers of styrene and divinylbenzene can be used in the entire pH range 1-14.

PRP-1 is a highly crosslinked (spherical material, contrary to XAD-2, with outstanding chemical resistance, is comparable with most of the organic solvents, and can be incorporated in LC systems. The major disadvantage of this material is its limited efficiency; however, the efficiency can be increased significantly by the addition of 1% dichloromethane to the eluting solvent [106]. Addition of dichloromethane causes some swelling of the material and thus an increase of pressure resistance. Except for the PRP-1 material the

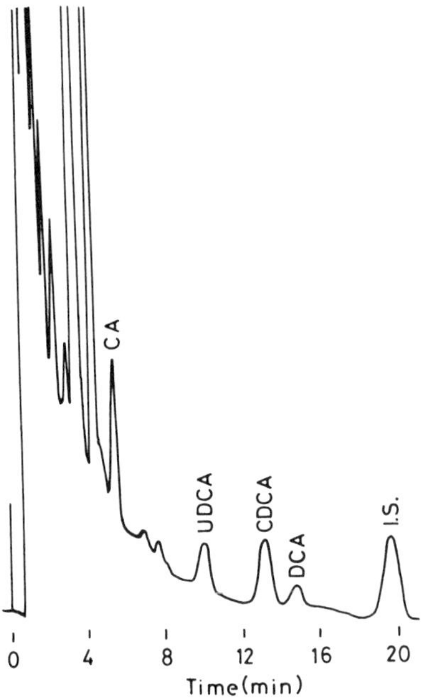


Figure 3.9 Chromatogram obtained with ion-exchange LC of free derivatized bile acids in serum after off-line SPI. CA, cholate; UDCA, urodeoxycholate; CDCA, chenodeoxycholate; DCA, deoxycholate; IS, deoxycholate-12-propionate. (From ref. 108.)

macroporous PLRPS material is very suitable in SPI cartridges. RP separations of polar, ionic, and ionizable compounds (carboxylic acids, sulfonamides, chlorophenols) which were troublesome with all types of silica-based materials, because of the residual silanol functions, are now feasible [107].

However, in the majority of the described sample pretreatment procedures, the hydrophobic chemically modified phases are applied (Fig. 3.9). The popularity of these phases can be explained by the following factors:

Reproducible results
Stability and relative inertness of the phases
Well-defined retention mechanisms
Availability of prefabricated cartridges
Large application area
Great number of applications already described

The sorbent (e.g., chemically modified silica) should be chosen in such a way that the attraction between the stationary phase (sorbent) and the analyte is stronger than the interaction between the analyte and the mobile phase because for a selective isolation of the analytes solvents must be applied with the weakest solvent strength possible in combination with the chosen sorbent. For instance, if a choice must be made between an octyl ($C_8$) or an octadecyl ($C_{18}$) modified phase, the octyl system must be preferred because this system can be used in combination with a solvent containing less organic modifier and more water; thus an eluent with a lower solvent strength. For the same reason an $C_8$-modified analytical column should be combined with a less hydrophobic precolumn (e.g., $C_2$) [18].

The recovery of the chemically bonded phase columns can be increased in some cases by washing the sorbent with ethylenediaminetetraacetic acid (EDTA) in order to remove cations from the silica material which can complexate with some of the analytes [107].

Because the principle of SPI is analog to normal LC the selectivity of the systems can be improved by the application of dynamic modified stationary phases. For instance, a chemically bonded hydrophobic phase is saturated with cetrimonium bromide to avoid irreversible adsorption of the analytes to the residual silanol functions. Subsequently this intermediary phase is saturated with a sulfonate to create an ion exchange column. The result is a new phase which gives very high recoveries for a number of cationic analytes [109].

Polar Sorbents. Polar sorbents (e.g., silica, alumina, magnesium silicate, diol, aminopropyl, cyanopropyl) are used for the pretreatment of medium polar and polar analytes. The cyano phase is a sorbent of medium polarity and can be applied in the NP as well as the RP mode.

The polar sorbent silica is the basis of a number of chemically bonded phases. Most of them are hydrophobic in nature, but the phenylboronic acid (Fig. 3.10) phase is a polar phase which reacts selectively with coplanar vicinal hydroxyl functions [98]. The sorbent can be used for the isolation of catecholamines. Alumina, one of the classical sorbents, can also be used for the selective isolation of catecholamines [110]. Both sorbents work on the same principle; at relatively high pH values the catecholamines are bound selectively to the sorbent and at relatively low pH values the analytes are eluted from the sorbent.

The cyanopropyl phase is a sorbent of medium polarity and can be used in the RP as well as the NP mode. For the pretreatment of vinca alkaloids and their metabolites in plasma and urine samples, SPI with cyanopropyl cartridges yielded the best results [111].

The aminopropyl phase can be used in NP as well as in RP, and as a weak anion exchanger, but it should not be used in combination

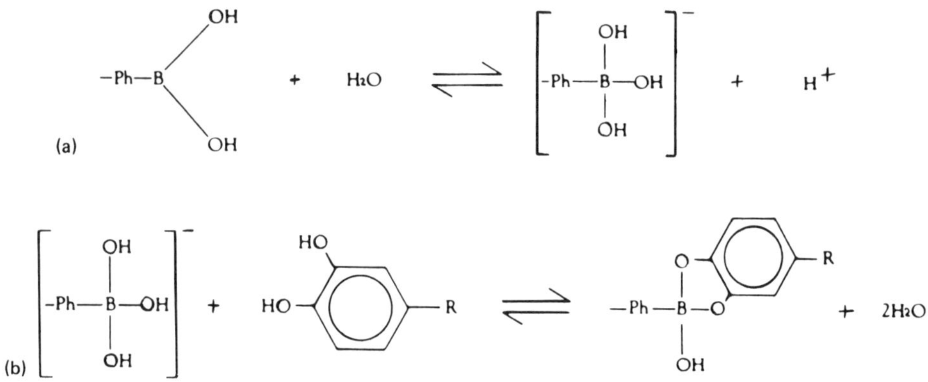

Figure 3.10 Adsorption of catecholamines onto phenylboronic acid phase. (a) activation of the sorbent and (b) adsorption of the analyte.

with aldehydes and ketones. Because of its mixed retention mechanism, isolations can be executed on this material that are not possible with the usual hydrophobic or ion exchange materials.

Last but not least, it is important to realize that in adsorption chromatography, with increasing solvent strength of the eluent the pure adsorption mechanism (dipole-dipole, dipole-induced dipole interactions) is replaced by a liquid-liquid partition equilibrium.

Ion Exchange Materials. Ion exchange (IE) materials are used for the pretreatment of ionic compounds. These sorbents allow the isolation of hydrophylic compounds, which is difficult or impossible with LLE. Ion exchange chromatography is a technique with high specificity, high selectivity, and high capacity. The applied stationary phase materials are permeable hydrophylic polymers or hydrophylic polymers bound to silica containing a fixed concentration of acidic and/or basic functions on the surface. In anion exchange chromatography the sorbent surface is covered with positively charged groups (aminopropyl, primary/secondary amine, quaternary ammonium) which bind reversibly to anionic solutes and in cation exchange chromatography the surface contains negatively charged functions (carboxylic, sulfonic) which bind reversibly to cationic solutes. The high selectivity of these materials is due to the great number of different IE materials, weak and strong exchangers, and the fact that the interaction of the solute with the stationary phase is strongly pH-dependent.

For instance, since most proteins are acidic at physiological pH values, anion exchange chromatography is the method of choice for the purification of samples of biomolecules [112,113].

For the metabolic profiling of organic acids, polar as well as less polar acids, DEAE-Sephadex anion exchange columns can be used. This material reacts rather specifically and high recoveries are obtained. The elution of the acids, however, should be performed with a 1.5 mol/liter solution of pyridine-acetate, and a lyophilization procedure is necessary before the samples can be further analyzed [112].

A severe drawback of the use of IE materials is that, because of the limited capacity of the resin-based IE materials, a precipitation step should be introduced in the procedure when the matrix under investigation contains a relatively high concentration of ionic components [114], and that relatively long equilibration times are needed in comparison with the use of the neutral chemically bonded phases.

Except for the chromatographic systems mentioned so far, which are based on the standard LC techniques for the analysis of small solutes, some other chromatographic techniques (gel filtration, hydrophobic interaction, affinity) can be applied for the purification and/or isolation of macromolecules (e.g., peptides, proteins) or for the separation of macromolecules from micromolecules.

Gel Filtration Chromatography. Gel filtration chromatography is a nondestructive clean-up procedure in which the separation is based on only one physical parameter: molecular size [115]. With this technique the separation of molecules with molecular weights ranging from several hundred to $10^7$ D is possible. The elution order depends on the molecular size. The larger molecules, unable to enter the pores, will elute first and the small molecules which can penetrate the pores will have the longer retention times. The main disadvantage of this technique is its limited resolving power. The technique is usually applied to the desalting of protein solutions on, for instance, Sephadex G-25 (bead-formed crosslinked dextran gel).

Hydrophobic Interaction Chromatography. Hydrophobic interaction chromatography is used for the purification and isolation of proteins in samples containing high concentrations of salts or other small hydrophilic molecules. The applied sorbents in hydrophobic interaction chromatography are similar to the sorbents used in RP-LC, but they contain a lower concentration of hydrophobic groups on their surface [116]. The samples are injected using a high salt concentration (e.g., 2.5 mol/liter ammonium sulfate) in the eluent (salting-out effect), and in this way only the macromolecules are adsorbed onto the sorbent. By decreasing the salt concentration in the eluent, the solutes are selectively eluted from the column [117]. The most hydrophobic proteins have the strongest retention on hydrophobic interaction phases. However, these high salt concentrations may be the reason for problems and therefore the application of polar bonded phases of the amide type with an adjustable hydrophobicity may be advantageous [118].

Because the retention on hydrophobic interaction columns is controlled by means of the ionic strength of the eluent, combination with an ion exchange analytical column allows an on-line combination of these two phases (see section on "(Semi)Automated Sample Pretreatment Procedures").

Affinity Chromatography. Affinity chromatography is an expensive and time-consuming technique which is based on the selective and reversible interaction between two molecules of which at least one is a macromolecule [119,120]. The high degree of specificity in affinity chromatography is introduced by the presence of special pairs of molecules. It is a chromatographic technique because one member of the pair (e.g., monoclonal-antibody, antigen, enzyme, inhibitor) is covalently bound to a sorbent (e.g., agarose). If a biological sample is injected on such a column, all of the injected compounds (e.g., proteins), except the other member of the affinity pair, fail to interact with the sorbent and are flushed through the column. After a washing step, the desorbed analyte is eluted from the column by means of a suitable competing ligand or by changing the pH or the ionic strength of the mobile phase. One of the problems with affinity techniques may be selectivity because closely related structures to the compound under investigation are retained as well.

General strategies for the development of affinity methods for the analysis of small molecules are described in the literature [121].

Affinity chromatography seems to be an elegant sample pretreatment procedure whenever small peptides should be analyzed in a protein-rich matrix. An example is shown by Kristjansson et al. [122]; with substance P (a peptide) antibodies were raised, harvested, purified, and immobilized on a rigid support to form an immunoaffinity sorbent, which is used for the clean-up of the sample.

An on-line procedure for the analysis of phenytoin in human plasma samples within 12 min using a phenytoin-specific polyclonal immunoglobulin which is attached to an organic silane derivative on silica is described by Johansson [123]. Although phenytoin is highly bound to plasma proteins, a dissociation and partition of phenytoin from the plasma proteins to a complete association with the immobilized antibodies is observed.

Hydroxyapatite. Hydroxyapatite (calcium phosphate) is a material which contains both negatively and positively charged groups at physiological pH values, and it can be considered as a special type of crystalline IE material. Strong IE materials have a very high binding capacity resulting in an irreversible binding of the protein. Therefore, weak IE materials may be advantageous [124]. Furthermore, for the hydroxyapatite as well as for the other IE materials, much higher capacities are obtained in comparison with the gel filtration materials. The sorbent offers special selectivities in sample pretreatment because it can be used for the isolation of negatively as well as positively charged proteins [125]. The sample is generally applied in a low ionic strength aqueous (phosphate) buffer and eluted with increasing concentration of the buffer. It is a highly discriminating interactive stationary phase material, but it is rather fragile (low flow rates), has a limited column lifetime, and shows a complex adsorption mechanism. The latter results, for instance, in a nonspecific irreversible binding of some components in biological materials.

In general, no general rules can be given for the selection of the most appropriate system for a certain problem, and therefore the most suitable system strongly depends on the individual samples: the combination of analyte and matrix, the LC system used, the applied detection system, and the required sensitivity.

Although in many cases the selectivity of one SPI column is sufficient to separate the analyte from the interfering matrix components, the combination of two or more columns can be used for a further increase of the selectivity. However, by using two or more cartridges in series, the application of automated systems is advisable to avoid lengthy pretreatment procedures.

An example is the selective isolation of basic solutes from a matrix containing a high concentration of ionic compounds. First, the analytes are retained on a hydrophobic phase column and after elution the second clean-up is performed on a cation exchange column. The advantage of this procedure is that the salts are removed during the first step of the clean-up so that they cannot influence the retention of analytes on the cation exchange column. In the analysis of tetracyclines in a complex matrix using a $C_{18}$ cartridge in combination with a weak IE cartridge, recoveries between 85 and 99% are obtained [126].

As mentioned in the beginning of this section, the combination of a derivatization procedure and a SPI method is a relatively new technique which is still developing. Rosenfeld et al. [93] impregnated XAD-2 with pentafluorobenzyl bromide, an alkylation reagent, and used it for the simultaneous extraction and derivatization of tetrahydrocannabinol carboxylic acid and phenobarbitone. Because the reaction time is about 1 hr at ambient temperature the procedure is executed in the off-line mode. However, the application of, for example, the fluorescence-introducing reagent OPA for the derivatization of primary amines allows an isolation-derivatization procedure which can be performed in an on-line even in an automated column-switching system. Certainly, in the near future, the combination of solid phase derivatization with solid phase pretreatment procedures will become more and more important [127].

The final remark of this section is that, although a division has been made between off-line and on-line sample pretreatment procedures, in terms of the use of SPI the majority of off-line procedures can be changed to on-line procedures, and vice versa.

## (Semi) Automated Sample Pretreatment Procedures

In LC isolation/concentration methods, by means of column-switching (CS) techniques, offer a new dimension to the sample preparation technique because in this way automation, throughput, and precision are strongly improved [128,129]. However, chromatographic sample clean-up and/or sample enrichment procedures preceding gas chromatographic analysis are normally performed off-line with respect to the gas chromatographic analysis.

Switching one or more valves enables the chromatographic eluent to be rerouted from one column to another. Combined with automated sample clean-up, the possibility of direct injection of plasma or urine sample into the LC system is more obvious than with the off-line procedures [130,131]. Another important application of CS is the on-line concentration of samples (e.g., river water, 24-hr urine) [132].

The (semi)automated procedures include continuous flow methods [133,134], CS [135], robotics, sample processors, and immobilized enzymes [136].

The use of immobilized enzymes for sample pretreatment is a relatively new technique. The "software" part (enzyme choice, immobilization procedure) is discussed in Chap. 6, whereas the "hardware" part is discussed in Chaps. 1 and 4.

### *Robotic Systems*

For the sample pretreatment of large numbers of samples the robotic systems are becoming more and more important the last few years (Fig. 3.11) [138,139]. Because in most laboratories 40-75% of the time is wasted by time-consuming manual steps such as weighing, sample and reagent transfer, dissolution, centrifugation, mixing, and solvent dispensing for extraction, robotic systems seem to be a welcome addition to the present range of sample pretreatment techniques [140]. The robotic system (pick-and-place system) is equipped with a three-dimensional (3-D) movable arm that can reach a number of attributes within a certain area and is able to transport the sample from one station to another and finally to transport it to an autosampler (Fig. 3.11). Furthermore, the throughput of the samples can be increased significantly, not because a robotic system is faster but because it works 24 hours a day without breaks [141].

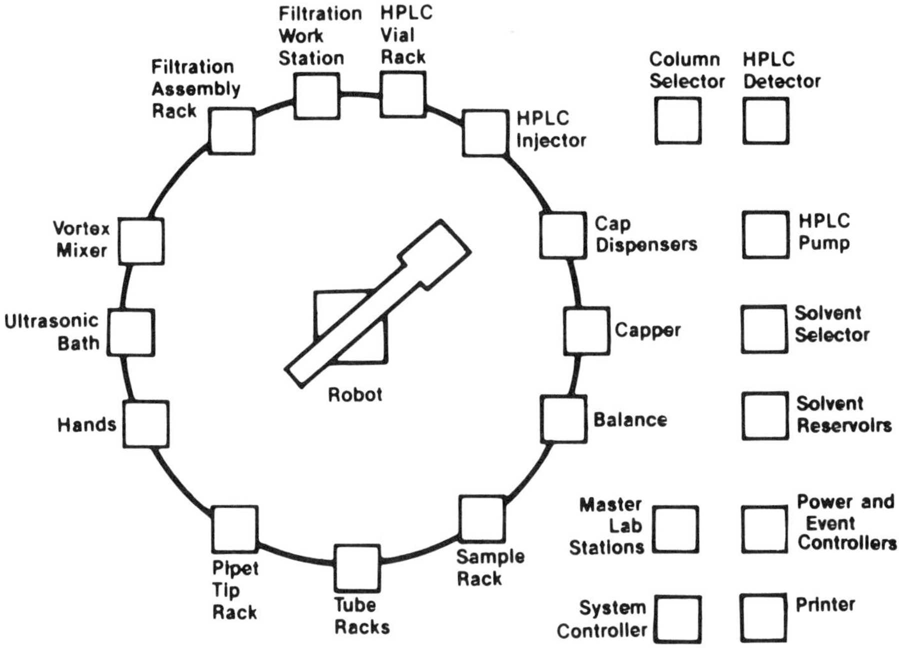


Figure 3.11 Schematic overview of robotic system. (From ref. 140.)

Robotic systems are able to weigh, dilute, mix and shake, filtrate, extract, and inject samples. Moreover, in the near future the possibilities with these systems will increase strongly.

The required software needed to guide the robotic systems is now under development, especially for dealing with the interface problems between the applied hardware and the robot controller unit. Software is already available for a number of applications such as the optimization of a spectrophotometric analysis, the analysis of samples in wastewater, and sample pretreatment prior to GC and LC determinations [142,143].

Robotic systems can be an excellent way to improve the precision of the results. For instance, a comparison of the relative standard deviations of the analysis of OPA-derivatized amino acids by LC gives a 5% standard deviation for manually treated samples, a standard deviation of 3% when sample processors are used, and a standard deviation of 2% when a robotic system is applied [141].

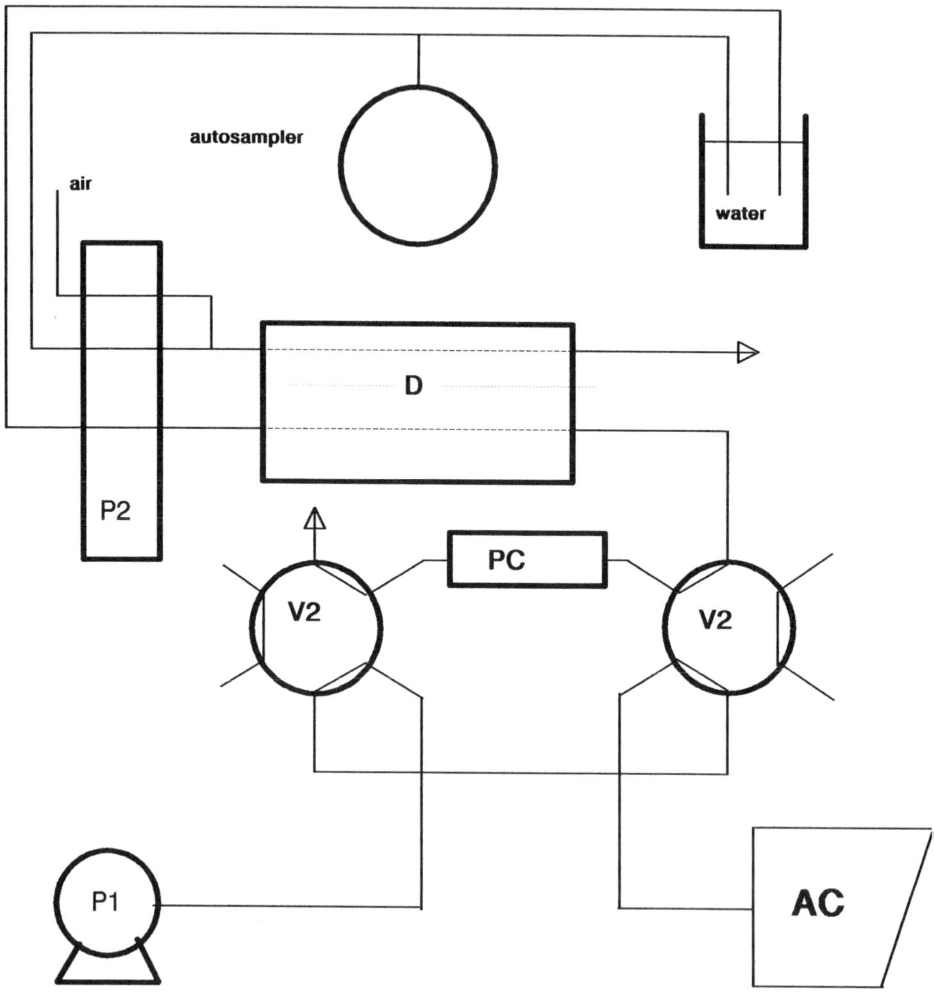


Figure 3.12 Continuous flow-based dialysis system combined with precolumn valve switching technology. AC, analytical column; D, dialysis block; $P_1$, low-pressure pump; $P_2$, high-pressure pump; PC, precolumn; $V_2$, six-port switching valves.

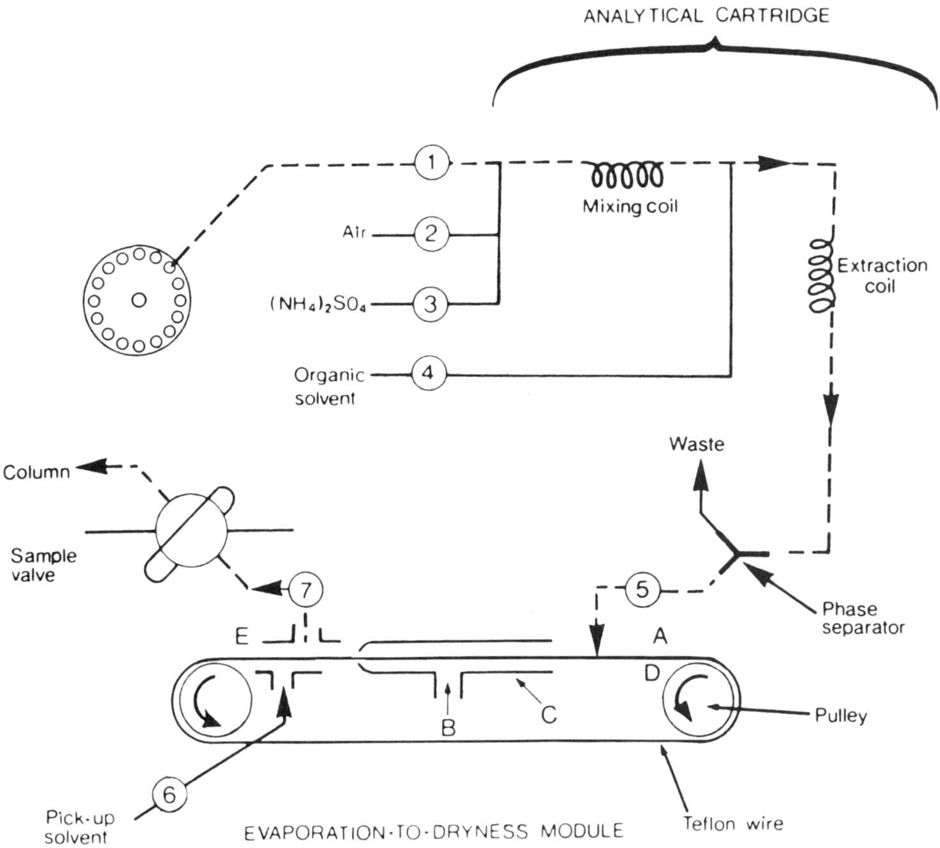


**Figure 3.13** Flow diagram of evaporation-to-dryness module. Sample pathway (----); other liquid streams (——). (From ref. 144.)

### *Continuous Flow Systems*

One of the fully automated sample pretreatment procedures is the use of continuous flow (CF) techniques such as the evaporation-to-dryness system [144] and a dialysis membrane (e.g., ASTED system) (Fig. 3.12) [145].

An example of a CF system for the analysis of theophylline in serum is based on the use of an automated clean-up in combination with an evaporation-to-dryness module (Fig. 3.13) [144]. Serum from an autosampler is air-segmented and mixed with an inorganic salt (e.g., ammonium sulfate) and an organic solvent. After equilibration in the extraction coil, the analyte-containing organic phase is transferred to the moving belt and evaporated by means of a

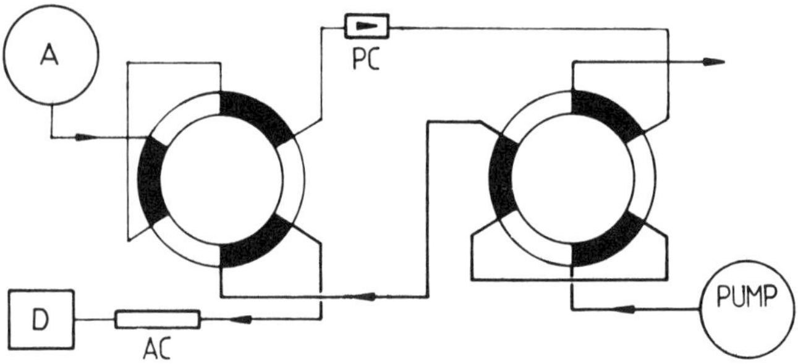


Figure 3.14 Valve switching unit for the analysis of mitomycin C samples applying CF. Loading of precolumn (PC) starts at A and follows the black blocks of the six-port valves. Elution of samples to the analytical column (AC) follows the white blocks of the valves. A, autosampler and dialysis block; D, detector. (From ref. 147.)

hot-air stream in the evaporation-to-dryness module. Applying this procedure a day-to-day reproducibility of about 5% can be obtained.

A second technique is the use of an on-line dialysis system. Normally techniques based on dialysis of samples are relatively slow. However, using a CF system a relatively high concentration gradient over the dialysis membrane is realized. This is due to the increased contact time between the donor and the acceptor streams. The automated dialysis systems (e.g., ASTED [145]) are based on the following sequence of steps: loading of sample, dialysis and concentration, injection, and regeneration [146]. The major disadvantage of this type of CF technique is the inherently low recovery of maximal 50% in the concurrent mode. However, applying the countercurrent mode recoveries of 100% can be obtained. The most pronounced advantages are high reproducibility, minimal sample pretreatment, and the possibility of developing a fully automated system.

A fully automated LC system is described for the analysis of mitomycin C in biological samples [147] after isolation of the analyte from the biological matrix (plasma, ascites, urine) with the CF system. The samples are concentrated on a RP precolumn and subsequently analyzed on RP analytical column applying a CS technique (Fig. 3.12). CS is applied to allow the combination of the relatively large volumes needed in CF techniques and the relatively small volumes that are required in LC (Fig. 3.14). Using the described system up to 100 samples a day can be analyzed with determination limits (absorption detection) in the order of 1 ng/ml and a linear dynamic range of at least three decades, with recoveries of about 25%, and with an improved reproducibility in comparison with other analytical methods.

A second application of a fully automated CF system is described for the screening of sulfonamide residues in milk, eggs, and meat using the CF system in combination with an XAD pretreatment and postchromatographic derivatization of the analytes [148].

A complication in the use of a dialysis membrane may be the degree of protein binding of the analyte, which may influence its recovery. In principle, only analytes that are non- or weakly bound to serum proteins can be analyzed without additional sample pretreatment with this type of CF procedure.

*Sample Processors*

One of the most important steps in automation is the use of autoinjection systems which can be combined with off-line as well as on-line sample pretreatment procedures [149].

Sample processors are cartesian programmable 3-D systems, in contrast to the robotic systems which are anthropomorphic systems, by the presence of the "programmable arm." However, the major disadvantage of the robotic systems is their limited speed.

The sample processors meet this disadvantage by performing autopipetting and mixing operations prior to automated sampling and injection procedures into the LC system. Sample processors equipped with one or two six-port valves can be used to [150]

- Inject into a second system in order to increase the throughput (e.g., a sample is injected in the first system, while the separation is performed on the second system)
- Collect fractions in (sealed) vials for automated collection and concentration
- Concentrate samples
- Clean up samples
- Program CS (a part of the sample is separated on the first column while another part is separated on the second column)
- Analyze a sample on two columns in series (two-column selection)
- Back-flush a guard column to remove a strongly retained material from the precolumn
- Back-flush the analytical column to remove a strongly retained material from the analytical column

All these procedures are programmable so as to be used in, say, automated precolumn derivatization of amino acids (see Chap. 1) [151].

Sample processors normally have a high accuracy and precision, allow a high throughput of samples, possess sufficient flexibility, and are far less expensive than robotic systems. The use of sample processors for the analysis of biological samples can be the reason for problems concerning the stability of the analytes. During extended storage of the samples at room temperature in the sample processor,

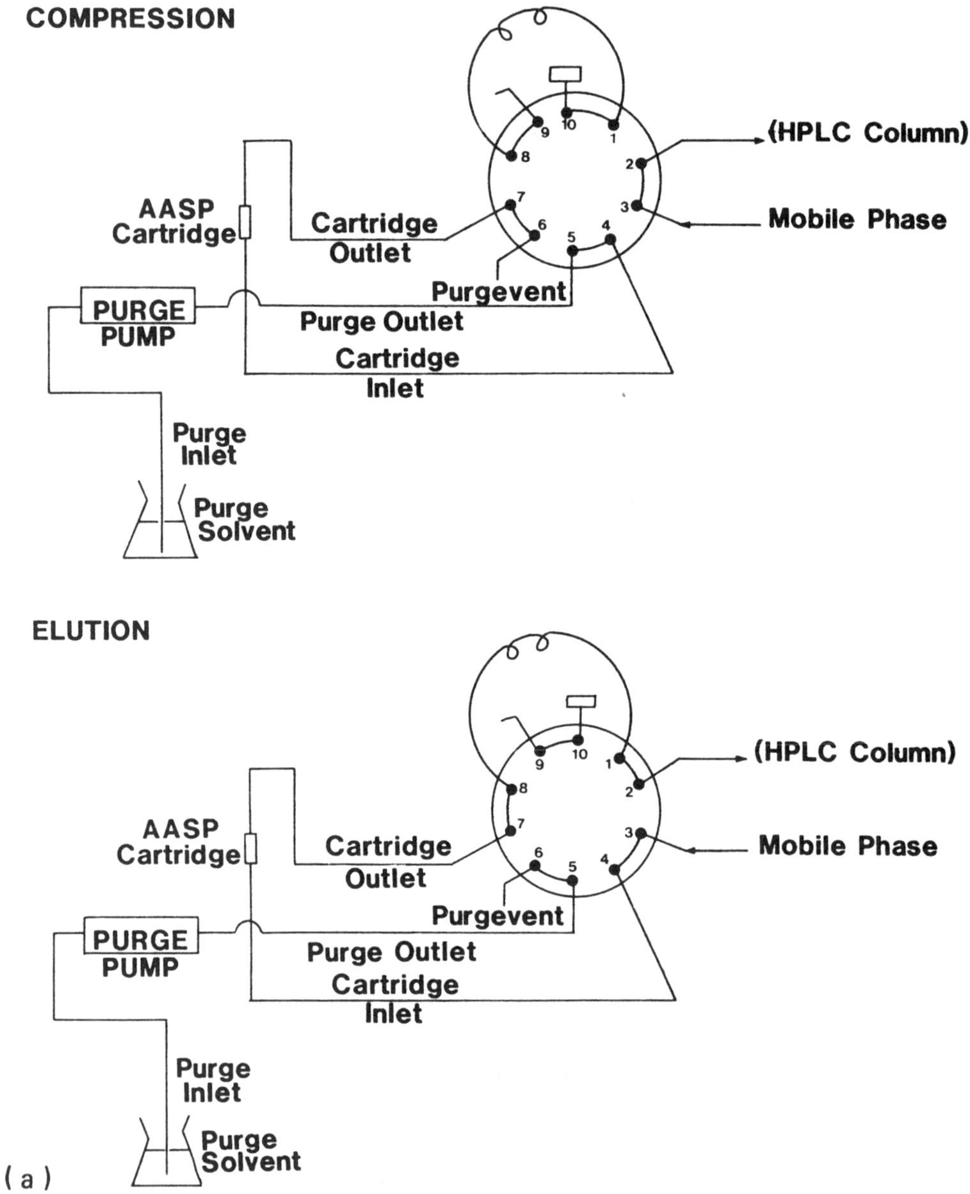


Figure 3.15 Schematic presentation of (a) AASP cassette and solvent pathways during washing and elution of the cartridges, and (b) detail of the cartridges. (From ref. 154.)

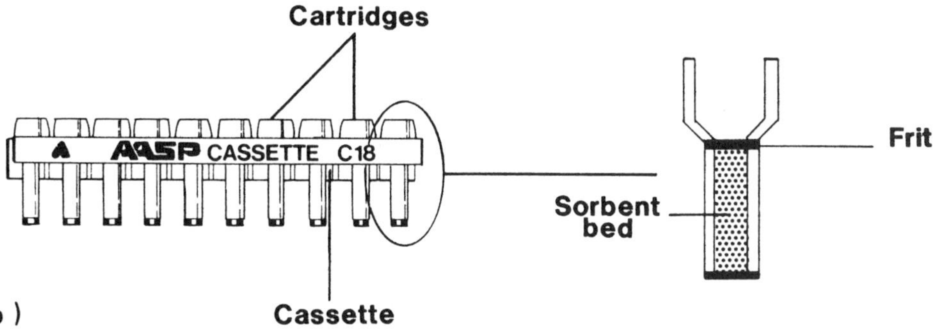


(b)

Figure 3.15 (continued)

decomposition of the analyte may occur. Refrigeration of the sample processor carousel, for instance, by placing the carousel in a refrigerator at 273 or 277 K, may be a necessity [152,153].

Off-line SPI sample pretreatment procedures can be partially automated by using, for instance, the AASP (advanced automated sample processor) [154-156]. The AASP system combines SPI with syringeless automated injection of the samples. The samples are prepared off-line (on the bench) in a cartridge holder (containing 10 cartridges). After loading the samples, the columns are washed and the cassettes containing the cartridges are loaded into the AASP. Prior to the automated injection the cartridges can be washed (Fig. 3.15), and subsequently the analytes are on-line flushed to the analytical column and analyzed.

However, one of the disadvantages of this system is incompatibility between the relatively large volumes of the elution solvent and the relatively small volumes of the injection solvent, which often results in band-broadening effects on the analytical column.

Quantification of an inotropic agent in plasma with a fully automated procedure using a combination of a sample processor, and the AASP is described by Murkitt et al. [157]. The sample processor prepares the samples by addition of several solvents and mixing them in a loop, and transfers these solutions to the AASP cassette. With this system at least 100 samples per day can be analyzed.

The main advantage of the AASP system in comparison with direct injection with CS techniques is that small quantities of the analyte (e.g., antiepileptic drugs) can be analyzed with a recovery of nearly 100% and a day-to-day precision of about 3% [158]. In principle the precision of the AASP system is comparable or even better than that of off-line solid phase cartridges.

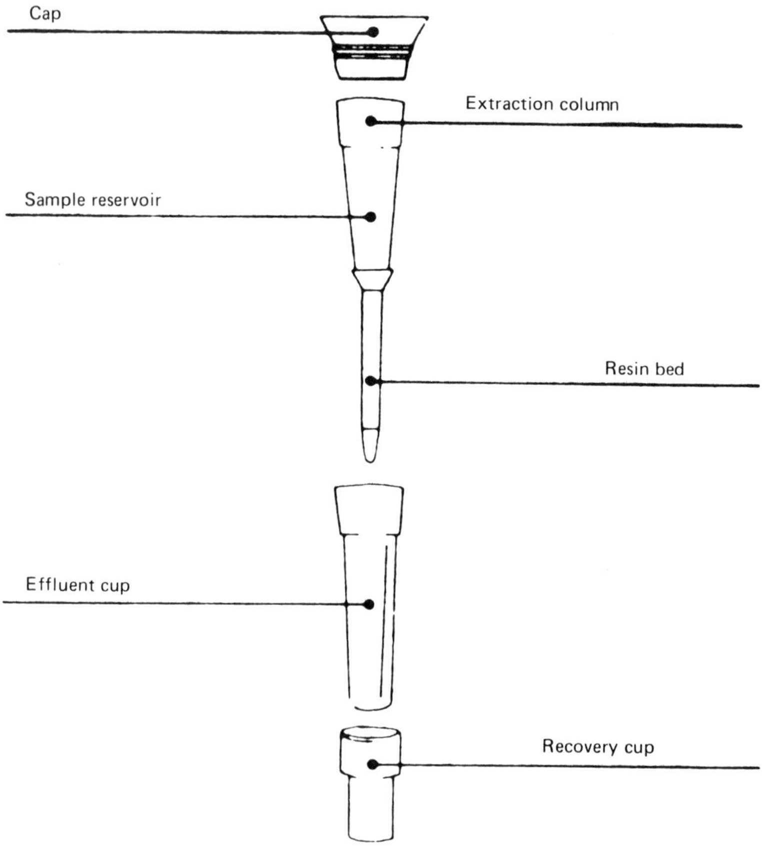


Figure 3.16 The Du Pont "Prep" rotor system during extraction and elution of the samples. (From ref. 67.)

Another semiautomated sample preparation system is the DuPont "Prep" system [159]. It is based on a bidirectional centrifuge that combines preparation and concentration of the sample in the same instrument. The rotor system allows the simultaneous analysis of 12 samples (Fig. 3.16). The disadvantage of this system is that the balance of the rotor is very important, meaning that the maximum number of samples must be pretreated every time. The advantage, of course, is that once the experimental conditions have been

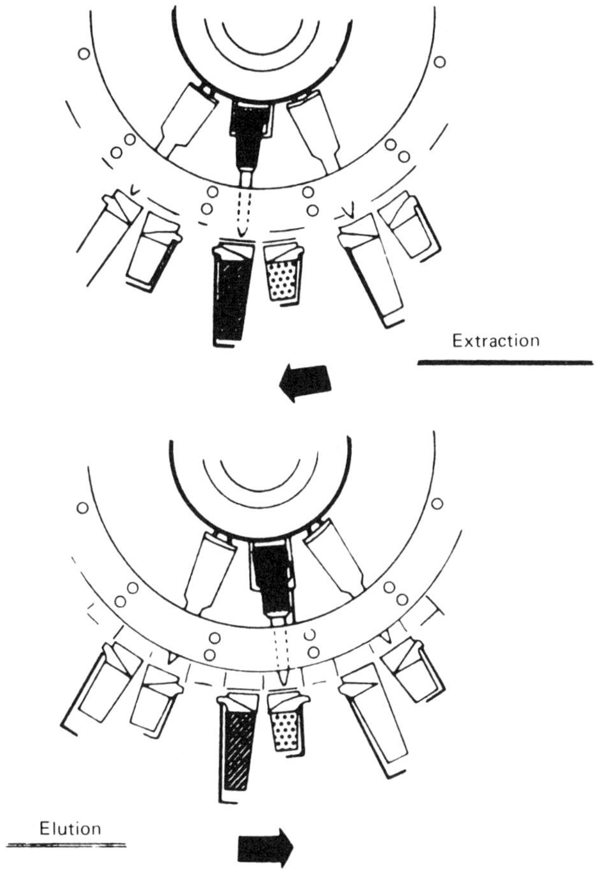


Figure 3.16 (continued)

established the process is fully automated, apart from the loading of the extraction cartridges into the instrument. Furthermore, accuracy and reproducibility are better than in off-line procedures [160,161].

An example of using a sample processor in combination with gel filtration is described for the analysis of pesticides in soils and food, which is of great importance in environmental studies. The necessary clean-up of the samples before the LC analysis can be processed using a gel filtration method in which the larger interfering compounds are eluted first and the analytes, being the smaller compounds, later [162].

An example of real on-line partial automated sample pretreatment would be the use of sample processors which are able to execute a limited number of sample pretreatment procedures such as dilution and addition of solvents in, for instance, the loop of the injection valve due to the presence of a six-port valve CS technique, introducing the possibility of sample concentration and sample clean-up with SPI precolumns.

*On-line Solid Phase Isolations*

In routine bioanalysis complex samples must often be analyzed. Except for proteins and fats, biological samples contain compounds of widely different polarities demanding long analysis times if isocratic elutions are performed. Although gradient elution can decrease the analysis time, the throughput of the samples is not proportionally enhanced due to the required regeneration time of the column.

Therefore, especially in trace and bioanalysis, the application of CS techniques is very useful because in this way concentration and clean-up becomes a part of the analytical procedure and sample manipulation is minimized [91,163]. Normally CS procedures are more reproducible than gradient elution systems. Furthermore, CS techniques can be used for the same reasons as described earlier for the use of sample processors.

Applying on-line SPI techniques, the samples are normally flushed by an additional pump, other than the one used for the analytical column, through the precolumn, and the elution of the analyte is performed by switching of one of the six-port valves (Fig. 3.17).

In CS several possibilities exist such as back-flushing [130] (elution in opposite direction as sample loading), forward-flushing [164] (elution in the same direction as sample loading), heart-cut [165], front-cut, and tail-cut, and the possibility of using a number of different columns in an on-line mode. Furthermore, the CS unit can be combined with several specific detection systems in a parallel mode or in series. The result is that the selective analysis of an analyte in a complex matrix is far easier to achieve by the application of CS methods. The principle of the use of precolumns and of CS techniques is discussed in Chaps. 1 and 4 of this book.

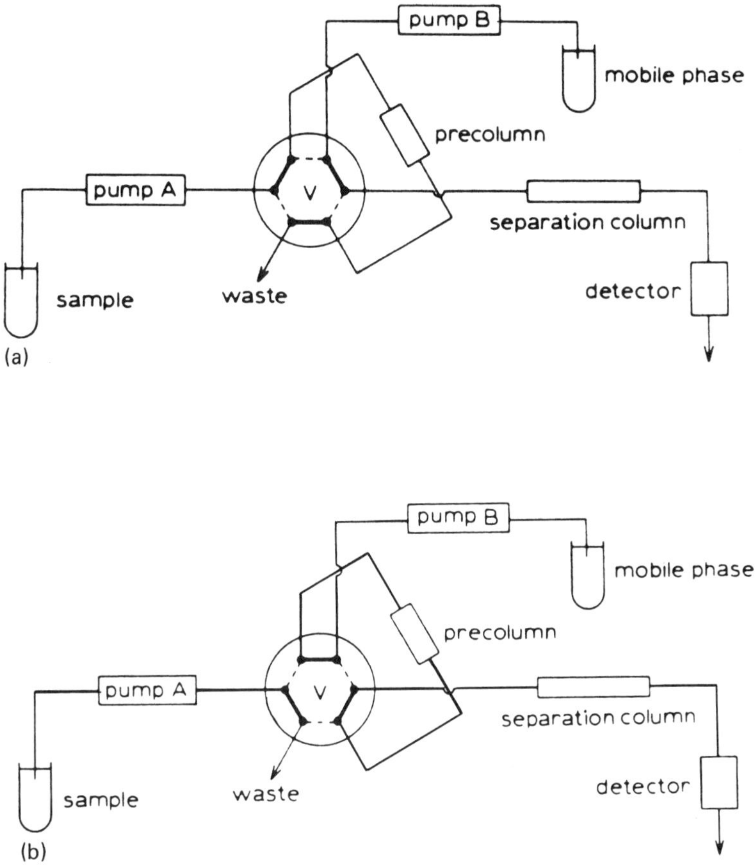


Figure 3.17 Configuration of on-line sample preconcentration, using one precolumn, one analytical column, and one six-port valve. (a) concentration mode; (b) separation mode; V, six-port switching valve.

Nowadays the application for routine analysis of on-line precolumns in automated SPI systems, prior to an LC separation, is a well-known technique [67,166]. In principle small, in comparison with the analytical column, precolumns (length: 2-10 mm; internal diameter, 2-4.6 mm) packed with relatively large particles (25-60 µm) and closed with screens instead of the normally applied frits are used. The best performance of the on-line precolumns is obtained when packed with porous 30-µm particles instead of 30-µm pellicular particles or with smaller porous particles [167]. Applying these rules using a boxcar configuration of the precolumn and the analytical column, about 2000 serum samples can be injected before the analytical

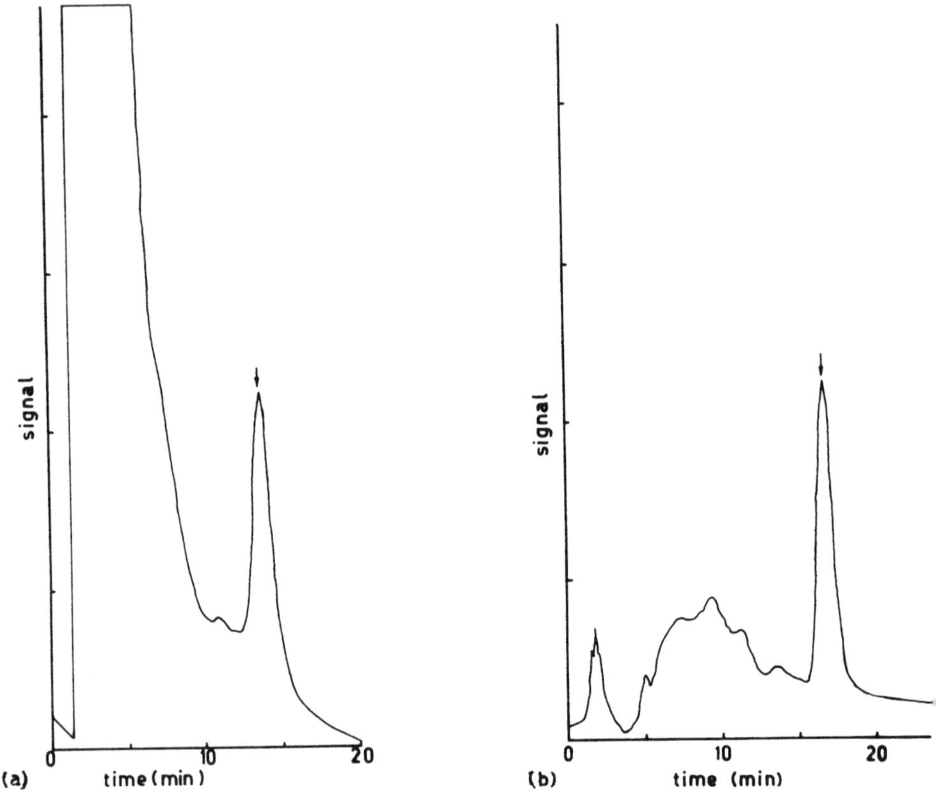


Figure 3.18 Chromatograms obtained with reversed phase LC of derivatized fluvoxamine in plasma samples using a CS system with a $C_2$ precolumn and an $C_8$ analytical column. (a) Direct elution of the precolumn and (b) after flushing of the precolumn. (From ref. 18.)

column should be replaced. The precolumns must be small because the k' of the analyte in the precolumn must be small in comparison with the k' of the analytical column to avoid additional band broadening in the system [168,169]. This means, for instance, that a $C_{18}$-modified analytical column cannot be used in combination with the more hydrophobic PRP-1 precolumn if for both columns the same eluent is used. However, additional band broadening can be limited by applying a different eluent with concentrating properties for the second column.

As mentioned before, it is possible to use a combination of several precolumns in series, but the sequence of the precolumns is

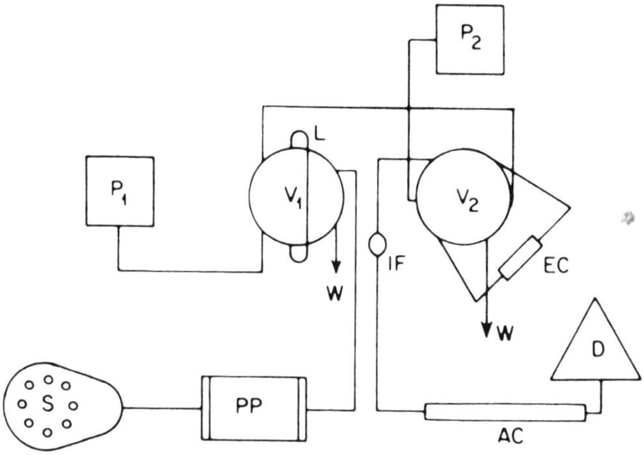


Figure 3.19 Configuration of the boxcar LC system. S, autosampler; PP, peristaltic pump; L, sample loop, $V_1$, $V_2$, and $V_3$, six-port switching valves; W, waste; $P_2$ and $P_3$, mobile phase pumps; $P_1$, wash pump; EC, extraction column; SC, short column; AC, analytical column; IF, in-line filter; D, detector. (From ref. 167.)

very important. The column with the weakest interactions should be placed before the precolumn with the stronger interactions (Fig. 3.18) (e.g., $C_{18}$ before PRP-1 and PRP-1 before ion exchange).

The advantages of CS methods over off-line SPI and the automated sample processors loaded with SPI cartridges are the ease of automation of the entire analytical procedure and excellent reproducibility and precision of the system. The most pronounced disadvantage, however, is the multiuse of the SPI columns, which is especially troublesome when some compounds are bound nearly irreversibly to the sorbent. These interactions may be the reason for severe memory effects. Another disadvantage is lack of flexibility of most of the applied CS systems. Furthermore, the lifetime of the precolumns is limited. After a certain number of injections the backpressure of the system increases and the reproducibility of the system decreases dramatically.

A special form of CS is boxcar configuration chromatography [135,167]. This technique simultaneously places multiple samples in an analytical column and thus increases the sample throughput (Fig. 3.19).

In boxcar chromatography a zone (heart cut) of the precolumn is transferred to one of two analytical columns. In this case the sample throughput can be doubled because as soon as the first sample reaches the first analytical column a second sample can be injected onto the second analytical column. However, in order to avoid contamination of the analytical columns back-flushing of the precolumn is a necessity in this case [170]. Boxcar chromatography is possible because often the precolumn can be cleaned and reequilibrated before elution from the analytical column is complete.

In the next part of this section a number of general examples of the application of CS techniques in sample pretreatment procedures will be given.

The direct analysis of plasma samples and other protein-containing samples is advantageous because it reduces analysis time and improves the accuracy and precision of the analytical method. The direct analysis onto a LC system is possible when a CS technique is applied with at least one precolumn (Fig. 3.17). Serum, plasma, urine, and saliva samples are denaturated with acetonitrile to eliminate the protein binding of the analytes and/or to block enzyme activity. Tissue homogenates and feces are homogenized in a suitable buffer, solubilized, and denaturated with acetonitrile. The resulting solutions are injected onto a precolumn packed with wide-pore RP material, allowing the pretreatment of samples up to 50 ml. Elution of the analytes from the precolumn to the analytical column can best be performed in the back-flush mode [15,171]. About 100-150 ml can be injected onto one precolumn before it must be replaced by another dry-packed column.

Almost the same principle (precolumn and back-flushing) is applied to the analysis of tripelennamine in milk [172]. However, to increase the lifetime of the columns it may be advisable to centrifuge and filter (50 μm) the rough samples (e.g., milk, plasma) before they are injected onto the CS system [173].

Normally, silica-based precolumns are applied. However, because these phases should be used at pH values of 8 or less [174, 175], the nonpolar adsorbents XAD-2 and PRP-1, which are compatible with all pH values between 1 and 14, are sometimes advantageous [176].

On-line deproteination of serum samples for LC analysis is also possible applying an adsorbent gel such as Butyl ToyoPearl 650 M [177]. This gel adsorbs proteins in a given medium by means of mild hydrophobic interactions by the butyl and hydroxyl functions on the surface. For medium hydrophilic as well as for hydrophobic analytes quantitative results are obtained regardless of the degree of protein binding. Butyl ToyoPearl is used for the analysis of gentamicin and it is found that 36 mg of bovine serum albumin can be adsorbed by 1 ml of the wet gel [178].

The precolumn venting plug (segmenting) technique is an improved simple CS system for the direct injection of nonpretreated plasma samples [179,180]. The precolumn is again used for concentration and clean-up of the samples, but before the sample is applied

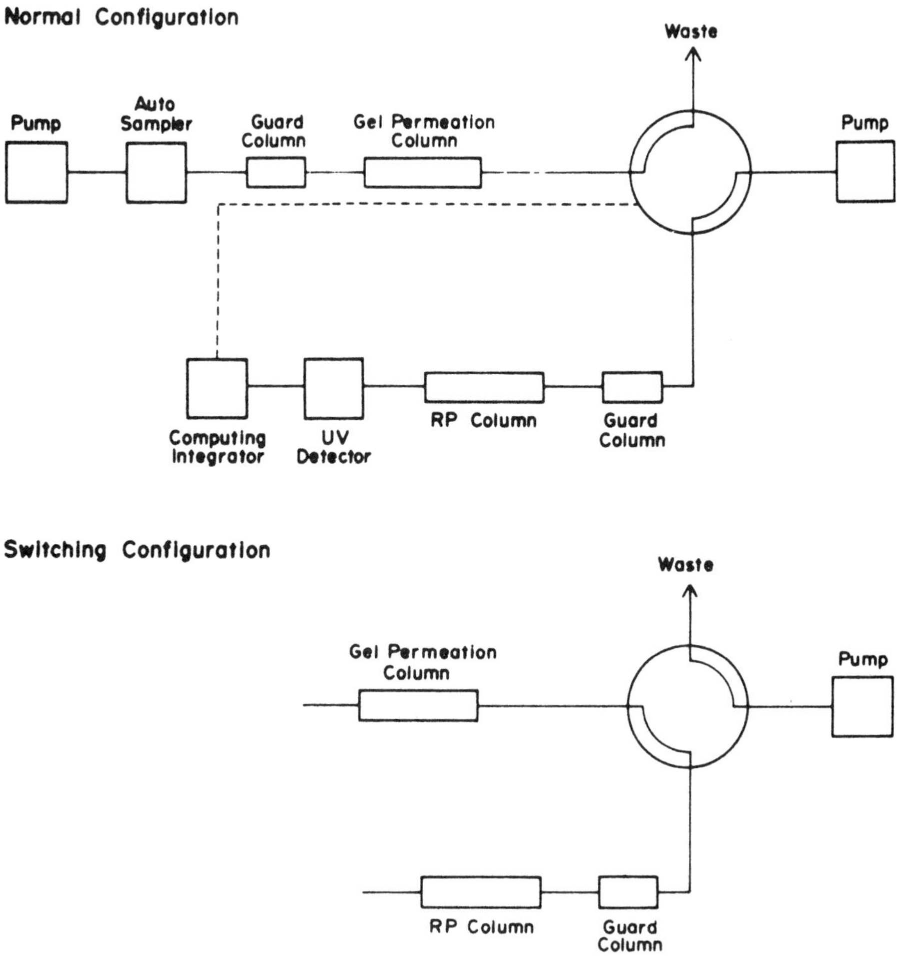


Figure 3.20 CS scheme of the analysis of theophylline in plasma samples using a gel filtration precolumn. (From ref. 181.)

to the precolumn aqueous plugs are introduced on both sides. The result is a stable system because contact between plasma proteins and organic solvent is prevented in this way. The proteins are washed to waste with the aqueous plug solvent. The advantage of this plug-venting technique is that the main fraction of the plasma protein does not reach the analytical column. The system is less time consuming and more stable than the earlier described back-flush [171] and forward-flush [164] methods.

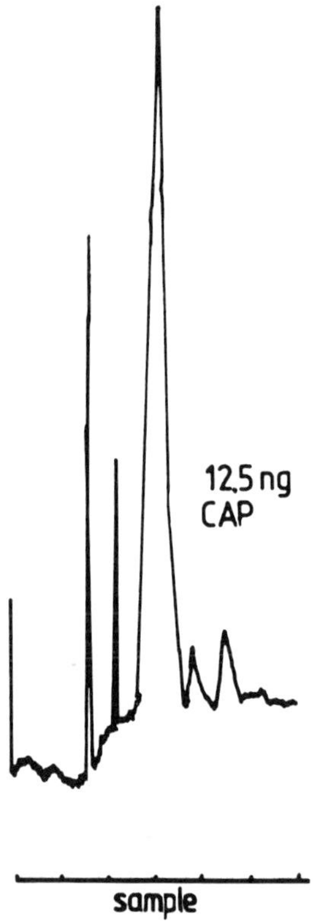


Figure 3.21 Chromatogram obtained with reversed-phase analysis of chloramphenicol in kidney tissue homogenates applying the heart-cut technique. (From ref. 165.)

Another possibility is the use of a gel filtration precolumn (Fig. 3.20). A fully automated system is described for the analysis of theophylline in 30-μl plasma samples. With this system 70 samples a day can be analyzed [181].

The application of micellar systems [182,183] is another possibility to prevent clogging of the proteins on the analytical column. Using a micellar mobile phase (e.g., aqueous sodium dodecyl sulfate) for the precolumn the precipitation problem is eliminated, due to the solubilizing power of the micelle. A disadvantage of this technique is the relative inefficiency of micellar chromatography; however, an advantage is that relatively large volumes (0.01-0.5 ml) can be injected. This system is applied to the bioanalysis of morphine, codeine, quinine, quinidine, and propranolol [184].

To increase the throughput of samples during the direct analysis of plasma samples, a fully automated system with some six-port switching valves and two or more alternating precolumns can be used [131]. For instance, antiepileptic drugs can be analyzed very rapidly in serum samples using the two alternating $C_{18}$ precolumns [164], or by using the box-car configuration consisting of a fast separation column prior to a second analytical column [129]. The first column rapidly separates a desired fraction which is transferred subsequently to the second column. Prior to elution of the sample fraction, fractions from other samples are placed on the latter column. The cut fractions thus migrate adjacent to one another in the analytical column.

The use of heart-cut techniques, an extension of the back-flush principle, can significantly improve the selectivity of the system just mentioned [165]. With this technique a fraction (heart cut) of the eluate of the first column is, for example, intercepted in a capillary loop or directly transferred to the second column. Two configurations are possible with this chromatographic techniques; two different eluents with one type of stationary phase, or two different stationary phases with one eluent. The heart-cut technique offers a high degree in selectivity in the sample pretreatment procedure (Fig. 3.21).

A disadvantage of on-line coupled columns in series may be the dilution of the samples. However, when, for instance, an injection solvent with a weaker elution strength (e.g., nonpolar solvent in combination with RP-LC is used, the samples can be concentrated on the top of the analytical column and in this way much larger volumes can be introduced onto the analytical column.

The highest selectivity can be obtained using precolumn technology based on a specific interacting sorbent. An automated procedure is described for the pretreatment and derivatization of trace levels (low ng/ml range) of ethynyl steroids in urine samples, applying a selective on-line preconcentration with small precolumns loaded with hydrophobic and silver(I)-loaded sorbents [185,186]. However, the application of this type of column to urine samples is rather critical because of the large number of interfering compounds. Therefore, prior to the silver-loaded column a hydrophobic PRP-1 column is placed to remove the inorganic and water-soluble compounds (Fig. 3.22).

Other ligand exchange systems are suitable for the clean-up of aniline and thiol-containing samples [188,189]. The mercury(II)-2-amino-1-cyclopentene-1-dithiocarboxylic acid (ACDA) phase can be used for the pretreatment of thiol-containing samples, while platinum(IV)-ACDA is suitable for aniline-containing samples. The retention of the analytes of the metal-loaded phases is governed by a number of different mechanisms [186]. The two major interactions

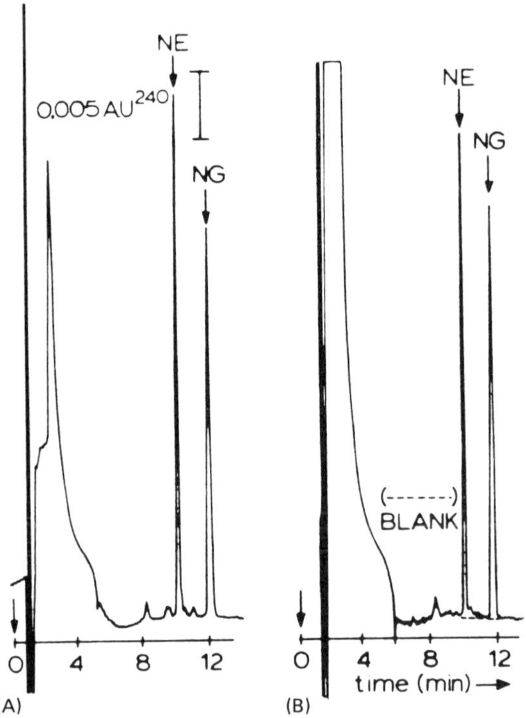


Figure 3.22 Chromatograms obtained with reversed-phase LC of norethindrone (NE) and norgestrel (NG) in academic (A) and urine (B) samples using a PRP-1 and a silver-oxime precolumn. (From ref. 187.)

are strong complexation of the metal ion and the functional group of the analyte, and the hydrophobic interactions between the π systems of the sorbent and the analyte. The desorption of the analytes from the precolumn is accomplished with an excess of a competing thiol (e.g., cysteine) or aniline. The disadvantages of this type of selective interacting phases are that regeneration of the columns is laborious and the breakthrough volumes are relatively small.

Strongly basic anion exchange resins can be used for pretreatment and preconcentration of phenols in wastewater [190] but, due to high concentrations of ionic compounds in wastewater samples, the phenols are first adsorbed onto a hydrophobic column while the anions are flushed to waste. In the next step the phenol is sorbed in a very small zone on the anion exchange column. Almost the same

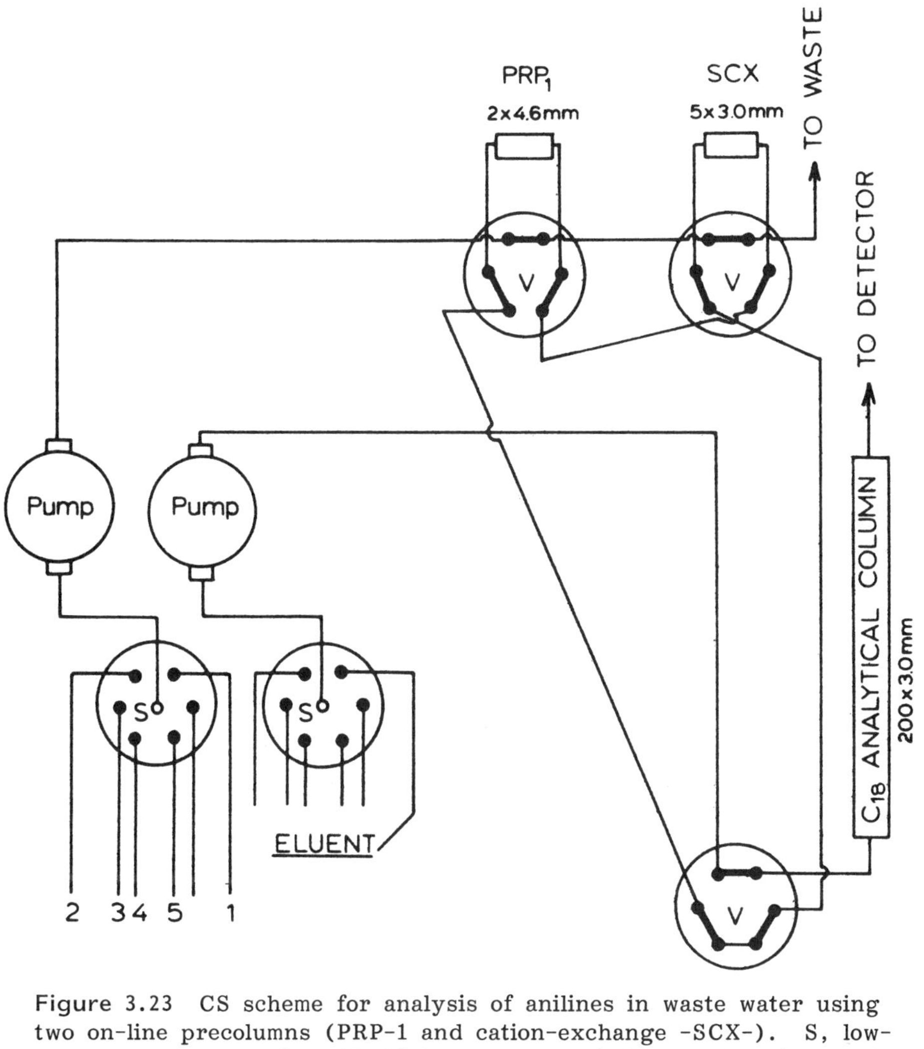


Figure 3.23 CS scheme for analysis of anilines in waste water using two on-line precolumns (PRP-1 and cation-exchange -SCX-). S, low-pressure selector valve; V, high-pressure switching valve; 1, sample; 2, water; 3, methanol; 4, 0.02 M perchloric acid; 5, 0.001 M perchloric acid. (From ref. 114.)

procedure is used for the analysis of anilines in environmental samples (Fig. 3.23) [114]. However, in this case a strong cation exchanger is applied instead of an anion exchanger. For the desorption of the anilines from the cation exchanger a relatively high

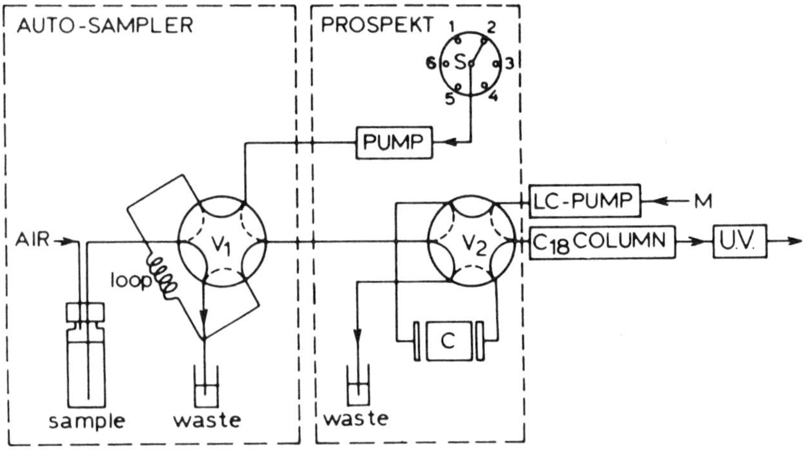


Figure 3.24 CS scheme of the automated sample handling system (PROSPEKT). $V_1$, injection valve; $V_2$-$V_4$, high-pressure switching valve; S, six-port solvent selection valve; M, mobile phase; C, disposable cartridges. (From ref. 192.)

concentration of aqueous potassium phosphate should be used to avoid band broadening.

Instead of using a special column for the sample pretreatment, the analyte can also be selectively modified before the pretreatment and subsequently cleaned on a standard hydrophobic column [191].

The main disadvantage of the use of on-line SPI techniques is the possibility of memory effects and clogging of the system. Furthermore, it has been observed that small amounts of analyte can be irreversibly adsorbed onto the sorbent, especially basic drugs (amines). One of the solutions may be the use of forward flushing instead of the normally applied back flushing. However, in this case short precolumns (5-mm length) should be applied [164].

Another solution may be the use of disposable cartridges which can be applied in an on-line system. Therefore, the Prospekt is under development to combine the possibilities of automation by means of CS techniques with disposable precolumns (Fig. 3.24).

The system consists of an autosampler, a microprocessor-controlled precolumn exchange module, a washing pump, an injection valve, and some high-pressure switching valves. The procedure is as follows: the cartridge exchange module is filled with the desired type of cartridges. First of all, the cartridge is automatically wetted and conditioned and subsequently the sample is applied to the top of the precolumn. By switching the valves ($V_2$ and $V_3$) several washing and elution steps can be applied before the analyte is transferred to the analytical column or to a second precolumn (e.g., with another sorbent) whenever further pretreatment is necessary [192].

System efficiency is not influenced by the cartridge system and no extra band broadening is observed, contrary to the AASP system in which a loss of 30% in the efficiency is observed. However, it is essential that short precolumns be applied in combination with low-volume connecting capillaries.

Applications are already described for the analysis of drugs in serum, anticancer drugs in plasma, barbiturates in urine, herbicides in river water, and caffeine in soft drinks [192].

## CONCLUSIONS

The emphasis in this chapter has been on the sample pretreatment of biological samples before the analytes are derivatized. However, the choice of sample pretreatment procedure must be compatible with the applied LC separation and detection system and with the used prechromatographic or postchromatographic derivatization procedure. Furthermore, the chromatographic system should not be damaged by introduction of aggresive chemicals from the derivatization mixture. Clean-up steps following the derivatization procedure are therefore frequently unavoidable. The removal of the excess reagent to prevent excessive detector response is another reason for additional sample pretreatment and has been mentioned before. In choosing a suitable clean-up procedure for the derivatization mixture the analytical chemist must be aware of the physicochemical properties of the constituents of the mixture. Very little is needed to cause appreciable loss of the minute amount of derivative formed in the reaction. The stability of the derivatives is of paramount importance and should be assessed carefully.

To find the appropriate sample pretreatment procedure for the analysis of one or more analytes in a certain concentration in a particular matrix, especially when small nonrepetitive sample series should be analyzed, is sometimes very difficult. In Table 3.2 an overview is given of the described off-line sample clean-up procedures which can be used for the analysis of small sample numbers.

Often the weakest link in the whole analytical procedure is the sample pretreatment, and therefore methods to increase the throughput of the samples (e.g., automation procedures) and more selective methods compatible with the sensitive detection systems (e.g., laser-induced fluorescence detection [193]) will be the topics of the near future.

This explains the advance of the sample processors. In comparing the cartesian samplers (sample processors) with the anthropomorphic arm systems (robotic systems), the most pronounced difference

Table 3.2 Off-line Isolation Procedures for Organic Analytes

| Isolation technique | Application | Ref. |
|---|---|---|
| Chemical modification | Initial pretreatment of unstable analytes | 29 |
| Covalent labeling | Initial pretreatment of unstable analytes | 30 |
| Dehydration | Removing aqueous solvent before further pretreatment | 2 |
| (Vacuum) distillation | Analysis nonvolatile analytes in large sample volumes | 27 |
| Electrophoresis | Sample clean-up of ionic analytes | 65,66 |
| Enzyme inhibition | Stabilization analytes in enzyme-containing matrices before further pretreatment | 33-38 |
| Freezing (e.g., liquid $N_2$, dry acetone) | Blocking enzyme activity, stabilization analyte | 14 |
| Head space | Analysis of volatile and nonvolatile analytes in one matrix | 64 |
| Homogenization solubilization | Initial pretreatment of (solid) matrices | 12,13, 15 |
| Hydrolysis | Analysis of conjugates in biological matrices | 39-42 |
| Liquid-liquid extraction (LLE) | The sample clean-up procedure for small numbers of organic analytes in all kinds of matrices with or without combination with the sample/analyte stabilization techniques | 67,68, 75,81, 84,87, 88 |
| Lyophilization | Removal of water, stabilization of analyte before further pretreatment or storage | 32 |
| Microwave irradiation | Solubilization samples, denaturation proteins, thawing samples; can be used as single clean-up procedures | 12,31 |
| Precipitation | Denaturation of proteins in biological samples, normally in combination with LLE or SPI | 43,44, 48,57 |
| Solid phase isolation (SPI) | The sample clean-up procedure for small to medium numbers of organic and inorganic analytes in all kinds of matrices with or without combination with the sample/analyte stabilization techniques | 67,89 |

Table 3.2 (continued)

| | | |
|---|---|---|
| Soxhlet extraction | Solubilization of analytes, in relatively large or in solid matrices | 28 |
| Ultrafiltration | Separation and clean-up of analytes on basis of molecular weights | 9,10, 60 |

is that the necessary space for the cartesian samplers is limited whereas for the robotic systems far more space is necessary. Furthermore, the cycle time of the cartesian samplers is much shorter than the cycle time of the robotic systems. On the other hand, the sample processors are only suitable for the handling of liquid samples, while the robotic systems can also be applied for solid samples and other objects.

However, one of the present limitations in the use of robotic systems is the limited availability of software to link the robotic controller with the computer, but the combination of the capabilities of these two technologies will certainly improve very strongly in the near future, resulting in a potentially complete and powerful technique for sample pretreatment [143].

The choice between the application of an off-line or on-line sample pretreatment technique and between an automated or an nonautomated clean-up strongly depends on the complexity of the sample (Table 3.3). For the routine analysis of medium- to large-size sample series it is advisable to automate the pretreatment using on-line techniques, and for the routine analysis of small- to medium-size sample series the application of off-line automated systems (sample processors, robotics) is advantageous.

In the near future the popularity of the off-line SPI will increase even more, even for relatively small numbers of samples, because of the availability of a wide range of chemically bonded phases which can be applied to the selective isolation of analytes from complex matrices [194]. The SPI will not displace the LLE, but both systems are supplemental to each other.

The application of the on-line automated, disposable SPI cartridges will become even more popular than the off-line use of SPI, especially with the introduction of a number of very selective interaction sorbents or with the introduction of new applications for the well-known sorbents.

An important future trend will be the introduction of immobilized systems (solid phase systems) which allow derivatization and sample pretreatment simultaneously or which allow the application of on-line automated derivatization procedures [127,195,196]. The three basic approaches of these techniques will be the use of silica-supported

Table 3.3 Sample Pretreatment Choice

| Number of samples | | | | | | | Number of charges |
|---|---|---|---|---|---|---|---|
| | | | *Automated* | | | | |
| Large | On-line: | I. | Column switching | On-line: | I. | Column switching | Large |
| | | I. | Continuous flow | : | I. | Continuous flow | |
| | Off/on-line: | II. | Sample processors | Off/on-line: | III. | Sample processors | |
| | Off-line: | I. | Robotics | Off-line: | II. | Robotics | |
| | | | *Automated or nonautomated* | | | | |
| | On-line: | I. | Continuous flow | On-line: | I. | Column switching | |
| Medium | : | II. | Column switching | : | I. | Continuous flow | Medium |
| | Off/on-line: | I. | Sample processors | Off/on-line: | II. | Sample processors | |
| | Off-line: | II. | SPI | Off-line: | I. | Robotics | |
| | : | III. | Robotics | | | | |
| | Off-line: | I. | SPI | On-line: | II. | Column switching | |
| | | II. | LLE | | II. | Continuous flow | |
| | Off-line: | I. | LLE | Off/on-line: | I. | Sample processors | |
| Small | | II. | SPI | Off-line: | III. | SPI | Small |
| | | | *Nonautomated* | | | | |
| | Low | | Medium sample complexity | | | High | |

I, II, III, Priority sequence; LLE, liquid-liquid extraction; SPI, solid phase isolation.

or coated reagents, the application of silica or polymeric ionically attached reagents, and the use of polymeric covalently attached reagents with labile tagging groups.

Sample pretreatment techniques that will become very important within a time period of about 10 years are those based on affinity chromatography because of their potential selectivity, but until now this technique was hardly used in sample pretreatment due to the lack of selectivity of the majority of the affinity phases.

As an end to this chapter the "magic" words of sample pretreatment in the near future are reiterated:

Smaller disposable SPI (on-line/off-line) cartridges
Immobilized pretreatment/derivatization procedures
Improvement of CF techniques, especially dialysis systems
Introduction of new selective sorbents for SPI
Introduction of affinity techniques
Improvement of robotic systems

Although sample pretreatment at the moment may be considered the weakest link in the total analytical procedure, this will shift in the near future to the selectivity- and sensitivity-determining part of the analytical procedure.

## REFERENCES


1. Diem, K., and C. Lentner (eds.), *Scientific Tables*, ed. 7, Documenta Geigy, Ciba-Geigy Limited, Basel, 1972.
2. Vaart, F. J. van de, A. Hulshoff, and A. W. M. Indemans, *Pharm. Weekbl.*, *115*:429 (1980).
3. Mohammed, H. Y., and F. F. Cantwell, *Anal. Chem.*, *50*:491 (1978).
4. Nielen, M. W. F., R. C. A. Koordes, R. W. Frei, and U. A. Th. Brinkman, *J. Chromatogr.*, *330*:113 (1985).
5. Slais, K., M. W. F. Nielen, U. A. Th. Brinkman, and R. W. Frei, *J. Chromatogr.*, *393*:57 (1987).
6. Wal, Sj. van der, *J. Liq. Chromatogr.*, *9*:1815 (1986).
7. Silva, J. A. F. de, *J. Chromatogr.*, *340*:3 (1985).
8. Lohman, J. J. H. M., *Pharm. Weekbl.*, *Sci. Ed.*, *8*:302 (1986).
9. Tjaden, U. R., R. A. M. van der Hoeven, J. A. C. Bierman, H. J. E. M. Reeuwijk, H. Lingeman, and J. van der Greef, *Chromatographia*, *24*:597 (1987).
10. Green, P. J., and M. J. Yucis, *Clin. Chem.*, *28*:1237 (1982).
11. Sophianopoulos, J. A., S. J. Durham, A. J. Sophianopoulos, H. J. Ragsdale, and W. O. Cropper, *Arch. Biochem. Biophys.*, *187*:132 (1978).

12. Rubenstein, M., S. Stein, and S. Udenfriend (eds.), *Hormonal Proteins and Peptides*, Vol. 9, Academic Press, New York, 1980.
13. Deyl, Z., J. Hyanek, and M. Horakova, *J. Chromatogr.*, *379*: 177 (1986).
14. Stolk, L. M. L., A. Fruijtier, and R. Umans, *Pharm. Weekbl., Sci. Ed.*, *8*:286 (1986).
15. Nielsen, H., *J. Chromatogr.*, *381*:63 (1986).
16. Hearn, M. T. W., *Adv. Chromatogr.*, *20*:4 (1982).
17. Szczerba, T. J., L. J. Glunz, J. D. Rateike, S. R. Patel, and J. A. Perry, *Chromatogr. Int.*, *19*:26 (1986).
18. Jong, G. J. de, *J. Chromatogr.*, *183*:203 (1981).
19. Hux, R. A., H. Y. Mohammed, and F. F. Cantwell, *Anal. Chem.*, *54*:113 (1982).
20. Hagestam, I. H., and T. C. Pinkerton, *Anal. Chem.*, *57*:1757 (1985).
21. Yoshida, H., I. Morita, T. Masujima, and H. Imai, *Chem. Pharm. Bull.*, *30*:2287 (1982).
22. Maurer, H., and K. Pfleger, *Fres. Z. Anal. Chem.*, *314*:586 (1983).
23. Min, B. H., J. Pao, W. A. Garland, J. A. F. de Silva, and M. Parsonnet, *J. Chromatogr.*, *183*:411 (1980).
24. Alric, R., M. Cociglio, J. P. Blayac, and R. Puech, *J. Chromatogr.*, *224*:289 (1981).
25. Kawasaki, T., M. Maeda, and A. Tsuji, *J. Chromatogr.*, *163*:143 (1979).
26. Lee, S. H., L. R. Field, W. N. Howald, and W. F. Trager, *Anal. Chem.*, *53*:467 (1981).
27. Greenhoff, K., and R. E. Wheeler, *J. Inst. Brew.*, *87*:35 (1981).
28. Ganzler, K., A. Salgo, and K. Valko, *J. Chromatogr.*, *371*:299 (1986).
29. Chamberlain, J., in *Methodological Developments in Biochemistry*, Vol. 5 (E. Reid, ed.), Elsevier, Amsterdam, 1976, p. 125.
30. Boxenbaum, H. G., and S. Riegelman, *J. Pharm. Sci.*, *63*: 1191 (1974).
31. Keusters, L., L. M. L. Stolk, R. Umans, and P. van Asten, *Pharm. Weekbl., Sci. Ed.*, *8*:194 (1986).
32. Chalmers, R. A., in *Methodological Developments in Biochemistry*, Vol. 5 (E. Reid, ed.), Elsevier, Amsterdam, 1976, p. 121.
33. Osselton, M. D., M. D. Hammond, and P. J. Twitchett, *J. Pharm. Pharmacol.*, *29*:460 (1977).
34. Osselton, M. D., I. C. Shaw, and H. M. Stevens, *Analyst*, *103*:1160 (1978).
35. Holzbecher, M., and H. A. Ellenberger, *J. Anal. Toxicol.*, *5*:62 (1981).
36. Allender, W. J., A. W. Archer, and A. G. Dawson, *J. Anal. Toxicol.*, *7*:203 (1983).

37. Hammond, M. D., and A. C. Moffat, *J. Forens. Sci. Soc.*, *22*:293 (1982).
38. Werkhoven-Goewie, C. E., U. A. Th. Brinkman, R. W. Frei, C. de Ruiter, and J. de Vries, *J. Chromatogr.*, *276*:349 (1983).
39. Kawasaki, T., M. Maeda, and A. Tsuji, *J. Chromatogr.*, *233*:61 (1982).
40. Takashi, U., A. Shigeru, and K. Kazuaki, *J. Chromatogr.*, *277*:308 (1983).
41. Lingeman, H., G. W. M. Meussen, C. van der Zouwen, W. J. M. Underberg, and A. Hulshoff, in *Bioactive Analytes: Including CNS Drugs, Peptides, and Enantiomers* (E. Reid, B. Scales, and I. D. Wilson, eds.), Plenum Press, New York, 1986, p. 343.
42. Leroy, P., S. Chakir, and A. Nicolas, *J. Chromatogr.*, *351*: 267 (1986).
43. Smith, R. V., and J. T. Stewart (eds.), *Textbook of Biopharmaceutic Analysis*, Lea & Febiger, Philadelphia, 1981.
44. Kabra, P. M., D. M. McDonald, and L. J. Marton, *J. Anal. Toxicol.*, *2*:127 (1978).
45. Neels, H. M., J. A. Totte, R. M. Verkerk, A. J. Vlietinck, and S. L. Scharpe, *J. Clin. Chem. Clin. Biochem.*, *21*:295 (1983).
46. Szabo, G. K., and T. R. Browne, *Clin. Chem.*, *28*:100 (1982).
47. Buice, R. G., B. C. Veit, S. E. McAlpin, B. J. Gurley, and P. Sidhu, *Res. Commun. Chem. Pathol. Pharmacol.*, *44*:401 (1984).
48. Brown, P. R., and R. E. Parkes, *Anal. Chem.*, *45*:94 (1973).
49. Kraak, J. C., F. Smedes, and J. W. A. Meijer, *Chromatographia*, *13*:673 (1980).
50. Solem, E., D. P. Agarwal, and H. W. Goedde, *Clin. Chim. Acta*, *59*:203 (1975).
51. Valor, P., *Ind. Eng. Chem. (Anal. Edn.)*, *8*:456 (1946).
52. Ings, R. M. J., J. A. McFadzean, and W. E. Ormerod, *Xenobiotica*, *5*:223 (1975).
53. Nicholls, L. C. (ed.), *Toxicology in the Scientific Investigation of Crime*, Butterworth, London, 1956.
54. Stevens, H. M., *J. Forens. Sci. Soc.*, *7*:184 (1967).
55. Breuer, J., G. Kaese, and H. Breuer, *Z. Klin. Chem.*, *6*:298 (1943).
56. Dubost, P., and S. Pascal, *Ann. Pharm. Fr.*, *11*:615 (1953).
57. Wiedmeier, V. T., S. P. Porterfield, and C. E. Hendrich, *J. Chromatogr.*, *231*:410 (1982).
58. Cooper, E. H., R. Turner, J. R. Webb, H. Lindblom, and L. Fagerstam, *J. Chromatogr.*, *327*:269 (1985).
59. Vlahos, I., *Clin. Chem.*, *28*:2286 (1982).

60. Porter, M. C., in *Handbook of Separation Techniques for Chemical Engineers* (P. A. Schweitzer, ed.), McGraw-Hill, New York, 1979.
61. Achari, R., *J. Chromatogr. Sci., 21*:278 (1983).
62. Blanchard, J., *J. Chromatogr., 226*:455 (1981).
63. Taylor, E. H., and B. H. Ackerman, *J. Liq. Chromatogr., 10*:323 (1987).
64. Giesselmann, G., and A. Hartman, *Abstract 10th Int. Symp. Column Liq. Chromatogr.*, San Francisco, 1986.
65. Everaerts, F. M., J. L. Beckers, and Th. P. Verheggen (eds.), *Isotachophoresis*. J. Chromatogr. Libr. Vol. 6, Elsevier, Amsterdam, 1976.
66. Kok, W. Th., K. P. Hupe, and R. W. Frei, *Abstract (Th-L4), 11th Int. Symp. Column Liq. Chromatogr.*, Amsterdam, 1987.
67. McDowall, R. D., J. C. Pearce, and G. S. Murkitt, *J. Pharm. Biomed. Anal., 4*:3 (1986).
68. Schill, G., in *Separation Methods for Drugs and Related Organic Compounds*, Swedish Academy of Pharmaceutical Sciences, Stockholm, 1978.
69. Wester, R., P. Noonan, C. Markos, R. Bible, W. Aksamit, and J. Hribar, *J. Chromatogr., 209*:463 (1981).
70. Cone, E. J., W. F. Buchwald, and W. D. Darwin, *Drug Metab. Disposition, 10*:561 (1982).
71. Hulshoff, A., and A. D. Förch, *J. Chromatogr., 220*:275 (1981).
72. Lingeman, H., *Pharm. Weekbl., Sci. Ed., 9*:183 (1987).
73. Furängen, A., *J. Chromatogr., 353*:259 (1986).
74. Yamada, K., E. Kayama, Y. Aizawa, K. Oka, and S. Hara, *J. Chromatogr., 223*:176 (1981).
75. Schill, G., in *Methodological Developments in Biochemistry*, Vol. 5 (E. Reid, ed.), Elsevier, Amsterdam, 1976, p. 87.
76. Fransson, B., and G. Schill, *Acta Pharm. Suec., 12*:107 (1975).
77. Roseboom, H., and A. Hulshoff, *J. Chromatogr., 173*:65 (1979).
78. Hulshoff, A., O. A. G. J. van der Houwen, D. M. Barends, and H. B. Kostenbauder, *Anal. Chim. Acta, 105*:139 (1979).
79. Langenberg, J. P., Thesis, Leiden University, Leiden, 1985.
80. Langenberg, J. P., and U. R. Tjaden, *J. Chromatogr., 305*:61 (1984).
81. Breiter, J., *Clin. Chem. Biochem., 14*:46 (1976).
82. Breiter, J., and R. Helger, *Z. Klin. Chem. Biochem., 13*:254 (1975).
83. Dickson, S. J., W. T. Cleary, A. W. Missen, E. A. Querée, and S. M. Shaw, *J. Anal. Toxicol., 4*:74 (1980).
84. Breiter, J., R. Helgar, and H. Lang, *Forensic Sci., 7*:131 (1976).
85. Breiter, J., *Kontakte, 3*:9 (1977).
86. Riedmann, M., Hewlett-Packard Appl. Note, 1981.

87. Cordonnier, J., M. van den Heede, and A. Heyndrickx, *Int. Analyst, 1*:28 (1987).
88. Sample Preparation Accessory (SPA), Milton Roy. LDC Division, Riviera Beach, FL, USA.
89. Tippins, B., *Int. Lab., 17*:28 (1987).
90. Majors, R. E., H. G. Barth, and C. H. Lochmüller, *Anal. Chem., 56*:300R (1984).
91. Frei, R. W., M. W. F. Nielen, and U. A. Th. Brinkman, *Int. J. Environ. Anal. Chem., 25*:3 (1986).
92. Mallet, V. N., G. L. Brun, R. N. McDonald, and K. Berkane, *J. Chromatogr., 160*:81 (1978).
93. Rosenfeld, M., M. Mareika-Russell, and A. Phatak, *J. Chromatogr., 283*:127 (1984).
94. Application Guide, Vols. 1 and 2, J. T. Baker, Philipsburg, NJ, USA, 1984.
95. Sorbent Extraction Technology Handbook, Analytichem International Inc., Harbor City, CA, USA, 1985.
96. Werkhoven-Goewie, C. E., U. A. Th. Brinkman, and R. W. Frei, *Anal. Chem., 53*:2072 (1981).
97. Shoup, R. E., and G. S. Mayer, *Anal. Chem., 54*:1164 (1982).
98. Smedes, F., J. C. Kraak, and H. Poppe, *J. Chromatogr., 231*:25 (1982).
99. Sjövall, J., and M. Axelson, *J. Pharm. Biomed. Anal., 2*:265 (1984).
100. Meala, R., and P. Vanko, *Clin. Chem., 20*:184 (1974).
101. Adams, R. F., and F. L. Vandemark, *Clin. Chem., 22*:25 (1976).
102. Kinberger, B., P. Wahrgren, and A. Holman, *Anal. Lett., 14*:1419 (1981).
103. Fischer, C., B. Heuer, K. Heuck, and M. Eichelbaum, *Biomed. Environ. Mass Spectrom., 13*:645 (1986).
104. Gustafson, R. L., R. L. Albright, J. Heisler, J. A. Lirio, and O. T. Reid, *I&EC Product Res. Dev., 7*:107 (1968).
105. Kullberg, M. P., and C. W. Gorodetzky, *Clin. Chem., 20*:177 (1974).
106. Reeuwijk, H. J. E. M., and U. R. Tjaden, *J. Chromatogr., 353*:339 (1986).
107. Lingeman, H., H. A. van Munster, J. H. Beynen, W. J. M. Underberg, and A. Hulshoff, *J. Chromatogr., 352*:261 (1986).
108. Goto, J., M. Saito, T. Chikai, N. Goto, and T. Nambara, *J. Chromatogr., 276*:289 (1983).
109. Gill, R., S. C. Qua, and A. C. Moffat, *J. Chromatogr., 255*:483 (1983).
110. Jong, J. de, A. J. F. Point, U. R. Tjaden, S. Beeksma, and J. C. Kraak, *J. Chromatogr., 414*:285 (1987).

111. Vendrig, D. E. M. M., J. Teeuwsen, J. J. M. Holthuis, and A. Hulshoff, *Abstract (Mo-P-118) 11th Int. Symp. Column Liq. Chromatogr.*, Amsterdam, 1987.
112. Chalmers, R. A., and R. W. Watts, *Analyst*, *97*:958 (1972).
113. Lankmayr, E. P., K. W. Budna, K. Müller, F. Nachtmann, and F. Rainer, *J. Chromatogr.*, *222*:249 (1981).
114. Nielen, M. W. F., R. W. Frei, and U. A. Th. Brinkman, *J. Chromatogr.*, *317*:557 (1984).
115. Petz, M., and U. Meetschen, *Z. Lebensm.-Unters. Forsch.*, *184*:85 (1987).
116. Ingraham, R. H., S. Y. M. Lau, A. K. Taneja, and R. S. Hodges, *J. Chromatogr.*, *327*:77 (1985).
117. Green, A. A., and W. L. Hughes, in *Methods in Enzymology*, Vol. 1 (S. P. Colowick and N. O. Kaplan, eds.), Academic Press, New York, 1955, p. 69.
118. Czok, M., and H. Engelhardt, *Abstract (Mo-P-37) 11th Int. Symp. Column Liq. Chromatogr.*, Amsterdam, 1987.
119. Ohlson, S., L. Hansson, P. O. Larsson, and K. Mosbach, *FEBS Lett.*, *93*:5 (1978).
120. Ey, P. L., S. J. Prowse, and C. R. Jenkin, *Immunochemistry*, *15*:429 (1978).
121. Butler, V. P. Jr., in *Methodological Developments in Biochemistry*, Vol. 5 (E. Reid, ed.), Elsevier, Amsterdam, 1976, p. 87.
122. Kristjansson, F., J. F. Stobaugh, and J. C. Brown, *Abstract (Mo-P-66) 11th Int. Symp. Column Liq. Chromatogr.*, Amsterdam, 1987.
123. Johansson, B., *J. Chromatogr.*, *381*:107 (1986).
124. Vos, A. J., *LAB/ABC*, *5(April)*:24 (1984).
125. Brooks, T. L., and A. Stevens, *Am. Lab.*, *(October)*:54 (1985).
126. Oka, H., Y. Ikai, N. Kawamura, K. Uno, M. Yamada, K.-I. Harada, M. Uchiyama, H. Asukabe, Y. Mori, and M. Suzuki, *J. Chromatogr.*, *389*:417 (1987).
127. Krull, I. S. (ed.), *Reaction Detection in Liquid Chromatography*, Marcel Dekker, New York, 1986.
128. Little, C. J., and F. Berdat, *Int. Labmate*, *10*:37 (1985).
129. Karger, B. L., R. W. Giese, and L. R. Snyder, *Trends Anal. Chem.*, *2*:106 (1983).
130. Little, C. J., D. Tompkins, O. Stahel, R. W. Frei, and C. E. Werkhoven-Goewie, *J. Chromatogr.*, *264*:183 (1983).
131. Roth, W., K. Beschke, R. Jauch, A. Zimmer, and F. W. Koss, *J. Chromatogr.*, *222*:13 (1981).
132. Werkhoven-Goewie, C. E., W. M. Boon, A. J. J. Praat, R. W. Frei, U. A. Th. Brinkman, and C. J. Little, *Chromatographia*, *16*:53 (1982).
133. Dolan, J. W., and L. R. Snyder, *J. Chromatogr.*, *185*:57 (1979).

134. Nordmeyer, F. R., and L. D. Hansen, *Anal. Chem.*, *54*:2605 (1982).
135. Snyder, L. R., J. W. Dolan, and Sj. van der Wal, *J. Chromatogr.*, *203*:3 (1981).
136. Erni, F., H. P. Keller, C. Morin, and M. Schmidt, *J. Chromatogr.*, *204*:65 (1981).
137. Bowers, L. D., and P. R. Johnson, *Clin. Chem.*, *27*:1554 (1981).
138. Vivilecchia, R. V., *HPLC Magazine*, *4*:94 (1986).
139. Strimaitis, J. R., *Int. Lab.*, *16*(9):56 (1986).
140. Venteicher, R. F., and J. van Antwerp, in *Advances in Laboratory Automation—Robotics* (G. L. Hawk and Y. R. Strimaitis, eds.), Zymark Corporation, Hopkinton, MA, USA, 1984.
141. Salit, M. L., and M. G. Cirillo, *Int. Lab.*, *16*(9):72 (1986).
142. Lochmüller, C. H., A. Colborn, and K. Cousins, Zymate Laboratory Automation Newsletter, Bd. 1, Nr. 1, Zymark Corporation, Hopkinton, MA, USA, 1984.
143. Brosemer, J., and J. Liscouski, *Int. Lab.*, *16*(9):82 (1986).
144. Dolan, J. W., Sj. van der Wal, S. J. Bannister, and L. R. Snyder, *Clin. Chem.*, *26*:871 (1980).
145. Automatic Sequential Trace Enrichment of Dialysates (ASTED), Gilson, Villiers-le-Bel, France.
146. Turnell, D. C., and J. D. H. Cooper, *J. Automatic Chem.*, *8*: 151 (1986).
147. Tjaden, U. R., E. A. de Bruijn, R. A. M. van der Hoeven, C. Jol. J. van der Greef, and H. Lingeman, *J. Chromatogr.*, *420*:53 (1987).
148. Aerts, M. M. L., W. M. J. Beek, C. A. Kan, and J. F. M. Nouws, *Archiv. Lebensm. Hyg.*, *37*:142 (1986).
149. Bell, J. P., R. A. Simpson, and R. L. Cunico, *Int. Lab.* *16*(9):62 (1986).
150. Internal data, Gilson Medical Electronics, Middleton, Wisconsin.
151. Einarsson, S., B. Josefsson, and S. Lagerkvist, *J. Chromatogr.*, *282*:609 (1983).
152. Simpson, R. C., and P. R. Brown, *LC*, *3*:537 (1985).
153. Bio-Radiations, Bio-Rad Laboratories, Richmond, March, 1984.
154. Pearce, J. C., J. A. Kelly, K. A. Fernandes, W. J. Leavens, and R. D. McDowall, *J. Chromatogr.*, *353*:371 (1986).
155. Korte, E., *LABO (December)*: 42 (1986).
156. Yago, L., *Am. Lab.*, *17*:118 (1985).
157. Murkitt, G. S., J. C. Pearce, and R. D. McDowall, *Abstract (Mo-P-116) 11th Int. Symp. Column Liq. Chromatogr.*, Amsterdam, 1987.
158. Juergens, U., *J. Chromatogr.*, *371*:307 (1986).
159. Williams, R. C., *Int. J. Environ. Anal. Chem.*, *18*:37 (1984).
160. Snider, B. G., L. J. Beaubien, D. J. Sears, and P. D. Rahn, *J. Pharm. Sci.*, *70*:1347 (1981).

161. Balkon, J., D. Prendes, and J. Viola, *J. Anal. Toxicol.*, *6*:228 (1982).
162. Verillon, F., and R. Glandian, *Int. Lab. (October)* (1985).
163. Frei, R. W., *Anal. Proc.*, *17*:519 (1980).
164. Juergens, U., *J. Chromatogr.*, *310*:97 (1984).
165. Tjaden, U. R., D. S. Stegehuis, H. J. E. M. Reeuwijk, H. Lingeman, and J. van der Greef, *Analyst*, *113*:171 (1988).
166. Zech, K., and R. Huber, *J. Chromatogr.*, *353*:351 (1986).
167. Nazareth, A., L. Jaramillo, B. L. Karger, R. W. Giese, and L. R. Snyder, *J. Chromatogr.*, *309*:357 (1984).
168. Nondek, L., and V. Chvalovsky, *J. Chromatogr.*, *312*:303 (1984).
169. Nondek, L., and V. Chvalovsky, *J. Chromatogr.*, *268*:395 (1983).
170. Verhagen, L. C., J. Strating, and U. R. Tjaden, *J. Chromatogr.*, *393*:95 (1987).
171. Voelter, W., T. Kronbach, K. Zech, and R. Huber, *J. Chromatogr.*, *239*:475 (1982).
172. Dadger, D., and A. Power, *Abstract (Mo-P-107) 11th Int. Symp. Column Liq. Chromatogr.*, Amsterdam, 1987.
173. Koziol, Th. R., and E. J. Woolf, *Abstract (Mo-P-113) 11th Int. Symp. Column Liq. Chromatogr.*, Amsterdam, 1987.
174. Lankelma, J., and H. Poppe, *J. Chromatogr.*, *149*:587 (1978).
175. Gfeller, J. C., and M. Stockmeyer, *J. Chromatogr.*, *198*:162 (1980).
176. Ishii, D., K. Hibi, K. Asai, M. Nagaya, K. Mochizuki, and Y. Mochida, *J. Chromatogr.*, *156*:173 (1978).
177. Tamai, G., H. Yoshida, and H. Imai, *Anal. Sci.*, *2*:481 (1986).
178. Tamai, G., H. Imai, and H. Yoshida, *Chromatographia*, *21*:519 (1986).
179. Daoud, N., T. Arvidsson, and K.-G. Wahlund, *J. Chromatogr.*, *385*:311 (1987).
180 Arvidsson, T., K.-G. Wahlund, and N. Daoud, *J. Chromatogr.*, *317*:213 (1984).
181. Chang, S. F., T. M. Welscher, and R. E. Ober, *J. Pharm. Sci.*, *72*:236 (1983).
182. Posluszny, J. V., D. Parriott, and R. Weinberger, *Abstract (Th-L-3) 11th Int. Symp. Column Liq. Chromatogr.*, Amsterdam, 1987.
183. Koenigbauer, M. J., and M. A. Curtis, *Abstract (Th-P-106) 11th Int. Symp. Column Liq. Chromatogr.*, Amsterdam, 1987.
184. Arunyanart, M., and L. J. Cline Love, *J. Chromatogr.*, *342*:293 (1983).
185. Tetsuo, M., M. Axelson, and J. Sjövall, *J. Steroid Biochem.*, *19*:175 (1983).

186. Nielen, M. W. F., H. E. van Ingen, A. J. Valk, R. W. Frei, and U. A. Th. Brinkman, *J. Liq. Chromatogr.* *10*:617 (1987).
187. Nielen, M. W. F., Thesis, Free University Amsterdam, Amsterdam, 1987.
188. Davankov, V. A., and A. V. Semechkin, *J. Chromatogr.*, *141*:313 (1977).
189. Veuthey, J.-L., M.-A. Bagnoud, and W. Haerdi, *J. Chromatogr.*, *393*:51 (1987).
190. Nielen, M. W. F., J. de Jong, R. W. Frei, and U. A. Th. Brinkman, *Int. J. Environ. Anal. Chem.*, *25*:37 (1986).
191. Irth, H., G. J. de Jong, U. A. Th. Brinkman, and R. W. Frei, *Abstract (Th-P-50) 11th Int. Symp. Column Liq. Chromatogr.* (1987).
192. Nielen, M. W. F., A. J. Valk, R. W. Frei, U. A. Th. Brinkman, Ph. Mussche, R. de Nijs, B. Ooms, and W. Smink, *J. Chromatogr.*, *393*:69 (1987).
193. Beld, C. M. B. van den, H. Lingeman, G. J. van Ringen, U. R. Tjaden, and J. van der Greef, *Anal. Chim. Acta*, *205*:15 (1988).
194. Elsohly, M. A., H. N. Elsohly, A. B. Jones, P. A. Dimson, and K. E. Wells, *J. Anal. Toxicol.*, *7*:262 (1983).
195. Colgan, S. T., I. S. Krull, C. Dorschel, and B. Bidlingmeyer, *Anal. Chem.*, *58*:2366 (1986).
196. Chou, T.-Y., S. T. Colgan, D. M. Kao, I. S. Krull, C. Dorschel, and B. Bidlingmeyer, *J. Chromatogr.*, *367*:335 (1986).

# 4

# Postchromatographic Reaction Detection

G. J. de JONG, UDO A. TH. BRINKMAN, and R. W. FREI
*Free University, Amsterdam, The Netherlands*

## INTRODUCTION

In recent years, column liquid chromatography (LC) has become a powerful tool for the analysis of a wide variety of samples. It is a disadvantage of LC, however, that it does not always meet the demands of modern trace level analysis with regard to sensitivity and/ or selectivity. One way to overcome this problem is to use chemical derivatization or related conversion techniques. Derivatization can also be used to improve the chromatographic properties of the analytes of interest and to improve their efficient trace enrichment.

In principle, derivatization or conversion of the analytes can be carried out by prechromatographic or postchromatographic* techniques. The main advantages of postchromatographic reactions are as follows:

- The analytes of interest are separated in their original form, which often permits adoption of existing separation procedures.
- Artifact formation plays a minor role, as opposed to derivatization prior to the separation step.
- The reaction does not need to be complete and does not have to be fully defined; the only requirement is reproducibility.

---

*In this chapter pre- or postchromatographic may be translated by pre- or postcolumn.

Disadvantages of postchromatographic reaction detection are the need to add reagent solution(s) which often require the use of additional pumps, the possibility of band broadening in the reactor which will diminish the chromatographic resolution, and the presence of excess reagent which may interfere with the signal of the reaction product(s). Interesting possibilities of reducing these disadvantages will be included in the next sections.

A variety of reactions have been used for on-line coupling with a LC separation via a postchromatographic technique. In this chapter the theoretical aspects of various postchromatographic reactors and reaction types will be discussed. Special attention will be paid to the hardware used for the reactions. Modern developments such as miniaturization and solid state reactors will be stressed. Examples of promising new reaction systems will be shown and some applications of more conventional postchromatographic reactors described.

## THEORETICAL ASPECTS

### Conventional Reactors

Conventional reactors are tubular, bed, and segmented-stream. These reactor types have often been described in books and review papers [1-5]. Therefore, only a summary of the theoretical and other aspects of these reactors will be presented.

#### *Open-Tubular Reactors*

The simplest type of reactor consists of straight or helically coiled tubes constructed from different materials such as glass, PTFE, or stainless steel. Generally, laminar flow patterns prevail. The combination of a parabolic velocity profile and molecular diffusion leads to a well-defined dispersion of the analyte zone. The variance of the dispersion is:

$$\sigma_t^2 = \frac{2D_m t^3}{L^2} + \frac{d_t^2 t}{96D_m} \qquad (1)$$

where $D_m$ is the molecular diffusion coefficient of the analyte, t is the mean residence time, L is the length of the tube, and $d_t$ is the inner diameter of the tube. For flow rates higher than about 0.1 ml/min, only the second term on the right-hand side of Eq. 1 is important (Taylor equation).

In order to reduce peak dispersion, a radial secondary flow is preferable. This can be obtained by a geometric deformation of the straight tube, in the easiest way by coiling. Owing to centrifugal forces acting on the flow pattern, secondary flows perpendicular to

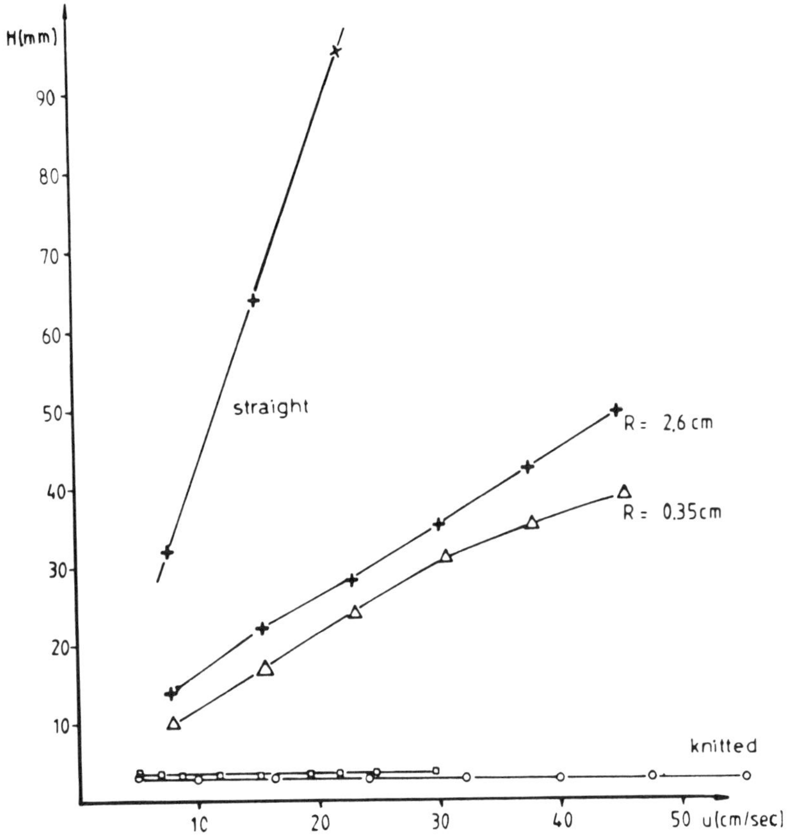


Figure 4.1 Peak dispersion in straight, tightly coiled, and knitted open tubes. (From ref. 5.)

the main flow are produced. As a result, better radial mixing and, consequently, a flattening of the laminar flow profile and a reduction of dispersion occurs. The variance of a coiled reactor is given by:

$$\sigma_t^2 = \frac{kd_t^2 t}{96D_m} \tag{2}$$

where k reflects the reduced axial dispersion. Still more optimally deformed capillaries can be obtained by knitting the tube. The reduction in dispersion resulting from coiling and knitting open tubes is shown in Fig. 4.1. Turbulent flow conditions cannot be achieved in conventional LC (flow rates of a few ml/min and tube diameters of 0.2-2 mm).

The pressure drop of a tubular reactor is given by the Poiseuille equation:

$$\Delta p = \frac{512\eta\Phi^2 t}{\pi^2 d_t^6} \tag{3}$$

where $\eta$ is the viscosity and $\Phi$ is the volumetric flow rate. Tubular reactors are preferably chosen for reactions with fast kinetics, i.e., residence times below about 30 sec.

*Packed Bed Reactors*

Packed bed reactors seem favorable for slower reactions. The variance of a bed reactor, a tube filled with inert particles, is:

$$\sigma_t^2 = \frac{2\gamma D_m t^3}{L^2} + \frac{A t^{1.67} d_p^{1.33}}{L^{0.67} D_m^{0.33}} \tag{4}$$

where $\gamma$ is the tortuosity factor, A is a constant depending on the bed geometry, and $d_p$ is the particle size. The pressure drop of a packed bed reactor can be calculated from the well-known Darcy equation:

$$\Delta p = \frac{u\eta L}{K_o d_p^2} \tag{5}$$

with u being the linear flow rate and $K_o$ the permeability constant. The decrease in peak dispersion achieved by decreasing the particle diameter must be paid for by an increase in pressure drop. Consequently, the reduction of the particle size and also the increase of the reactor bed length are limited by the available pressure drop. Ideally, adsorption of the analyte(s) and/or reaction product(s) to the packed bed should not take place.

*Segmented Stream Reactors*

The concept of segmentation of a flowing liquid stream with air or other gas bubbles can effectively suppress dispersion of sample plugs during extended residence times. The goal of segmentation is to reduce axial diffusion of the sample zones. Generally, the dispersion in a segmented flow system is brought about by wetting of the tubular wall and, hence, transfer of the analyte(s) from one segment to another. Snyder et al. [6,7] presented the following semiempirical expression for the band broadening in such a system:

$$\sigma_t^2 = \left[\frac{0.2d_t^{0.67}(\Phi + \eta V_G)^{1.67}\eta^{0.67}}{\sigma^{0.67}D_R\Phi} + \frac{1}{n}\right] \times \left[\frac{2.35(\Phi + \eta V_G)^{1.67}\eta^{0.67}t}{\sigma^{0.67}\Phi d_t^{1.33}}\right] \quad (6)$$

where n is the segmentation frequency, $V_G$ the volume of a gas bubble, and $\sigma$ its surface tension; $D_R$ is the dispersion coefficient for the analyte in water. Important terms for the optimization of a segmented stream reactor are the tube diameter $d_t$, the volumetric flow rate $\Phi$, and the segmentation frequency n.

An interesting alternative to air segmentation is solvent segmentation. This principle is based on, for example, segmentation of the LC eluate with organic nonmiscible solvent plugs. Liquid segmentation has the advantage that the system is less dependent on pressure, flow rate, and temperature fluctuations because of the lower compressibility of liquids. Moreover, the solvent that is added for segmentation can also extract analyte(s) or derivative(s), e.g., ion pairs [5] and hence reduce the background signal of the reagent.

The mathematical treatment of solvent-segmented systems is still more complex because dispersion is strongly influenced by the wetting of the tube by the liquids, as has been shown by Lawrence et al. [8] for a model system of postchromatographic fluorescent ion pair formation and extraction. For glass coils, band broadening was found to be almost independent of the flow rate of the aqueous phase, whereas it strongly increased with decreasing flow rate in the case of PTFE coils. Since the ion pair to be detected is rapidly extracted into the organic phase, band broadening is higher in PTFE tubes with strong wetting of the organic segments on the wall.

In practice, the band broadening in a segmented stream reaction system is mainly caused by the debubbler (air segmentation) or the phase separator (solvent segmentation) which is placed in front of the detector. A low dispersion requires debubblers and phase separators with a very small volume and optimal geometry. The easiest way to remove air is by a simple T piece with one upward opening. The air passes through the opening because of a density difference. Some designs of phase separators for physical desegmentation will be described in the section on "Practical Aspects and Applications." An alternative to physical desegmentation is electronic desegmentation [9].

Generally, segmented stream reactors are preferred for slow reactions (> about 5 min). For relatively rapid reactions the contribution of the debubbler or phase separator to the band broadening is

Table 4.1 Observed Band Broadening and Pressure Drop for Various Reactors

| Reactor type | $\sigma_t$ (sec) | $\Delta p$ (MPa) |
|---|---|---|
| Coiled tubular, 25 m × 0.25 mm | 2.3 | 5 |
| Coiled tubular, 6 m × 0.5 mm | 4.5 | 0.3 |
| Knitted tubular, 6 m × 0.5 mm | 4.0 | 0.3 |
| Packed bed, 20 cm × 4.6 mm, $d_p$ = 40 μm | 1.7 | 0.3 |
| Packed bed, 20 cm × 4.6 mm, $d_p$ = 17 μm | 1.2 | 1.6 |
| Segmented flow, 2.0 mm I.D. + debubbler[a] | 2.2 | <0.1 |
| Segmented flow, 1.1 mm I.D., no debubbler | 1.0 | <0.1 |

[a]F = 25 $mm^3$/sec.

*Source:* Ref. 10

so important that the use of tubular or bed reactors is more attractive. Moreover, these two reactors are simpler to construct and operate. Deelder et al. [10] presented an excellent comparison of the three reactors for the reaction of urea with *p*-dimethylaminobenzaldehyde (see Table 4.1). They conclude that the lowest dispersion is obtained for packed-bed reactors and for segmented stream systems. However, the debubbler caused a twofold increase in the band broadening, $\sigma_t$, of the reactor. For a tubular reactor a similar dispersion can only be obtained with a tube with a low internal diameter. However, the back pressure of such a reactor is then relatively high. For a segmented stream reactor a relatively high internal diameter can be chosen and, therefore, the back pressure is very low in comparison with the other two reactor types.

Recently, Shih and Carr [11] studied the influence of the flow rate in postchromatographic reactors on the peak height and peak area. They conclude that postchromatographic reaction detectors constitute a distinct class of detectors which differ from mass-flow-sensitive and concentration-sensitive detectors. If the rate of the reaction is low, the peak height will increase linearly with a decrease in the flow rate and the peak area will quadruple if the flow rate is reduced to one-half. Moreover, a simple mathematical model in which dispersion and chemical reaction are treated as independent processes shows that there is an optimum reactor length for any given set of conditions, i.e., reaction rate and band broadening.

### Solid State Reactors

Almost all the postchromatographic reactions are homogeneous reactions. Until now, little work has been done in the field of heterogeneous derivatization. In these systems, the derivatization reagent is present as a finely divided solid or as a solid which is physically adsorbed or electrostatically or covalently bonded to the support. Another possibility is the use of a packing material which acts as a catalyst. Solid state reactions have some important advantages over homogeneous reactions. A pump and a mixing tee for addition of the reagent are not needed. Provided that no bleeding occurs there is no derivatization reagent present in the LC eluate that can increase the background and noise level—advantages which are shared with the photochemical reactor (see below). The reagent cost is often reduced considerably and the problems encountered with toxic reagents are now less. Often reagents can be used which are not sufficiently soluble in the mobile phase for homogeneous reactions. A special type of reactor is a catalytic reactor, which does not have to be reloaded periodically. The reactor should have a reasonably long lifetime and should not irreversibly adsorb the analyte(s) and/or reaction products. The LC mobile phase should be fully compatible with both the separation and the reaction conditions.

The use of solid state reactions causes a specific contribution to band broadening. Contrary to the more classical packed bed reactor, a certain retention of the analyte(s) is required for an efficient reaction and a separation of reactant and product(s) can take place. Therefore, band-broadening phenomena are rather complex in solid state reactors. Using the mathematical model of Langer and Patton [12], Nondek et al. [13] developed a quantitative relationship for the reaction band broadening on a solid state catalyst in addition to the normal band broadening of a packed bed:

$$\sigma_r = \frac{\sqrt{\ln 2}}{8} \frac{(1 - k'_p/k'_R)}{k_r} \tag{7}$$

where $k'_p$ and $k'_R$ are the capacity factors of the product and the reactant, respectively, and $k_r$ is the rate constant of the first-order reaction. If $k'_p$ and $k'_R$ are equal, it is clear that no reaction band broadening will occur. If $k_r$ approaches infinity, the pulse of the reactant is immediately converted to the pulse of the product and passes through the rest of the bed without any reaction band broadening. The results for the decomposition of diacetone alcohol on alumina (formation of acetone) and the catalytic hydrolysis of 1-naphthalyl-*N*-methylcarbamate on an anion exchanger (formation of methylamine which was selectively detected via a

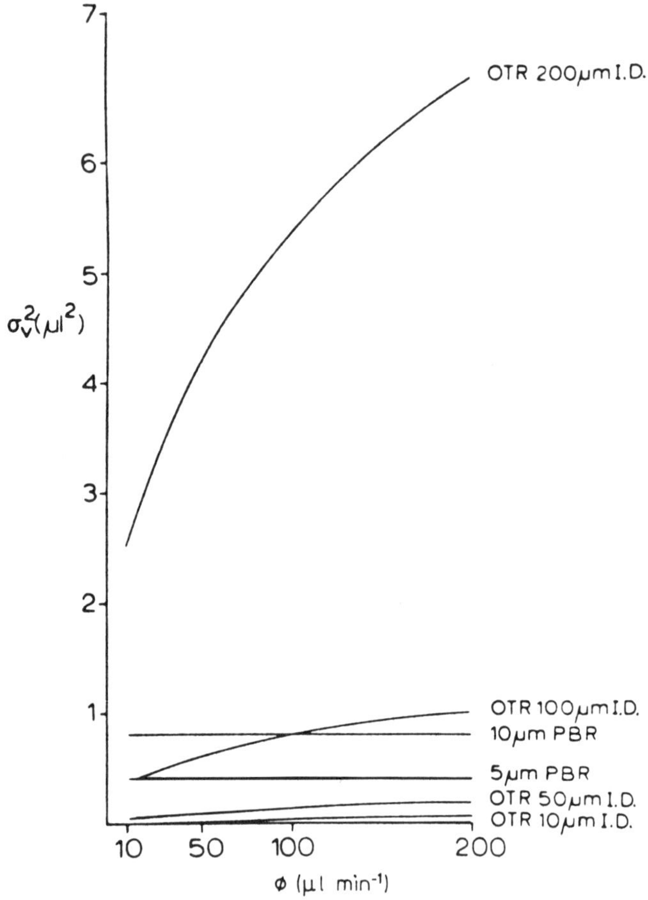


Figure 4.2 Calculated volume variance vs. flow rate for open-tubular reactors. Reactor void volume: 20 μl. Mobile phase: acetonitrile-water (80:20). Packed bed reactor data are experimental (HETP of packed beds is about 6 $d_p$). (From ref. 15.)

fluorogenic reaction with *o*-phthalaldehyde (OPA)) were in agreement with the mathematical model.

For enzyme reactors the kinetics of enzyme reactions should be considered [14]. The amount of product formed during passage through the reactor is given by the integrated form of the Michaelis-Menten equation:

$$k_r[E]t = ([S_o] - [S_L]) + K_M \ln \frac{[S_L]}{[S_o]} \tag{8}$$

where $k_r$ is the rate of dissociation of the enzyme-substrate complex, [E] the concentration of the enzyme, $[S_O]$ and $[S_L]$ the concentrations of the substrate at the entrance and exit of the reactor, respectively, and $K_M$ the Michaelis constant, i.e., the stability constant of the enzyme-substrate complex. In the region of analytical interest (low substrate concentration), the second term on the right-hand side of Eq. 8 predominates.

### Miniaturization

Recently, much attention has been paid to miniaturization of LC systems. The main advantages of such systems are low consumption of solvents and packing materials and high mass sensitivity, which is very useful if only small amounts of sample are available. However, the small peak volumes require that the extracolumn dead volume, including that of the detector, be considerably reduced relative to conventional LC systems. For microbore LC (column of about 1 mm I.D.) the dead volume should not exceed about 1 µl in order to avoid excessive peak dispersion and loss of resolution. For the use of still smaller internal diameters and, especially, for open-tubular LC columns, a considerably larger reduction of the dead volume is necessary. Apffel et al. [15] calculated the volume variance for some reactors with Eq. 2. The results for a reactor volume of 20 µl are shown in Fig. 4.2. In this figure the variance is also shown for some packed bed reactors with a reduced plate height of 6 (experimental value). Bed reactors packed with 5-µm particles and coiled open-tubular reactors with internal diameters of about 50 µm are optimal. The open-tubular reactor with an internal diameter of 10 µm is not considered in this comparison because of excessive back pressure. Figure 4.3 shows the pressure drop of the different reactors calculated with Eqs. 3 and 5. As for this aspect, packed bed reactors are superior to open-tubular reactors. For a maximum pressure of 100 bar, it is not possible to operate 50-µm open-tubular reactors with a reactor volume of 20 µl above 25 µl/min, whereas a bed reactor with 5-µm particles can be operated at flow rates higher than 250 µl/min. It should be noted that the mixing tee is often a major source of band broadening. This can be diminished by a high flow rate of the reagent but this also leads to a high dilution of the sample and, therefore, lower sensitivity.

In studies on external band broadening in open-tubular LC systems, Niessen et al. [16] found that most of the contributions to total peak variance give rise to exponential broadening of the Gaussian peak eluting from the column. The exponentially modified Gauss (EMG) concept has been generally accepted as the most useful model to take into account the asymmetrical nature of the peak. Peak deconvolution methods can be used to calculate the total external band broadening and the different contributions to this band broadening.

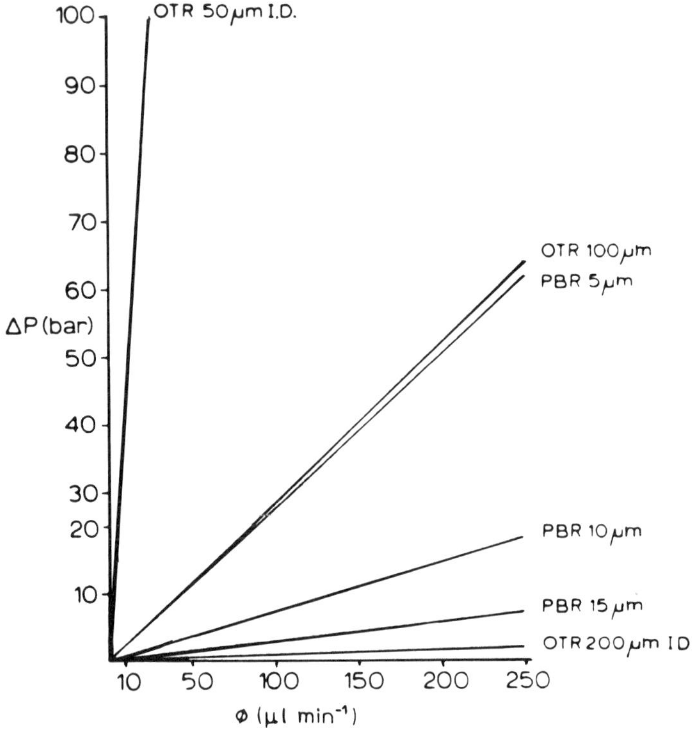


Figure 4.3 Calculated pressure drop vs. flow rate. Reactor void volume: 20 μl. Mobile phase: acetonitrile-water (80:20). (From ref. 15.)

## PRACTICAL ASPECTS AND APPLICATIONS

### Open-Tubular Reactors

In conventional reaction systems a reagent is added to the LC eluate via a mixing unit and after reaction in a capillary reactor or in a packed bed the product is detected. For fast reactions the contribution of the mixing unit to the total band broadening can be significant. Moreover, very efficient and fast mixing is important. Therefore proper construction of the mixing unit is a major condition for a good functioning of a reaction detector. Different mixing geometries are shown in Fig. 4.4. Generally, types III and IV are the best geometries because the relatively high turbulence in the mixing unit causes an efficient mixing with a low band broadening. Other efficient mixers are the cyclone mixer [17] and a rotating flow mixing device [18]. The cyclone mixer is shown in Fig. 4.5.

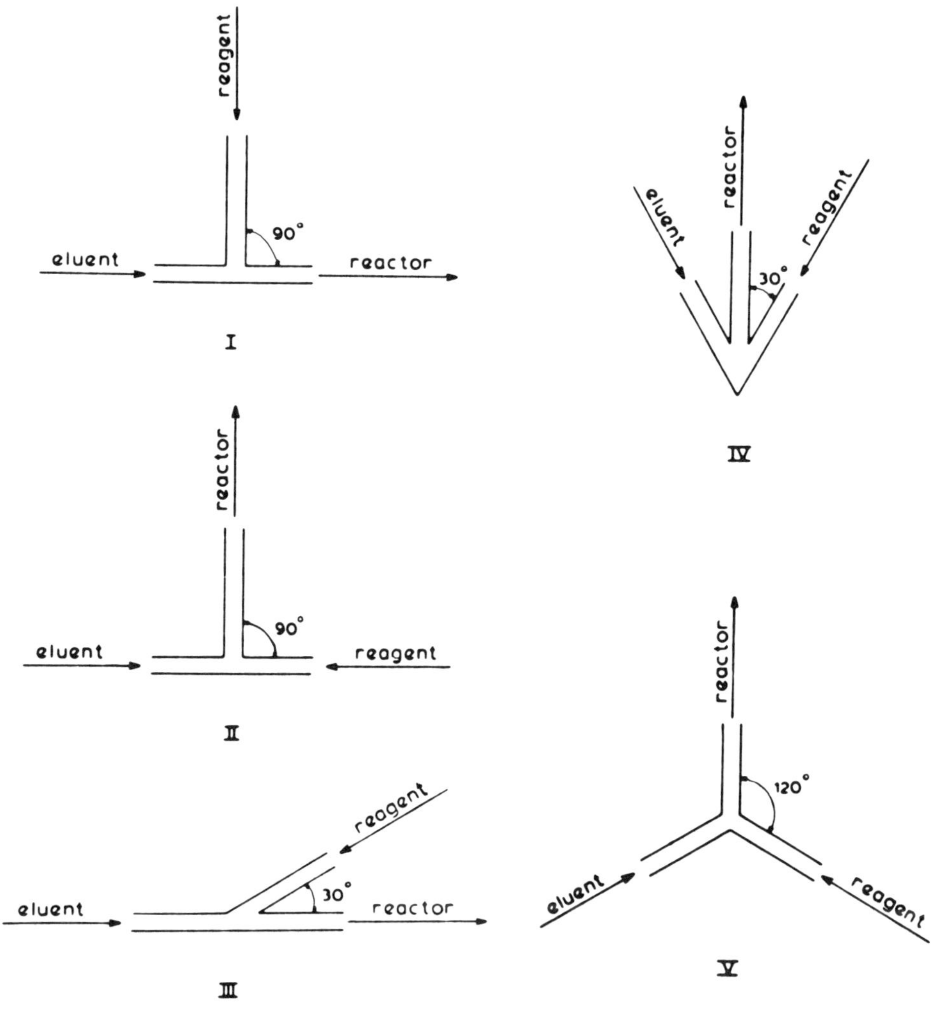


Figure 4.4 Different geometries for mixing units. (From ref. 4.)

The reagent is added tangentially in a small circular slit and mixes with the column effluent at the tip of a cone. The rotating flow mixer is shown in Fig. 4.6. The two streams are rotated in the same direction; the two inlets are in the bottom of the small vessel at opposite sides and an outlet is in the center of the top of the vessel. Recently, a sophisticated mixer has been introduced that contains 36 clockwise and anticlockwise spin chambers with a total volume of only 10 μl (Visco-Jet Micro Mixer, Lee Co., Westbrook, CT).

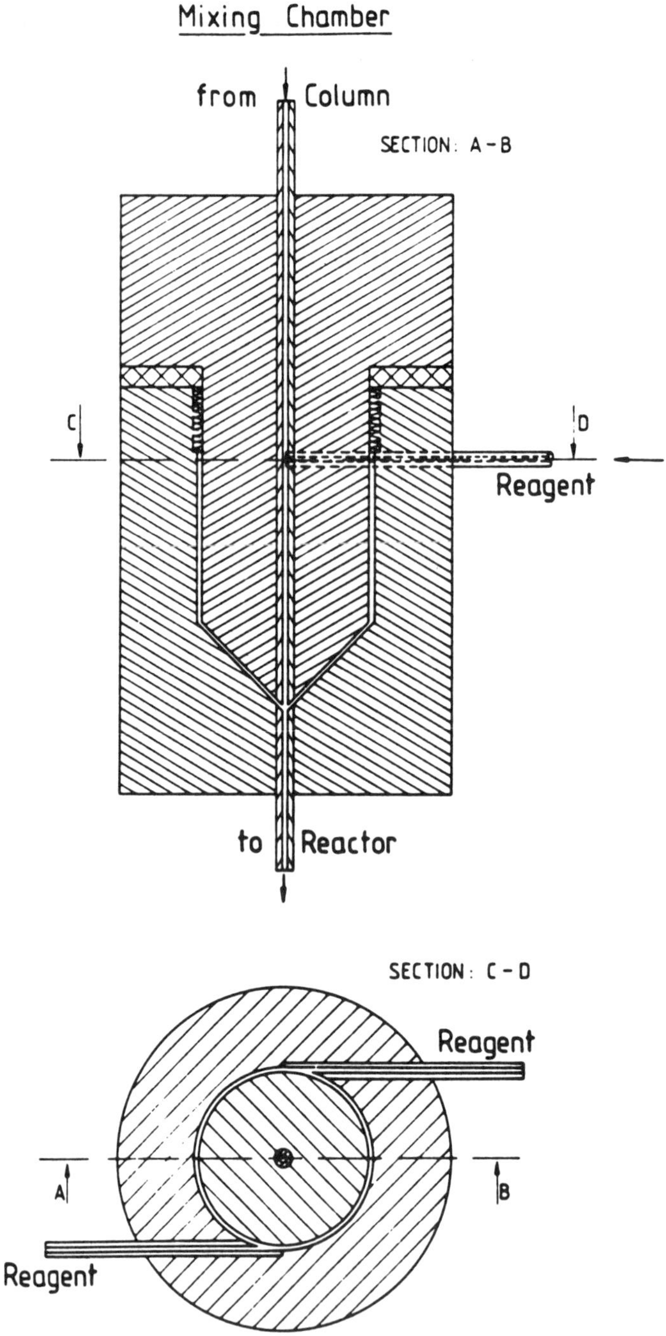


Figure 4.5 Scheme of a cyclone-type mixer. (From ref. 17.)

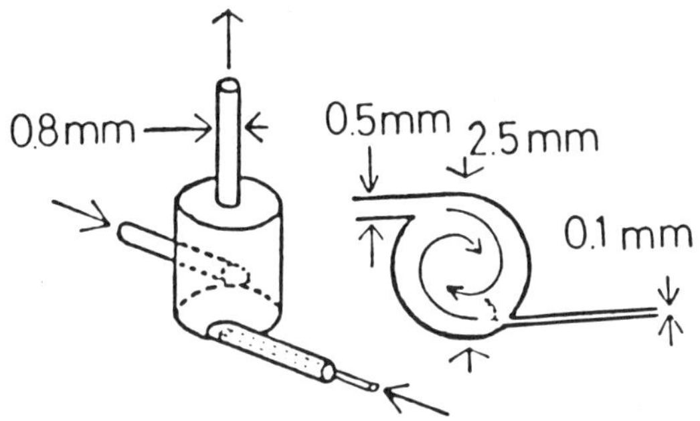


Figure 4.6 Rotating flow mixer. (From ref. 18.)

Porous membranes [19,20] and hollow-fiber membranes [21] have also been used for reagent addition. In this way, mixing units can be eliminated. A recent application is presented by Haginaka et al. [22]. The sulfonated hollow-fiber membrane was suspended in a sodium hydroxide solution and the analytes clavulanic acid and sublactam were degraded in alkaline solution to products with strong UV absorption at 270-280 nm. This elegant reactor seems very interesting since no pump is required for reagent addition.

For addition of a reagent by a pump, it is important that the reagent solution be delivered at a constant flow rate and without pulse. This helps to reduce drift and noise of the baseline. Usually, LC pumps with pulse dampers are used. However, pulseless syringe pumps are preferred. Since the back pressure is often only a few bars, low-pressure syringe pumps are very suitable. Aggressive reagents like sulfuric acid at high concentrations should not be delivered by the pump. The displacement of such reagents by an inert liquid will prolong the lifetime of the pump [17].

A fluorogenic reagent which is frequently used for labeling of primary amines is OPA. The reagent itself is not fluorescent and the reaction with primary amines (in the presence of 2-mercaptoethanol) can be carried out at room temperature and is often complete in less than 1 min. The isoindole products can be detected in the pg range. An interesting extension of this reaction in the prechromatographic mode is the use of *N-tert*butyloxycarbonyl-L-cysteine (Boc-L-cysteine) or *N*-acetyl-L-cysteine as thiol for the separation of amino acids enantiomers [23,24]. In some papers [25-28] the coupling of the OPA derivatization to an on-line hydrolysis of carbamates by a sodium hydroxide solution or an anion exchanger has been described and detection limits of femtomoles have been obtained. The system using an ion exchange resin and an illustrative example are shown in Figs. 4.7 and 4.8. This is an example of the use of two simple reactions in series in order to obtain a high sensitivity and selectivity.

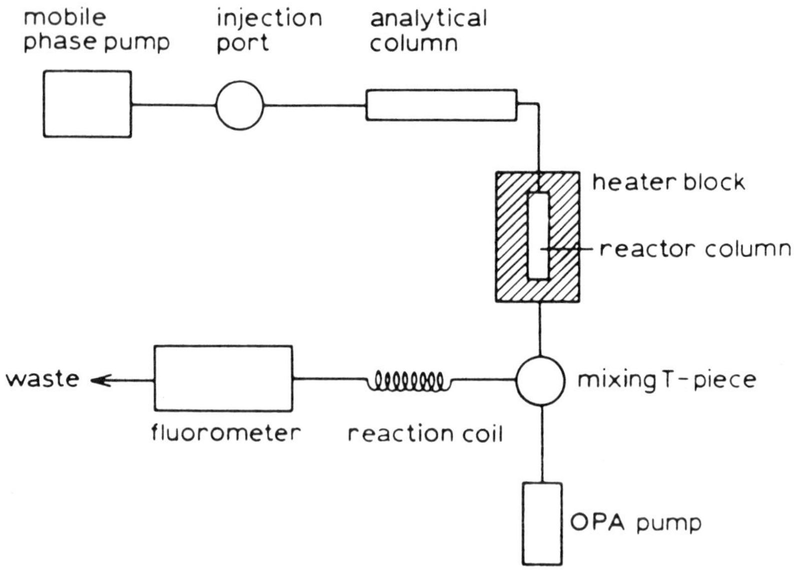


**Figure 4.7** Scheme of the system for separation and detection of carbamates. For experimental details, see legend to Fig. 4.8. (From ref. 27.)

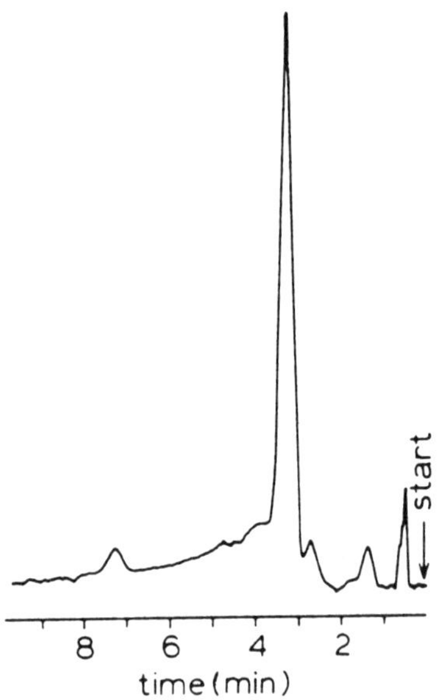

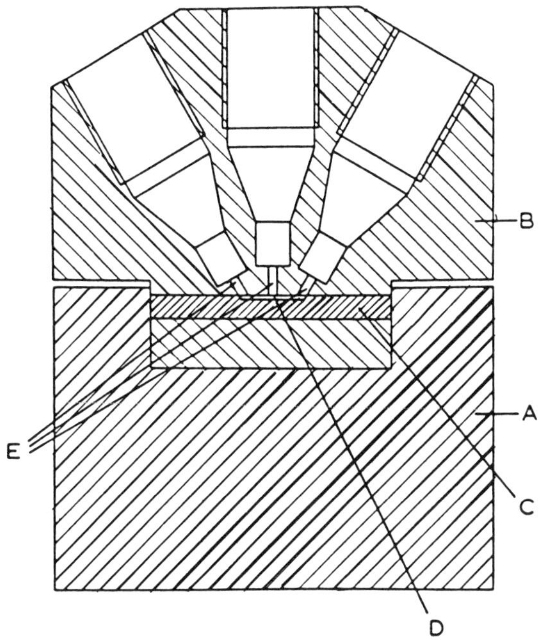


Figure 4.9 Scheme of a mixing T piece for miniaturized LC. (From ref. 29.)

The above system for the determination of carbamates has been successfully miniaturized [29]. For the addition of OPA a specially designed T piece was constructed (Fig. 4.9). The two parts, A and B, of the T piece are mounted together by screws. The lower part, A, contains a PTFE plate (C) in which a very small groove (D; volume, 60 nl) has been made. This groove is positioned just below three short capillaries (E) of 0.12 mm I.D. in the upper part. The capillaries on the left and right are the inlet tubes and the central capillary serves as the exit. A stainless steel capillary of

---

Figure 4.8 Chromatogram of polluted canal water spiked with 0.2 ppb carbaryl. Amount preconcentrated, 20 ml; 2 × 4.6 mm I.D. precolumn packed with 5 μm LiChrosorb RP-18; LC column packed with 5 μm LiChrosorb RP-18; mobile phase, methanol-water (3:2) at 0.6 ml/min; 40 × 4.6 mm I.D. reactor packed with Amberlite CG-400; OPA reagent flow-rate, 30 μl/min; detection with Perkin-Elmer 204A fluorescence spectrometer at $\lambda_{exc}$ = 340 nm and $\lambda_{em}$ = 455 nm. (From ref. 27.)

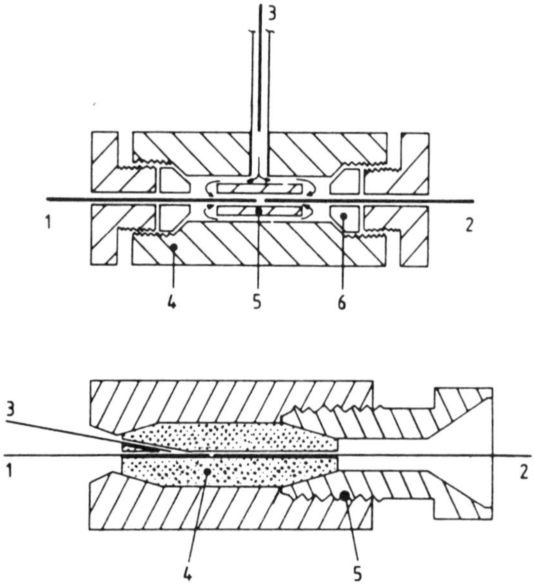


Based on Valco 1/16-in. zero dead volume connector; 4: connector body, 5: stainless steel liner, 6: teflon ferrules, arrows: reagent flow.

Based on Supelco butt-connector; 4: vespel ferrule, 5: connector body.

Figure 4.10 Design of mixing devices for open-tubular LC. (From ref. 30.)

0.17 m × 0.12 mm I.D. was used as the OPA reactor. The total band broadening in the postchromatographic reaction detector system was roughly the same as the value of about 5 s (statistical second moment) reported for a conventional LC system. Derivatization of amines with OPA has even been applied in open-tubular LC [30]. The mixing devices with apparent volumes of less than 10 nl are shown in Fig. 4.10.

### Segmented Stream Reactors

In the literature many air-segmented systems have been described [4] for relatively slow reactions. As was stated earlier, solvent segmentation is now often preferred because next to suppression of the band broadening an efficient separation of the reaction product and the high excess of the more polar reagent can be obtained. Ion pair formation combined with on-line extraction is an illustrative example of such a system [5]. The principle of this approach is that an acidic or basic solute which has little or no UV-Vis absorption or fluorescence can be detected sensitively if a suitable ion pair is formed and extracted before the detection. The detection is based on the absorbance or fluorescence of the counterion in the organic

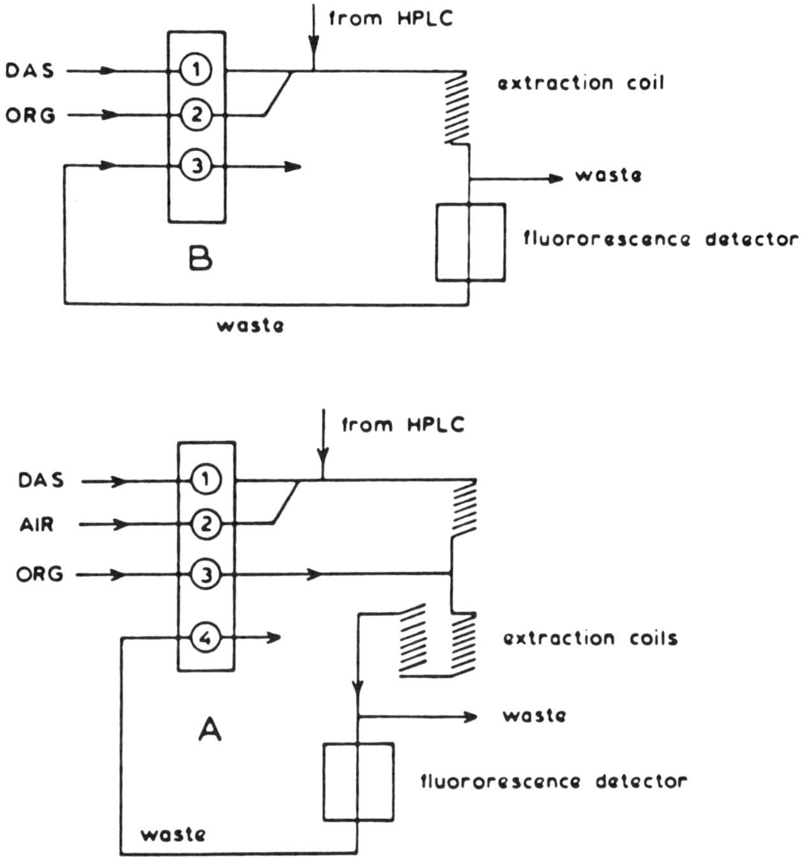


Figure 4.11 Schemes for ion pair extraction and fluorescence detection (A) with air segmentation and (B) with solvent segmentation. (From ref. 5.)

phase, the concentration of which is proportional to the concentration of the solute. The strongly fluorescent ion pair reagent 9,10-dimethoxyanthracene-2-sulfonate (DAS) is often used. A scheme of the extraction detection system is shown in Fig. 4.11. The ion pair reagent can also be added to the LC mobile phase and then influence the chromatographic selectivity. The organic modifier content of the aqueous mobile phase is very important since this determines the coextraction of the counterion into the organic phase and, therefore, the background signal. This technique can also be applied in normal phase chromatography. The LC eluent which is now

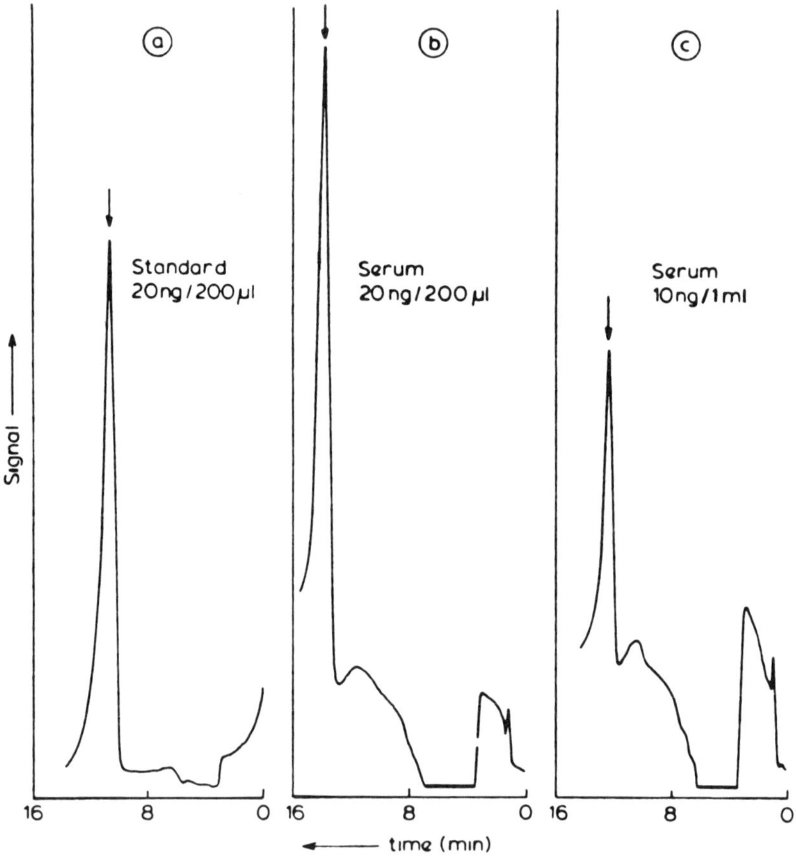


Figure 4.12 Chromatograms of secoverine-containing samples. (a) Aqueous standard solution of 100 ng/ml of secoverine; direct injection of 200 µl. (b) Spiked serum sample containing 100 ng/ml of secoverine; preconcentration from 200 µl. (c) Spiked serum sample containing 10 ng/ml of secoverine; preconcentration from 1 ml. Note that the amount of secoverine injected in chromatogram c is half that in chromatogram b. Detection system: see Fig. 4.11B. (From ref. 31.)

the organic phase, is segmented with an aqueous stream containing the ion-pairing reagent. In this case the concentration of the polar modifier, e.g., methanol, of the mobile phase has a strong effect on the background signal. The signal-to-noise ratio decreases considerably at higher methanol concentrations [8].

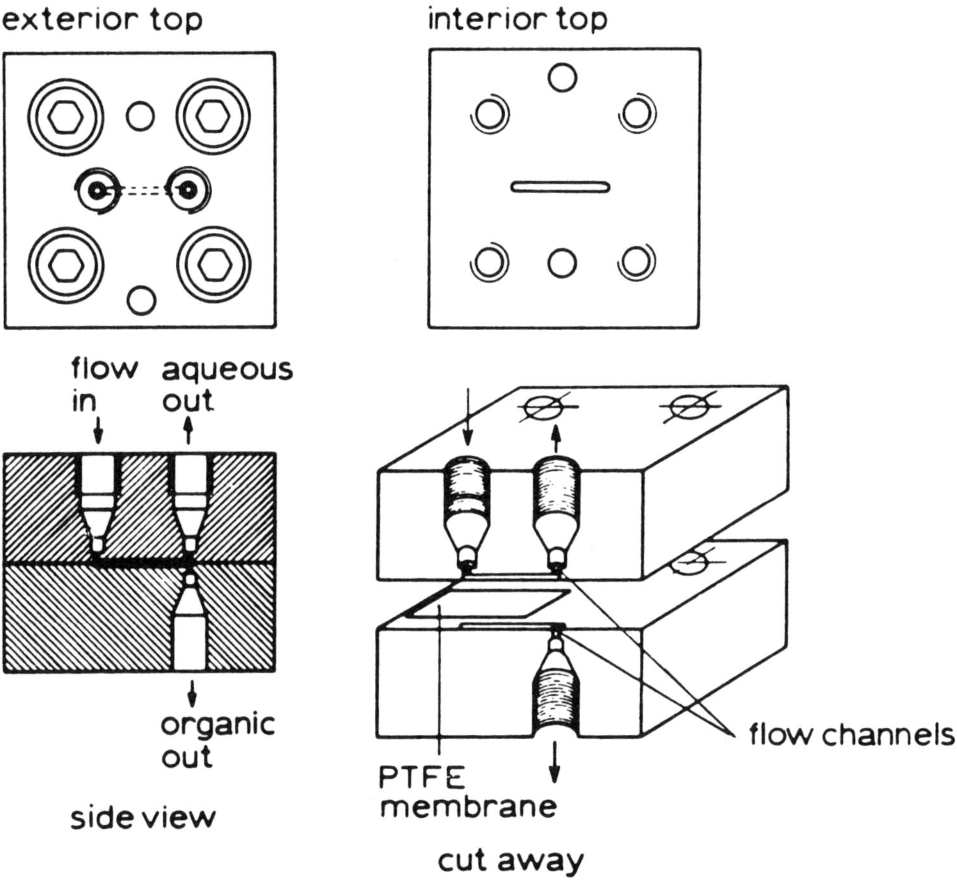


Figure 4.13 Scheme of membrane phase separator. (From ref. 33.)

One example of a reversed phase LC separation coupled to an extraction detection system is the determination of the tertiary amine-type drug secoverine in serum [31]. After enzymatic hydrolysis of the sample with subtilisin A [32] 1 ml of serum was preconcentrated on a short (2 mm length) cyano-bonded precolumn and, subsequently, secoverine was eluted to the cyano-bonded analytical column. This stationary phase has the advantage that only 10% dioxane as modifier is sufficient for a fast elution of secoverine. Therefore, the coextraction of DAS into the extraction solvent 1,2-dichloroethane is very low. Figure 4.12 shows that a 100% recovery is obtained from a 1-ml serum sample. The detection limit of secoverine (signal-to-noise ratio of 2) is about 20 pg and the selectivity of the system is very high.

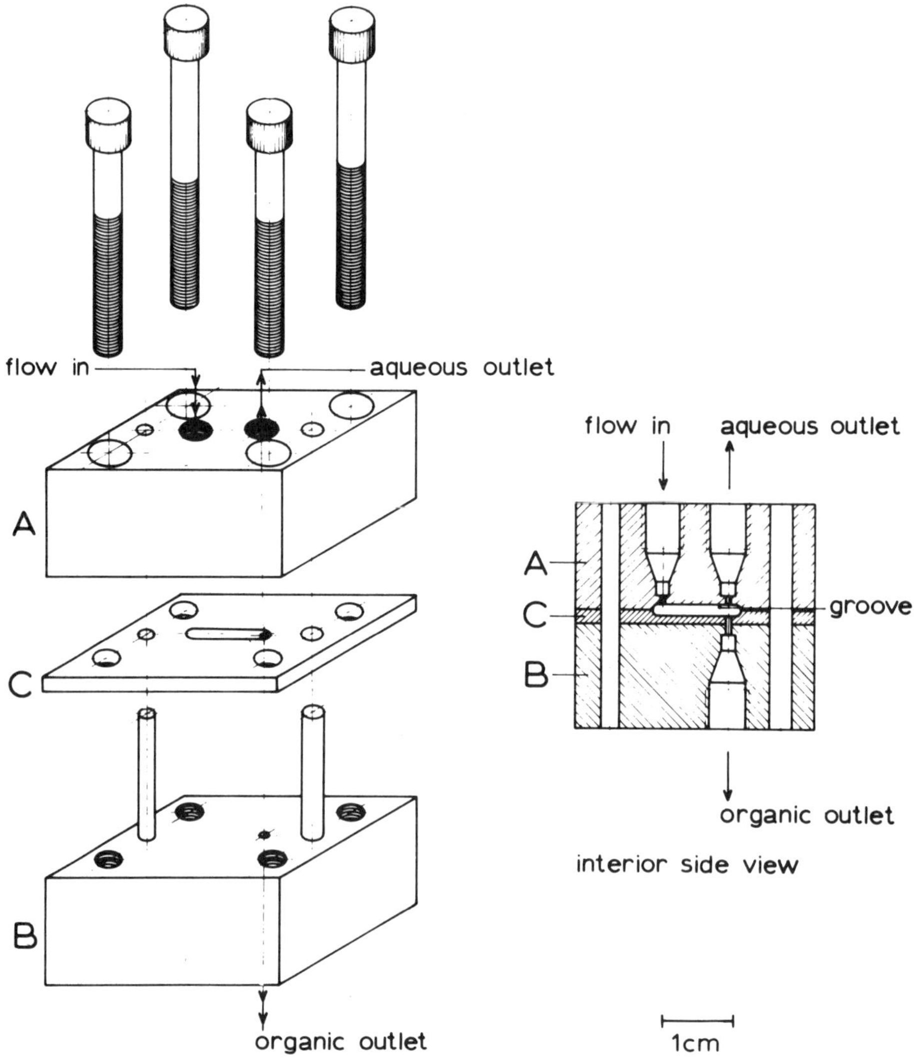


Figure 4.14 Design of the sandwich phase separator. (From ref. 34.)

As was mentioned above, the phase separator is a critical part of an extraction detection system. The major principle used for phase separation is wetting. The first designs were based on the insertion of PTFE tubing in a glass tubing. The different wettability of glass and PTFE creates separation of the aqueous and organic segments. Studies with membrane separators have shown a high potential for application to postchromatographic extractions, particularly for miniaturized systems if dead volume has to be reduced. Apffel et al. [33] designed a low dead-volume membrane separator for miniaturized LC using PTFE membranes with 0.5-μm pores. Figure 4.13 shows a scheme of the separator; the dead volume was only 0.5 μl. However, it was found that this type of separator is less suitable for the analysis of biological samples. Rapid clogging of the membrane occurred in the analysis of plasma samples. Therefore, De Ruiter et al. [34] developed so-called sandwich phase separators based solely on wetting. Figure 4.14 shows a scheme of such a phase separator which consists of two stainless steel blocks, one with (A) and one without (B) a groove and a PTFE (or another hydrophobic material) disc with a groove (C). These are held together by four screws. The inlets and outlets are machined into the blocks to accommodate 1/16-in. Valco male fittings. The aqueous phase is attracted to the polar stainless steel surface and the organic solvent flows along the PTFE surface. The aqueous outlet is connected to a microneedle valve to allow the application of back-pressure to adjust the organic flow through the detector. The band broadening of a sandwich phase separator with groove volumes between 8 and 43 μl is only 2.5-4 s (statistical second moment) and flow rates are not very critical, permitting total flow rates between 0.5 and 3 ml without seriously impeding the function of the 43-μl separator.

### Solid State Reactors

Most published LC reaction detection studies have involved the use of homogeneous reactions. The advantages of heterogeneous over homogeneous reactions have already been mentioned. A few types of solid phase reactions will be briefly discussed. Rather short solid-state reactors packed with ion exchange resins and, utilizing reaction temperatures of up to 100°C, have been used for the hydrolysis of carbamates [26, cf. above] and guanidines [35]. In another application nonreducing carbohydrates were hydrolyzed postchromatographically on a strong cation exchanger and the resulting reducing carbohydrates were reacted with various reagents. The characteristics of various cation exchangers were compared by Vratny et al. [36] and a commercial postchromatographic reaction system is now available from Kratos (Ramsey, NJ).

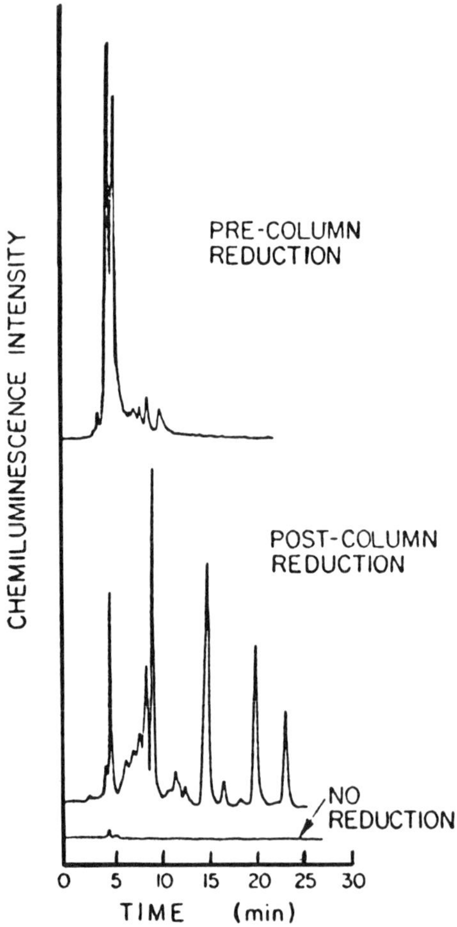


Figure 4.15 Chromatograms of a carbon black extract with chemiluminescence detection and pre- or postchromatographic reduction with a zinc reactor. (From ref. 41.)

Borohydride on silica, alumina, or polymeric supports can reduce a large variety of organic compounds [37]. The main advantage is a gain in selectivity and functional group analysis with LC. Studebaker [38] used a reactor consisting of a thiol-Sepharose that was reacted to form a mixed disulfide with a chromophore such as *N*-dinitrophenylcysteine or a fluorophore such as didansylcysteine. Thiols eluting from the LC column can form a disulfide which incorporates the chromophore or fluorophore. Disulfides are first cleaved in a reactor containing only thiol-Sepharose and form thiolate anions which then

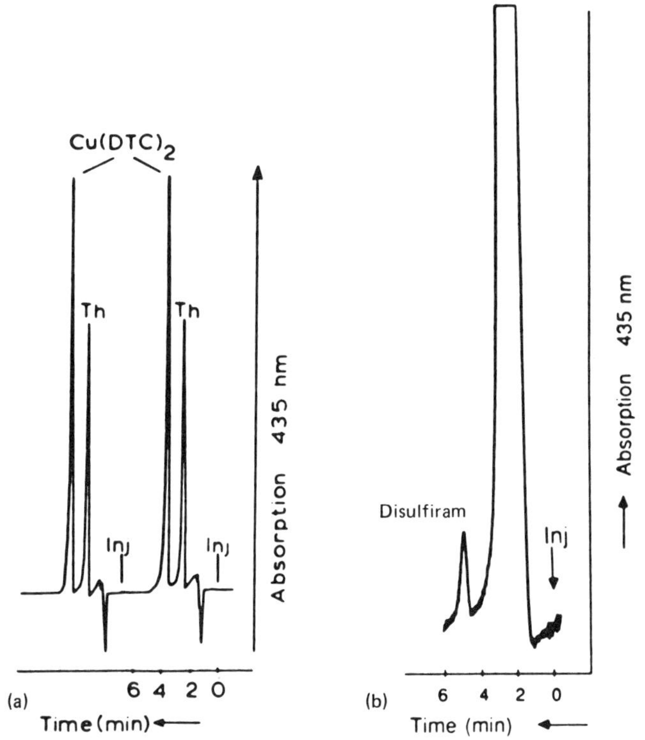


Figure 4.16 (a) Chromatogram of the duplicate injection of surface water spiked with 10 ppm of thiram and 20 ppm of $Cu(DTC)_2$. LC conditions: analytical column, 200 × 2.1 mm I.D. packed with 5 μm Hypersil ODS; copper reactor, 2.0 × 2.1 mm I.D.; mobile phase, acetonitrile-aqueous acetate buffer (10 mM, pH 5.0) (70:30); flow rate, 0.3 ml/min; injection volume, 20 μl; detection wavelength, 435 nm. (b) Chromatogram of urine spiked with 87 ppb of disulfiram (sample stabilized with 10 mM EDTA citrate). LC conditions: analytical column, 200 × 2.1 mm I.D. packed with 5 μm Hypersil ODS; precolumn, 4.0 × 2.1 mm I.D. packed with 5 μm LiChrosorb RP-18; copper reactor, 2.0 × 2.1 mm I.D.; mobile phase, acetonitrile-aqueous acetate buffer (10 mM, pH 5.0) (65:35); flow rate, 0.4 ml/min; preconcentration of 1.0 ml of urine; detection wavelength, 435 nm. (From ref. 42.)

react on a second reactor which contains the chromophore of fluorophore. Millot et al. [39] used the same principle with disulfide-silica supports for the detection of thiols. The advantage of this type of support is the lower band broadening compared with Sepharose.

Rüter et al. [40] described the postchromatographic oxidation of catecholamines with $PbO_2$/silica or $MnO_2$/silica reactors. The reaction product trihydroxyindole can be detected by fluorescence detection with a high sensitivity. Sigvardson et al. [41] applied a zinc reactor for the postchromatographic reduction of nitro polycyclic aromatic hydrocarbons to amino-substituted analogs. These compounds can be detected with extremely high sensitivity (0.1-10 pg) by a peroxyoxalate chemiluminescence detection system (see Fig. 4.15). Refilling of the reactor after about 8-10 hr is required. Irth et al. [42] used a reactor containing metallic copper for the complexation of dithiocarbamates which are then selectively detected at 435 nm. The reactor of only 2-4 × 2.1 mm I.D. gave 100% conversion at flow rates of 0.2-0.4 ml/min and the band broadening, $\sigma_t$, of the reactor was only about 2 s. The useful lifetime of the postchromatographic reactor is determined by the amount of analyte injected and by the presence of oxygen which causes passivation. With a reactor that contains about 8 mg of copper, over 200 analyses can be run. Figure 4.16 shows two applications of this selective detection system, i.e., the determination of thiram in surface water and the determination of disulfiram in urine.

Many other solid-state reactions appear to have a high potential for postchromatographic reaction detection in LC. Addition of a reagent has also been realized by a reactor mounted parallel to the separation column. The mobile phase is split upstream from the injection valve. In this way a relatively large reactor can be used because it does not contribute to band broadening. The principle has been successfully used for the postchromatographic addition of hydroxide ions released from an anion exchanger to the mobile phase [43]. In this way, barbiturates are converted to their anionic form and hence sensitive detection at 254 nm is possible in contrast to the less selective low-UV region (below 230 nm) which is required for barbiturates in nonalkaline medium. Another interesting example of a solid-state reactor parallel to the LC column is an iodine reactor. For the fluorescence detection of the aflatoxins $B_1$ and $G_1$, a 20-fold gain in fluorescence was found after reaction with iodine [44]. Labeling reactions with adsorbed reagents is another alternative as was demonstrated by Colgan et al. [45] in an off-line mode for the labeling of alkyl halides and epoxides with picric acid. One of the most promising types of solid-state reactors are immobilized enzymes, which are discussed separately in Chap. 6.

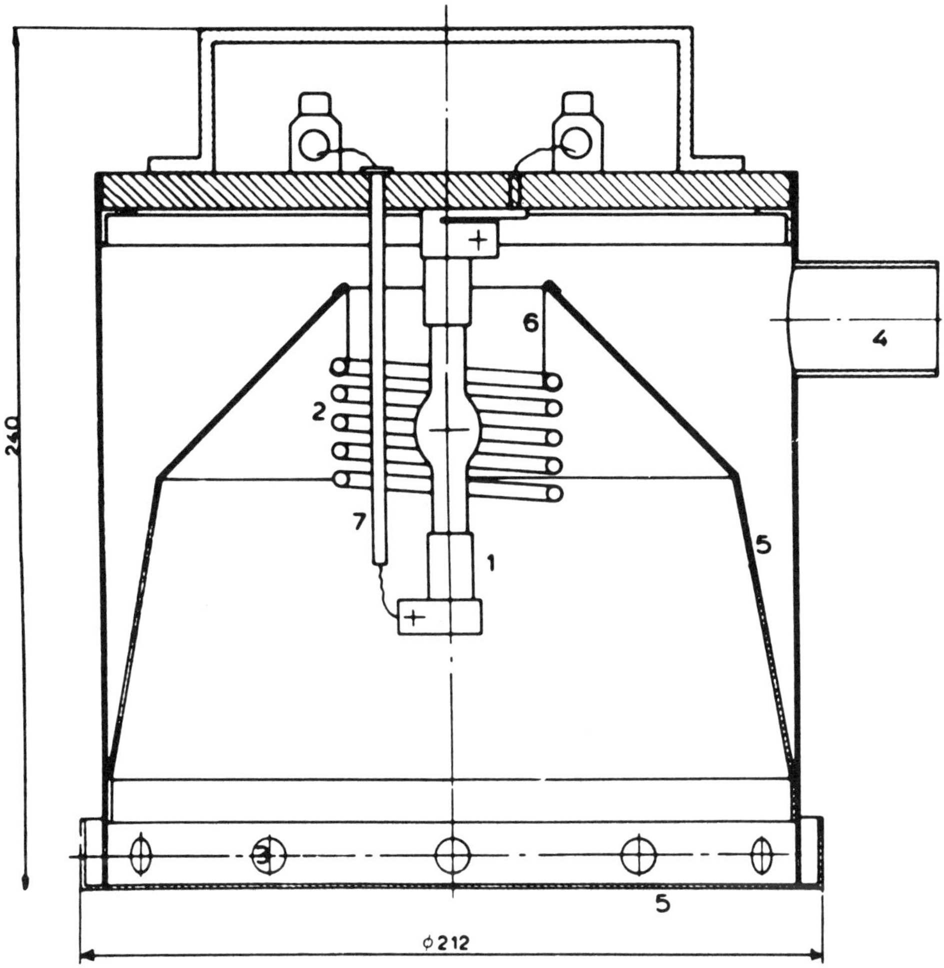


Figure 4.17 Photochemical reactor: (1) 200-W Xe-Hg high-pressure lamp; (2) quartz capillary, 0.5 mm I.D.; (3) inlet for air cooling; (4) outlet for air cooling; (5) aluminum reflection shield; (6) clip; (7) glass isolation. (From ref. 46.)

### Photochemical Reactors

Many photochemical reactions are useful for postchromatographic reaction detection in LC. The analyte is converted by UV irradiation to a product or products that have strongly improved detectability in fluorescence, electrochemical, or UV detection. Other possibilities are photoconductivity, photoionization, and photoelectrochemical detection. Photoconductivity reactions use light to generate new products that have different conductivity properties. Photoionization reactions form ions and electrons which are collected and detected. In a photoelectrochemical detector light impinges on the working electrode of an electrochemical detector in order to convert a nonelectroactive species to an electrophore.

Photochemical reactions are dependent on the composition of the mobile phase, the residence time of the analyte in the reactor, and, of course, the wavelength and intensity of the light source. All these variables should be considered for optimization of the sensitivity and selectivity of a photochemical reaction. An advantage of postchromatographic photochemical reactions over more conventional conversion reactions is the absence of pump(s) and mixing tee(s) for reagent addition.

Until recently, no commercial photochemical reactors for LC were available, and, therefore, different home-made reactors have been used. One example is shown in Fig. 4.17 [46]. The light source is a 200-W Xe-Hg lamp, the LC eluate passes through a quartz or PTFE capillary of 0.5 mm I.D., and air cooling takes place. An aluminum reflection shield encloses the lamp and the capillary and may improve the yield of the reaction. PTFE capillaries are preferred since in addition to providing good transparency in the 200- to 300-nm region PTFE tubing is inexpensive, is available in a wide variety of diameters, and is easier to handle and install than quartz. The temperature should be kept constant because temperature fluctuations will affect flow rates, reaction kinetics, and/or reaction pathways. Uihlein and Schwab [47] used a very elegant knitted reactor with the UV light source being mounted on a sledge in order to be easily moved further into or out of the reactor to vary the time of irradiation. The band broadening was reduced by knitting of the tube (see above). Recently, Nieder and Jaeger [48] presented a novel design of a photochemical reactor of which a schematic drawing is shown in Fig. 4.18. The reactor consists of a PTFE capillary coiled around two low-pressure mercury UV lamps. The band broadening was found to be clearly less in the nonhelical coil than in a helical coil. The potential of the novel design is demonstrated in Fig. 4.19 for the determination of the anticancer drug tamoxifen and one of its metabolites, *N*-desmethyltamoxifen, which show fluorescence gain factors of 30 after irradiation. Addition of a reagent to the mobile phase can be essential for the

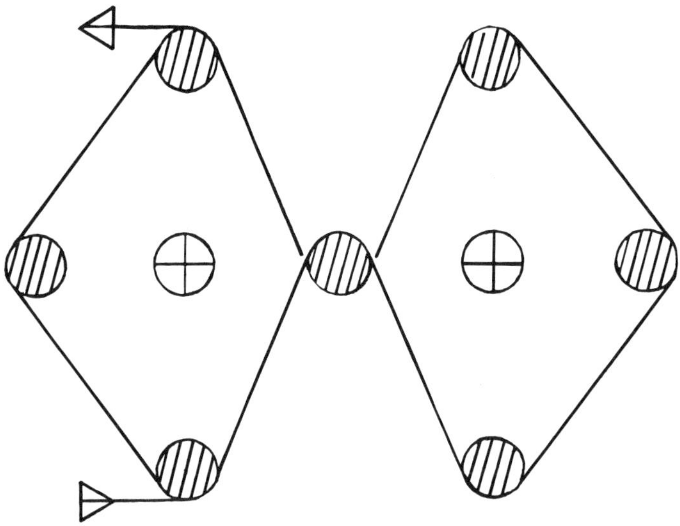

Figure 4.18 Scheme of a photochemical reactor with two UV lamps. The front view of the nonhelical coil is shown. Shaded circles are metal rods and the circles with a cross are UV lamps. (From ref. 48.)

proper mechanism of a photochemical reaction. This was illustrated by Lefevere et al. [49] for the photochemical reaction detection of vitamin K homologs. Many fluorescent reaction products may be formed by photolysis of these compounds. The formation of naphthohydroquinones with an intense fluorescence was promoted by adding ascorbic acid as stabilizer. A concentration of only 0.25 ng/ml in the LC mobile phase was sufficient for this effect.

A recent application is the determination of chlorophenols in surface water [50]. The analytes were separated as their dansyl derivatives with postchromatographic irradiation releasing the highly fluorescent dansyl-OH and dansyl-$OCH_3$. Fluorescence gain factors of up to 8000-fold were obtained for derivatives with a low native fluorescence (see Fig. 4.20). The system also seems to offer many possibilities for the detection of drugs and metabolites with phenolic groups. A number of applications of LC with on-line photochemical reactions and fluorescence detection is reported in Table 4.2. An example of the use of photochemical reactions followed by electrochemical detection is shown in Fig. 4.21 [51]. Some nitro compounds (explosives and drugs) release inorganic nitrite upon photolysis which can easily be detected at +1.0 V. The nitro compounds themselves can only be detected in the reductive mode, which is rather problematic.

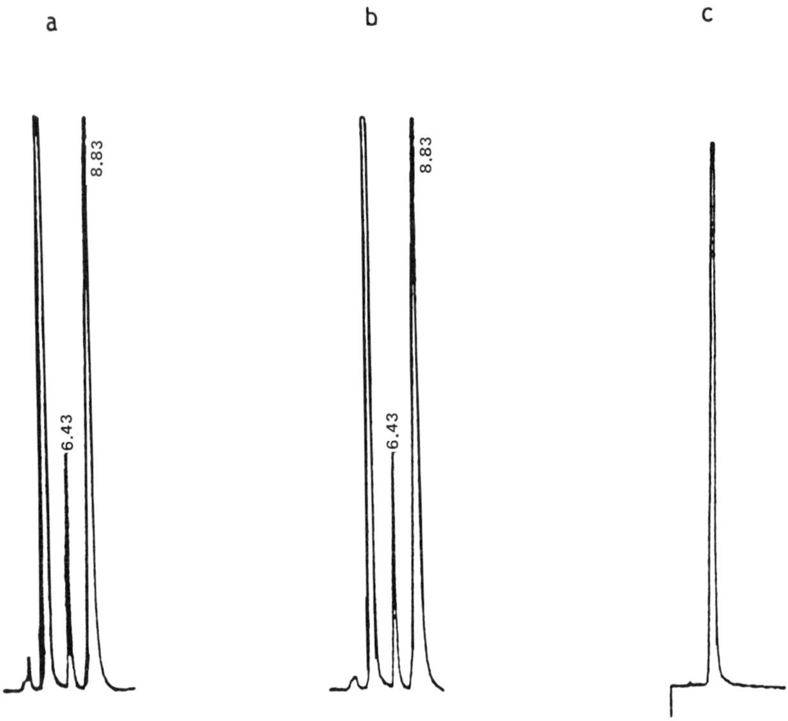


Figure 4.19 Evaluation of the photochemical yield. Injection of 1 ng of *N*-desmethyltamoxifen (peak at 6.43 min) and 5 ng of tamoxifen (peak at 8.83 min) directly on-column under different operating conditions of the photoreactor: (a) two lamps on; (b) one lamp on; (c) lamps off. Operation of only one lamp is sufficient for quantitative photochemical rearrangement. Without irradiation the native fluorescence of the unreacted analytes is too weak to be detected. (From ref. 48.)

Table 4.2 Applications of HPLC with On-line Photochemical Reaction Detection[a]

| | | |
|---|---|---|
| Cannabinol | Demoxepam | Methotrexate |
| Chlorophenols | Diethylstilbestrol | Phenothiazines |
| Clobazam | LSD | Tamoxifen |
| Clomiphene | Metal ions | Vitamin $K_1$ |

[a]For references, see Ref. 5. Most applications are for urine, blood, and/or surface water samples.

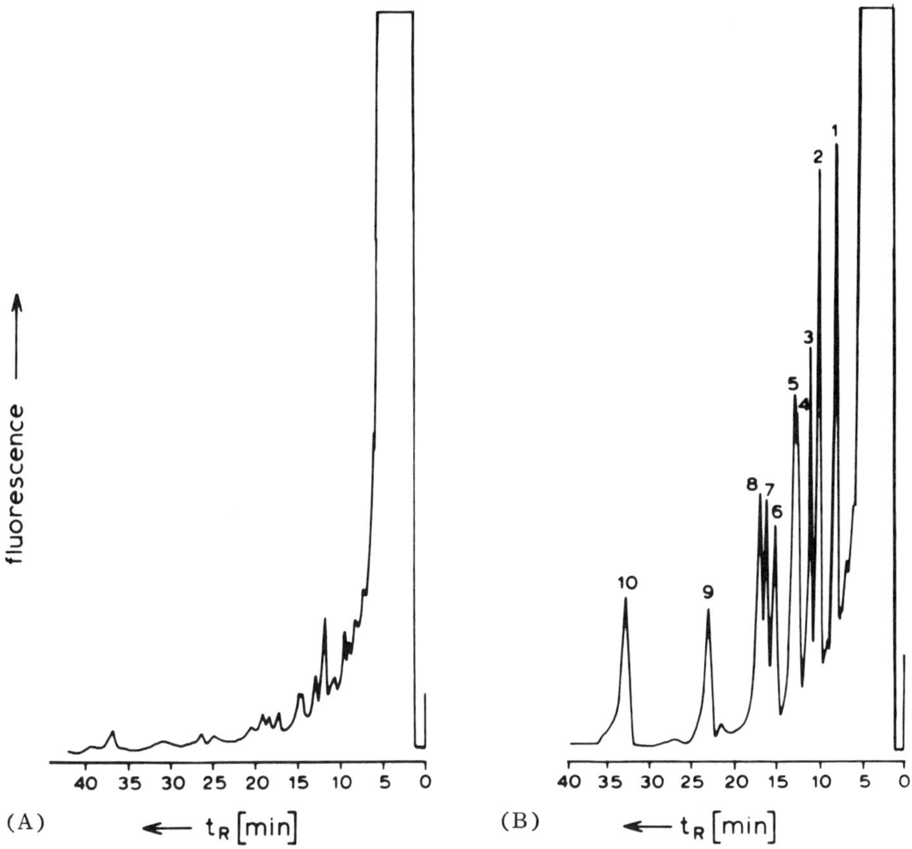


Figure 4.20 Analysis of river water. Column: 250 × 3.1 mm I.D. 5-μm LiChrosorb RP-18; mobile phase: methanol-0.025 M phosphate buffer pH 7.5 (82:18) at 1.0 ml/min; river water sample size: 100 ml; detector: Perkin-Elmer LS-4 set at $\lambda_{exc}$ = 330 nm (slit 15 nm), $\lambda_{em}$ = 470 nm (slit 20 mm); (A) unspiked river water; (B) river water spiked at a 1 ppb level with chlorophenols; peak identities: (1) 2,3,6-trichlorophenol, (2) 2,4,6-trichlorophenol, (3) 2,3,4-trichlorophenol, (4) 2,4,5-trichlorophenol, (5) 2,3,5-trichlorophenol, (6) 3,4,5-trichlorophenol, (7) 2,3,5,6-tetrachlorophenol, (8) 2,3,4,6-tetrachlorophenol, (9) 2,3,4,5-tetrachlorophenol, (10) pentachlorophenol. (From ref. 50.)

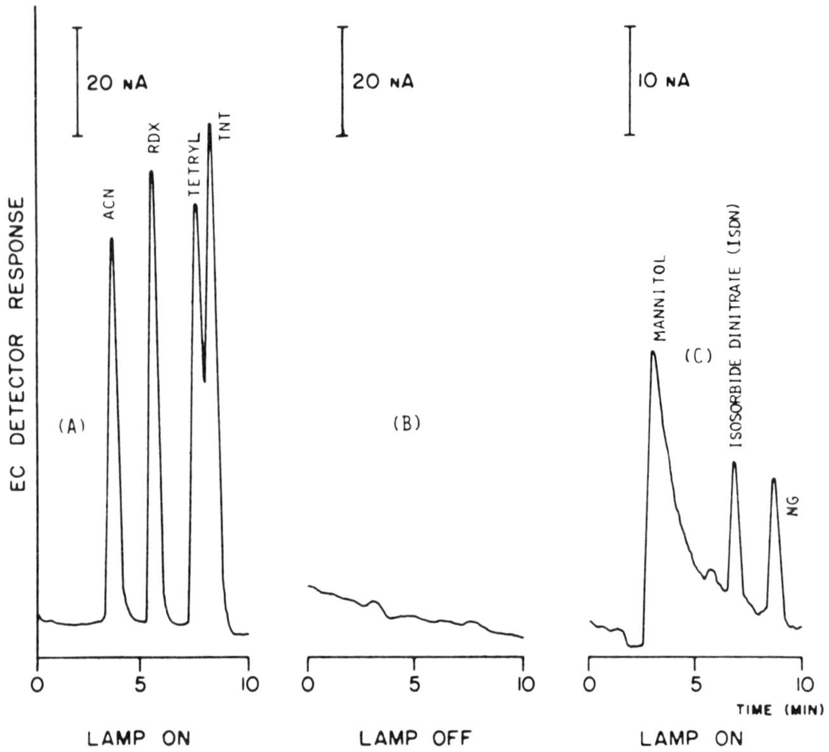


Figure 4.21 (A) LC-hν-EC single-electrode chromatograms of three explosives: RDX, TNT, and **TETRYL** standards using 10-μm RP-8, 25 cm × 4.1 mm I.D., with methanol-0.1 M NaCl (50:50) as mobile phase at 1.4 ml/min, glassy carbon electrode at 1.0 V; (B) lamp off; (C) lamp on with isosorbide dinitrate (ISDN) and nitroglycerin (NG) drug standards. (From ref. 51.)

The photoconductivity detector has been described in the literature for some years [52-55] and is commercially available from Tracor Instruments (Tracor Inc., Austin, TX). Upon irradiation nonionic compounds are converted into ionic species which increases the conductivity of an electrolyte solution. For example, halogenated compounds are photolytically decomposed to halogen acids and other products. This detector is also suitable for, e.g., *N*-nitrosamines, organothiophosphates, sulfonamides, and other drugs. Figure 4.22 shows a block diagram of the commercial photoconductivity detector and Fig. 4.23 demonstrates its high sensitivity and selectivity. The photoconductivity detector can only be used with aqueous mobile

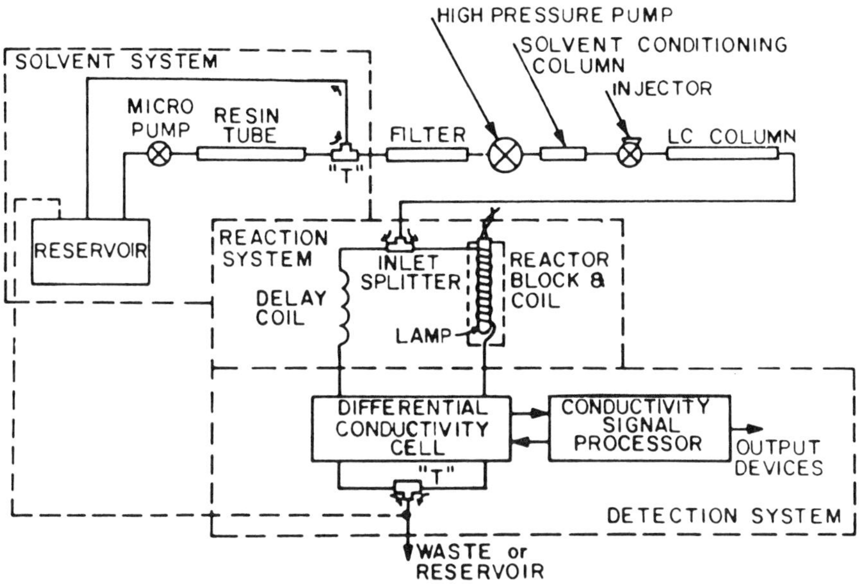


Figure 4.22 Scheme of the photoconductivity detector. (From ref. 52.)

phases and mobile phase constituents should not cause excessively high background levels. Photoionization and photoelectrochemical detection have been reviewed in Ref. 5.

## Electrochemical Detection with Postchromatographic Reactions

Only a few papers have appeared on the use of postchromatographic reactions followed by electrochemical detection. An elegant approach is postchromatographic electrochemical generation of bromine or iodine, reaction of the bromine or iodine with the analyte(s) of interest, and electrochemical detection of the excess of halogen. This system was successfully applied by King and Kissinger [56] for the detection of unsaturated fatty acids and prostaglandins and by Kok et al. [57,58] for the detection of thioethers. Detection limits in the picomole range have been obtained. The electrochemical cell for the production of bromine should have a small dead volume as it contributes to the band broadening. The cell developed by Kok et al. is shown in Fig. 4.24. The generated halogen can also be used for the production of strongly fluorescent species as was demonstrated by Kok et al. [59] for the reaction of phenothiazines and aflatoxins with bromine.

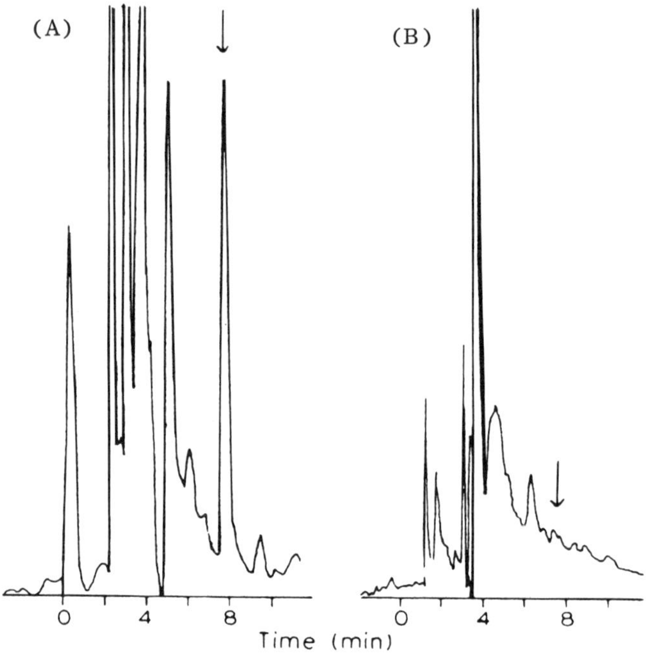


Figure 4.23 Chromatograms of extract of fresh strawberry sample containing incurred residue of 0.03 ppm of captan. (A) Photoconductivity detector; (B) UV detector at 220 nm and 0.01 aufs. Injection of 7 μl of extract equivalent to 486 mg of sample was made. Column: 25 cm × 4.6 mm I.D. Zorbax CN; mobile phase: methanol-water (55:45) at 1 ml/min. (From ref. 54.)

Recently, a method was described for the sub-ppm level detection of metal ions by postchromatographic complexation with a dithiocarbamate ligand and monitoring of the oxidation current of the ligand [60]. The oxidation current at the selected potential decreases if complexation with a metal ion takes place. The ligand (pyrrolidine-carbodithioate) is oxidized at a low anodic potential (+0.65 V) and therefore the detection system is very selective.

The coupling of enzyme reactions that produce hydrogen peroxide with electrochemical detection of the peroxide is also very interesting. Eva et al. [61] and Asano et al. [62] used this system for the determination of acetylcholine in brain tissues. The detection limit was less than 10 pmol.

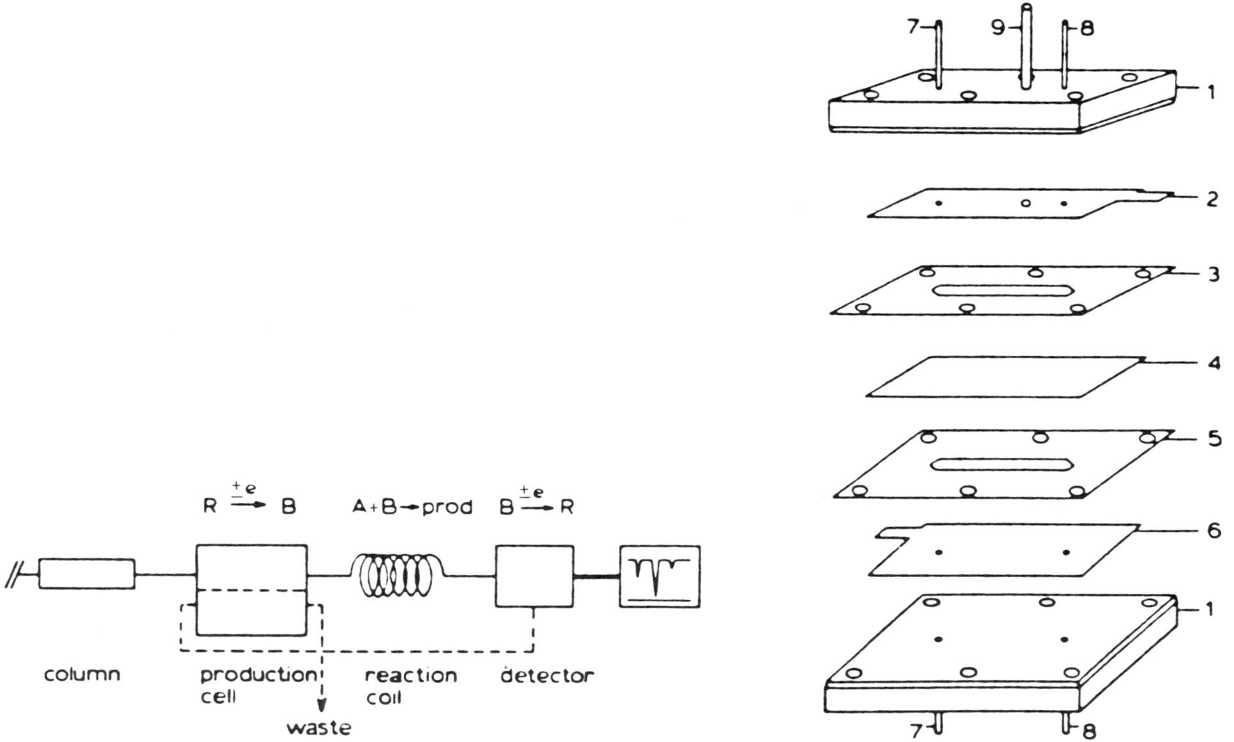


Figure 4.24 (Left) Scheme of on-line electrochemical reagent production and detection. (Right) Coulometric cell: (1) stainless steel support with PTFE insulation; (2) Pt working electrode; (3) spacer; (4) ion exchange membrane; (5) spacer; (6) stainless steel counterelectrode; (7) inlet; (8) outlet; (9) channel to reference electrode (optional). (From ref. 57.)

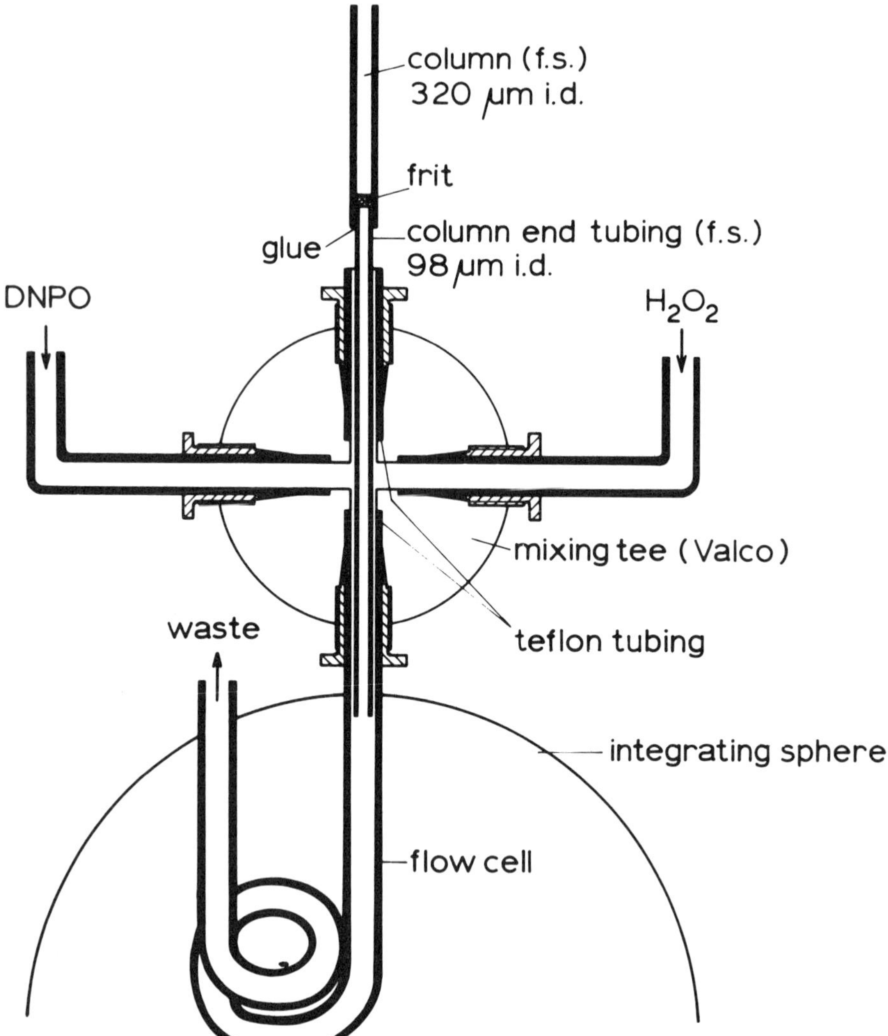


Figure 4.25 Scheme of the system for mixing of the LC eluate of a packed capillary column and the reagents for peroxyoxalate chemiluminescence detection. (From ref. 65.)

### Chemiluminescence Reactions

Chemiluminescence as a technique for LC detection is discussed in Chap. 10. Here, only two aspects which have serious consequences for a chemiluminescence reaction detection system will be emphasized. The emission signal of a chemiluminescence reaction displays an exponential decay. In a dynamic (LC) system, the half-life of the chemiluminescence signal is of prime importance. For given values of the flow rates of the mobile phase and the reagent streams, the dead volume between the mixing tee and the flow cell, and the volume of the flow cell itself, the chemiluminescence half-life determines the percentage of the emitted light that will be measured [63]. In actual practice, the decay of the signal can be so rapid that even a small dead time between the mixing of the reagent solution(s) with the LC effluent and the detection cell inlet will result in considerable loss of emitted light. This problem can be solved by carrying out the mixing within an integrating sphere which also efficiently collects (nearly) all the emitted light. It is noteworthy that even with a relatively large 450-μl glass coil as flow cell [63], additional band broadening in a LC system with a 1-mm-I.D. column is almost negligible if the chemiluminescence signal completely decays in a small part of this cell. For a full discussion of this so-called chemical band narrowing effect, one should consult Refs. 63 and 64. The same principle has been applied [65] to LC with 320-μm-I.D. packed capillary columns. The detection system is shown in Fig. 4.25 and explained in the legend. With several dansylated compounds, detection limits of about 400 attomol were obtained.

Secondly, effective light collection plays a dominant role in chemiluminescence detection. If a fluorescence detector is selected, the optical system should collect a large part of the light. The light source is not used and therefore the excitation part of the apparatus plays no role. The optical aperture of a commercial fluorescence detector is often too small. The Kratos (Schoeffel) FS-970 (Kratos, Ramsey, NJ) detector with the 2π steradian mirror in front of the flow cell is very suitable for chemiluminescence detection [64,66]. Positioning of the cell in very close proximity to the photomultiplier is also essential. As mentioned above and shown in Fig. 4.25, an integrating sphere can collect (nearly) all the emitted light. Gandelman and Birks [67] also described an interesting design, i.e., a flow cell (PTFE coil, 60 μl) placed within an ellipsoidal mirror. The cell is positioned near the first focus of the mirror and the second focus is placed on the photomultiplier cathode.

Van Zoonen et al. [68] developed a simple design for the chemiluminescence detection of hydrogen peroxide with the peroxyoxalate reaction (see Chap. 10). The reagents, bis(2,4,6-trichlorophenyl)-oxalate and 3-aminofluoranthene, were chemically bonded to glass beads and these beads were packed in the detector flow cell in front of the photomultiplier. The system was combined on-line with immobilized enzymes for detection of hydrogen peroxide which was produced by the enzyme [69] or with photochemical reactions [70]. This demonstrates that solid-state reactor principles can also be adapted to chemiluminescence detection.

## CONCLUSIONS

The field of postchromatographic reaction detection in LC is now expanding rapidly. A variety of reactors and reaction types have been developed, and the theoretical description of postchromatographic reactors is often sufficiently known to allow proper optimization of the systems. In order to achieve optimum performance, the selection of suitable hardware, i.e., pumps, mixing tees, phase separators, materials, and dimensions of reactors, is highly important. Especially with miniaturized systems, the reduction of the dead volume of the reaction system deserves much attention because the band broadening in the postchromatographic part of the system can give a considerable contribution to the total band broadening. The use of solid-state reactors seems very attractive in such cases. It is obvious that additional research in this area is required, so much the more because it also involves the use of immobilized enzymes (see Chap. 6) and the even newer area of immobilized antibodies with the extreme selectivity of antibody-antigen interactions. If such systems can be run using enzyme-labeled antigens and chemical amplification, excellent sensitivity can also be attained. However, a great deal of work remains to effectively convert immunoassay techniques into postchromatographic reaction systems. Also catalytic redox reactions, e.g., on metal-loaded surfaces, offer many possibilities. If necessary, low amounts of reagents can be added to the mobile phase without influencing chromatographic behavior. Other interesting pumpless reaction systems are photochemical reactions. Photochemical split-off of strongly fluorescent or chemiluminescent labels is especially promising [50]. As for segmented stream systems, two-phase reactions have opened a new area. For well-chosen reactions, the reaction products of the analytes are rapidly formed and transported in the organic phase. It seems worthwhile to investigate this type of reaction for various types of compounds and to develop good labels for various detectors.

Postchromatographic reaction detectors can be inserted into automatic analytical systems. In such a case the sensitivity and selectivity of the reaction itself will often permit a relatively simple sample preparation. For the analysis of highly complex mixtures, the introduction of on-line precolumn technology can further enhance the sensitivity and selectivity of the procedure. In this way, very powerful integrated systems can be obtained [4,32]. Here it should be noted that the selectivity needed in the sample pretreatment step is dependent on the selectivity of the detector and vice versa.

In the authors' opinion, there is a distinct need for both flexible general-purpose and tailor-made dedicated LC cum postchromatographic reaction detection systems. At present, the commercial availability of suitable hardware and accessories is still rather limited, and further marketing activities from the instrument companies will certainly be welcome.

## REFERENCES


1. Lawrence, J. F., and R. W. Frei, *Chemical Derivatization in Liquid Chromatography*, Elsevier, Amsterdam, 1976.
2. Blau, K., and G. S. King, *Handbook of Derivatives for Chromatography*, Heyden, London, 1977.
3. Schwedt, G., *Chemische Reaktionsdetektoren für die schnelle Flüssigkeitschromatographie*, Hüthig, Heidelberg, 1980.
4. Frei, R. W., and J. F. Lawrence (eds.), *Chemical Derivatization in Analytical Chemistry*, Vols. 1-2, Plenum Press, New York, 1981-1982.
5. Krull, I. S. (ed.), *Reaction Detection in Liquid Chromatography*, Marcel Dekker, New York, 1986.
6. Snyder, L. R., and H. J. Adler, *Anal. Chem.*, *48*:1017, 1022 (1976).
7. Snyder, L. R., *J. Chromatogr.*, *125*:287 (1976).
8. Lawrence, J. F., U. A. Th. Brinkman, and R. W. Frei, *J. Chromatogr.*, *185*:473 (1979).
9. Smythe, W. J., S. L. Bellinger, H. G. Diebler, and R. Pannewitz, U.S. Patent 3,826,615 (1974).
10. Deelder, R. S., A. T. J. M. Kuypers, and J. H. M. van den Berg, *J. Chromatogr.*, *255*:545 (1983).
11. Shih, Y. T., and P. W. Carr, *Anal. Chim. Acta*, *167*:137 (1985).
12. Langer, S. H., and J. E. Patton, *New Developments in Gas Chromatography*, John Wiley and Sons, New York, 1973, p. 294.
13. Nondek, L., U. A. Th. Brinkman, and R. W. Frei, *Anal. Chem.*, *55*:1466 (1983).
14. Laidler, K. J., and P. S. Bunting, *The Chemical Kinetics of Enzyme Action*, 2nd ed., Clarendon Press, Oxford, 1973.
15. Apffel, J. A., U. A. Th. Brinkman, and R. W. Frei, *Chromatographia*, *17*:125 (1983).
16. Niessen, W. M. A., H. P. M. van Vliet, and H. Poppe, *Chromatographia*, *20*:357 (1985).
17. Engelhardt, H., and U. D. Neue, *Chromatographia*, *15*:403 (1982).

18. Kobayashi, S., and K. Imai, *Anal. Chem.*, *52*:1548 (1980).
19. Pilosof, D., and T. A. Neiman, *Anal. Chem.*, *54*:1698 (1982).
20. Malavolti, N. L., D. Pilosof, and T. A. Neiman, *Anal. Chem.*, *56*:2191 (1984).
21. Davis, J. C., and D. P. Peterson, *Anal. Chem.*, *57*:768 (1985).
22. Haginaka, J., J. Wakai, and H. Yasuda, *Anal. Chem.*, *59*:324 (1987).
23. Buck, R. H., and K. Krummen, *J. Chromatogr.*, *315*:279 (1984).
24. Nimura, N., and T. Kinoshita, *J. Chromatogr.*, *352*:169 (1986).
25. Moye, H. A., S. J. Scherer, P. A. St. John, *Anal. Lett.*, *10*:1049 (1977).
26. Nondek, L., R. W. Frei, and U. A. Th. Brinkman, *J. Chromatogr.*, *282*:141 (1983).
27. Low, K. S., U. A. Th. Brinkman, and R. W. Frei, *Anal. Lett.*, *17*:915 (1984).
28. Engelhardt, H., and B. Lillig, *Chromatographia*, *21*:136 (1986).
29. Jansen, H., U. A. Th. Brinkman, and R. W. Frei, *Chromatographia*, *20*:453 (1985).
30. van Vliet, H. P. M., G. J. M. Bruin, J. C. Kraak, and H. Poppe, *J. Chromatogr.*, *363*:187 (1986).
31. Reddingius, R. J., G. J. de Jong, U. A. Th. Brinkman, and R. W. Frei, *J. Chromatogr.*, *205*:77 (1981).
32. Werkhoven-Goewie, C. E., C. de Ruiter, U. A. Th. Brinkman, R. W. Frei, G. J. de Jong, C. J. Little, and O. Stahel, *J. Chromatogr.*, *255*:79 (1983).
33. Apffel, J. A., U. A. Th. Brinkman, and R. W. Frei, *Chromatographia*, *18*:5 (1984).
34. de Ruiter, C., J. H. Wolf, U. A. Th. Brinkman, and R. W. Frei, *Anal. Chim. Acta*, *192*:267 (1987).
35. Jansen, H., E. G. van der Velde, U. A. Th. Brinkman, R. W. Frei, and H. Veening, *Anal. Chem.*, *58*:1380 (1986).
36. Vrátny, P., R. W. Frei, U. A. Th. Brinkman, and M. W. F. Nielen, *J. Chromatogr.*, *295*:355 (1984).
37. Krull, I. S., S. Colgan, K.-H. Xie, U. Neue, R. King, and B. Bidlingmeyer, *J. Liq. Chromatogr.*, *6*:1015 (1983).
38. Studebaker, J. F., *J. Chromatogr.*, *185*:497 (1979).
39. Millot, M. C., B. Sebille, and J. P. Mahieu, *J. Chromatogr.*, *354*:155 (1986).
40. Rüter, J., U. P. Kurz, and B. Neidhart, *J. Liq. Chromatogr.*, *8*:2475 (1985).
41. Sigvardson, K. W., and J. W. Birks, *J. Chromatogr.*, *316*:507 (1984).
42. Irth, H., G. J. de Jong, U. A. Th. Brinkman, and R. W. Frei, *J. Chromatogr.*, *370*:439 (1986).
43. Jansen, H., C. J. M. Vermunt, U. A. Th. Brinkman, and R. W. Frei, *J. Chromatogr.*, *366*:135 (1986).

44. Jansen, H., R. Jansen, U. A. Th. Brinkman, and R. W. Frei, *Chromatographia*, *24*:555 (1987).
45. Colgan, S. T., I. S. Krull, U. Neue, C. Dorschel, C. Stacey, A. Newhart, and B. Bidlingmeyer, *J. Chromatogr.*, *333*:349 (1985).
46. Scholten, A. H. M. T., U. A. Th. Brinkman, and R. W. Frei, *Anal. Chim. Acta*, *114*:137 (1980).
47. Uihlein, M., and E. Schwab, *Chromatographia*, *15*:140 (1982).
48. Nieder, M., and H. Jaeger, *J. Chromatogr.*, *413*:207 (1987).
49. Lefevere, M. F., R. W. Frei, A. H. M. T. Scholten, and U. A. Th. Brinkman, *Chromatographia*, *15*:459 (1982).
50. de Ruiter, C., J. Bohle, G. J. de Jong, U. A. Th. Brinkman, and R. W. Frei, *Anal. Chem.*, *160*:666 (1988).
51. Krull, I. S., C. Selavka, X.-D. Ding, K. Bratin, and G. Forcier, *Bioanaly. Systems Current Sep.*, *5*:57 (1984).
52. Popovich, D. J., J. B. Dixon, and B. J. Ehrlich, *J. Chromatogr. Sci.*, *17*:643 (1979).
53. McKinley, W. A., *J. Anal. Toxicol.*, *5*:209 (1981).
54. Walters, S. M., *J. Chromatogr.*, *259*:227 (1983).
55. Jasinski, J. S., *Anal. Chem.*, *56*:2214 (1984).
56. King, W. P., and P. T. Kissinger, *Clin. Chem.*, *26*:1484 (1980).
57. Kok, W. Th., U. A. Th. Brinkman, and R. W. Frei, *Anal. Chim. Acta*, *162*:19 (1984).
58. Kok, W. Th., J. J. Halvax, W. H. Voogt, U. A. Th. Brinkman, and R. W. Frei, *Anal. Chem.*, *57*:2580 (1985).
59. Kok, W. Th., W. H. Voogt, U. A. Th. Brinkman, and R. W. Frei, *J. Chromatogr.*, *354*:249 (1986).
60. Hojabri, H., A. G. Lavin, G. G. Wallace, and J. M. Rivrillo, *Anal. Chem.*, *59*:54 (1987).
61. Eva, C., M. Hadjiconstantinou, N. H. Neff, and J. L. Meek, *Anal. Biochem.*, *143*:320 (1984).
62. Asano, M., T. Miyauchi, T. Kato, K. Fujimori, and K. Yamamoto, *J. Liq. Chromatogr.*, *9*:199 (1986).
63. de Jong, G. J., N. Lammers, F. J. Spruit, R. W. Frei, and U. A. Th. Brinkman, *J. Chromatogr.*, *353*:249 (1986).
64. de Jong, G. J., N. Lammers, F. J. Spruit, U. A. Th. Brinkman, and R. W. Frei, *Chromatographia*, *18*:129 (1984).
65. de Jong, G. J., N. Lammers, F. J. Spruit, C. Dewaele, and M. Verzele, *Anal. Chem.*, *59*:1458 (1987).
66. Imai, K., and R. Weinberger, *Trends Anal. Chem.*, *4*:170 (1985).
67. Gandelman, M. S., and J. W. Birks, *J. Chromatogr.*, *242*:21 (1982).
68. van Zoonen, D. A. Kamminga, C. Gooijer, N. H. Velthorst, R. W. Frei, and G. Gübitz, *Anal. Chim. Acta*, *174*:151 (1985).

69. van Zoonen, P., C. Gooijer, N. H. Velthorst, R. W. Frei, J. H. Wolf, J. Gerrits, and F. Flentge, *J. Pharm. Biomed. Anal.*, *5*:485 (1987).
70. Poulsen, J. R., J. W. Birks, G. Gübitz, P. van Zoonen, C. Gooijer, N. H. Velthorst, and R. W. Frei, *J. Chromatogr.*, *360*:371 (1986).

# 5
# Enantiomeric Derivatization

SÁNDOR GÖRÖG *Chemical Works of Gedeon Richter, Ltd., Budapest, Hungary*

## INTRODUCTION: IMPORTANCE OF ENANTIOMERIC SEPARATIONS

The chromatographic separation of enantiomers is one of the most important topics in contemporary biomedical analysis and at the same time one of the most exciting branches of chromatographic research. Hundreds of papers have been published in the last decade dealing with the theoretical and practical aspects of this subject from the point of view of biochemical pharmacology as well as chromatography. The reasons for this enormous interest in the field of enantiomers are as follows:

(1) The differences between the biological-pharmacological activties of the antipodes of several racemic drugs and the differences in their fates in vivo are of interest to pharmacologists. However, in order to study these effects rapid and reliable methods for the separation and quantitation of the enantiomers, even in biological samples, are required. A recently published monograph is recommended [1] concerning the theoretical background of the biological side of these questions. In this section only a few, almost randomly chosen examples are listed with the aim of demonstrating the importance of the chromatographic separation of enantiomers. The problems mentioned in this section will be discussed in one or more of the following sections of this chapter.

Propranolol, a widely used beta-blocking agent, is used in clinical therapy as a racemic mixture in spite of the fact that its S enantiomer is 100 times as potent as the R form. At the same time the latter is metabolized more rapidly [2-8]. A similar situation exists with other beta blockers such as alprenolol and metoprolol [9], pindolol and oxprenolol [7], acebutolol and diacetol [10]. Mexiletine is a racemic antiarrhythmic drug and the pharmacological properties of the individual enantiomers are not known. Based on a separation method for the enantiomers the pharmacokinetic curves could be obtained [11].

In the case of racemic antiinflammatory 2-arylpropionic acids (e.g., ibuprofen), the S enantiomers are more active and an analytical separation method had to be developed for the investigation of the enantioselective biotransformation of R(-) to S(+) [12-14].

Furthermore, the necessity of an enantioselective separation of racemic alcohol derivatives such as warfarin [15] and proxyphylline [16] has been demonstrated.

(2) Several drugs are administered as pure enantiomers rather than as their racemic mixtures. For these analytes the optical purities of the bulk material and the pharmaceutical formulations have to be established. Of course, chromatographic methods supply much more reliable data on the optical conformation than the classical methods based on measurement of the optical rotatory power. Typical samples are the determination of an (*R*)-(-)-amphetamine impurity in (*S*)-(+)-amphetamine [17-19] and the estimation of (*R*)-(-) impurity in (*S*)-(+)-epinephrine [20].

(3) The rapidly increasing interest in asymmetrical syntheses has stimulated the development of new chromatographic techniques for the determination of the optical purity of the starting materials and intermediates of these syntheses. The most typical examples are peptide syntheses where the optical purity of the reaction products depends on the quality of the starting material and the effectivity of the prevention of racemization during the subsequent reaction steps. The first question can be answered directly by checking the optical purity of the amino acids and their protected derivatives while an answer for the second question can be obtained by checking the purity again after hydrolysis of the peptides [21-33].

## POSSIBILITIES FOR CHROMATOGRAPHIC SEPARATION OF ENANTIOMERS

In general, the basis for chromatographic separation of two or more compounds on a given stationary phase is the difference in the binding forces (e.g., dipole-dipole, charge transfer, hydrophobic interactions, Coulomb forces, hydrogen bonding) between the stationary phase and the various compounds to be separated. If only the commonly used achiral stationary phases are taken into consideration as selectors, and the enantiomeric pairs of chiral compounds as the selectants (analytes), it can be stated that, taking into account only the above listed interactions, the two enantiomers are absolutely equivalent and for this reason no separation can be achieved. However, if the goal of an analytical procedure is the resolution of an enantiomeric pair of compounds, this procedure should contain at least one chiral interaction step.

In general, there are three methods for achieving this as discussed below:

(1) Transformation of the enantiomers to a pair of diastereomeric derivatives using a chiral reagent.

| (R,S) | + R' | ⟶ R-R' + S-R' | |
|---|---|---|---|
| racemate to be solved | chiral reagent | diastereomeric derivatives | (1) |

In this reaction (Eq. 1), covalently bound diastereomers are formed. These diastereomers are no longer equivalent regarding the binding forces to the stationary phase and hence they can be separated even on an achiral stationary phase. The derivatization reaction is carried out before chromatography is performed and for the separation of the diastereomers both gas chromatographic (GC) and liquid chromatographic (LC) techniques can be applied. This principle is applicable in those circumstances where the molecule of the selectant contains easily derivatizable functional groups (primary or secondary amino, hydroxyl, or carboxyl groups). Accordingly, the chiral reagents are usually strong acylating agents such as acid chlorides, isocyanates and isothiocyanates, and esterifying agents.

(2) Transformation of the enantiomers to a pair of diastereomeric complexes using a chiral reagent dissolved in the mobile phase.

| (R,S) | + R' | ⟶ R- - -R' + S- - -R' | |
|---|---|---|---|
| racemate to be solved | chiral mobile phase additive | diastereomeric adducts or complexes | (2) |

Just as with the covalently bound derivatives, in this case achiral stationary phases can also be used for the separation of the diastereomeric derivatives. The main differences between reactions 2 and 1:

a. The chiral reagents in reaction 2 are not reactive materials to form covalently bound diastereomeric derivatives but chiral substances capable of forming various kinds of adducts or complexes with the enantiomers to be separated. The main types of these reagents are as follows:

Metal chelates with chiral ligands where the basis of the separation is the ligand exchange equilibrium between the reagent and the enantiomers to be separated

Chiral cations or anions which form diastereomeric ion pairs with the charged selectants

Chiral molecules with strong hydrogen-bonding capabilities (at least three sites) which form a pair of diastereomeric adducts with the racemate to be resolved

Chiral macromolecules such as proteins

Macromolecules with chiral cavities which form inclusion complexes such as cyclodextrins and chiral crown ethers

b. The reagents must be dissolved in the mobile phase.

c. Reaction 2 leads to an equilibrium which exists during the chromatographic analysis and is involved in its separation mechanism.

Because these interactions only can take place in the liquid phase, GC cannot be used, in contrast to the methods discussed under 1 and 3; the application of the principles mentioned above is restricted to LC techniques and in practice mainly to high-performance liquid chromatography (HPLC).

(3) In this group of methods the enantiomeric mixture is separated directly on a chiral stationary phase, meaning that no derivatization reaction is necessary before or during the chromatographic separation.

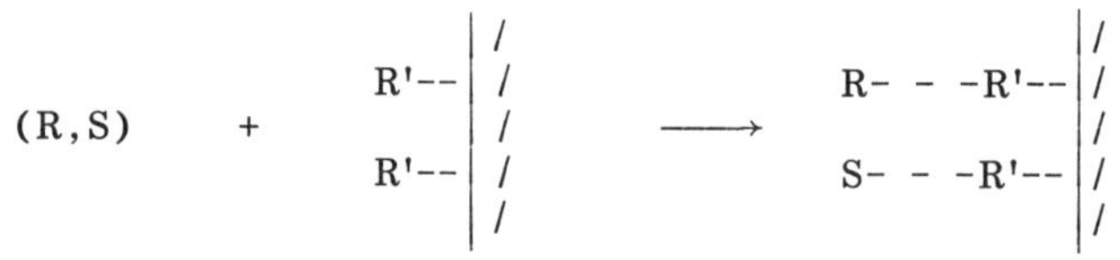

(3)

racemate to be solved — chemically bonded chiral reagent — enantiomers bound to the chiral stationary phase

The basis for the separation is that the stability constants of the diastereomeric adducts formed on the surface of the chiral stationary phase (adsorbent or the liquid phase) are different.

$$K_{R---R'--|/} > K_{S---R'--|/} \qquad (4)$$

This principle can be used in GC as well as in LC. In the overwhelming majority of analytical separations the principal stationary phases are in fact achiral, polysiloxanes in the case of GC and silica gel in the case of LC, but they are chirally modified by covalently bonded chiral moieties (e.g., amino acids, peptides, and their derivatives including metal chelates, proteins, cyclodextrins).

In accordance with the aims and scope of the present handbook, of the above outlined possibilities in this chapter only those enantiomeric separations will be discussed in detail where the LC separation is based on the formation of covalently bonded diastereomeric derivatives and the emphasis is on the selection of the chiral reagent and on the derivatization reaction itself (LC aspects of Introduction section). As from the vast number of papers and reviews on other branches of enantiomeric derivatizations and separation and the whole field in general, a recently published volume [34] and two excellent chapters by Lindner [33] and Lindner and Petterson [36] are mentioned.

## PREREQUISITES FOR DERIVATIZING AGENTS AND REACTIONS

To enable effective separation of the enantiomers and to solve a number of practical problems, the derivatizing agent and the derivatization reaction should fulfill several prerequisites. As a consequence of the delicate nature of the problems, these requirements must be dealt with more seriously in comparison with achiral derivatizations. The main prerequisites are listed below:

(1) Structural features of the reagent should enable well-separable diastereomers to be formed. The greater the conformational differences that exist between the diastereomeric reaction products of the derivatization reaction, the better the separation that can be expected. This depends on

The distance between the two chiral centers
The bulkiness of the groups in the vicinity of the chiral centers
The presence of polar or polarizable groups enabling the formation of hydrogen bonding or other intramolecular interactions stabilizing certain conformations

Several papers deal with the theoretical and empirical aspects of these questions [15,37-42]. Generally speaking, it is difficult to set up general rules because of the entirely different separation mechanisms in the various systems; as can be seen in Tables 5.1-5.3, both RP and NP separations are used in the chromatographic steps. In this chapter only a few practical examples are shown based on the separation methods collected in the above-mentioned tables.

The optimal distance between the two chiral centers is two to four atoms. This requirement is fulfilled in almost all methods discussed in this chapter.

The bulkiness of the substituents in the vicinity of one of the chiral centers favorably influences the conformational rigidity of the diastereomeric derivatives and thus their resolution. This is certainly the reason for what can be easily seen from the structures of reagents in Tables 5.1-5.3, namely, that the overwhelming majority

Table 5.1 Chiral Derivatization of Primary and Secondary Amines

| Compounds to be separated | Reagent(s) | Stationary phase[a] | Detection[b] | Ref. |
|---|---|---|---|---|
| Amino acid esters | (-)-α-Methoxy-α-methyl-1-naphthalene acetic acid | NP | UV (280) | 28 |
| α-Phenylethylamine | *O*-Methylmandelyl chloride | NP | UV (255) | 49 |
| Oxfenicine | *R*-(+)-1-methoxy-1-trifluoromethylphenylacetic anhydride | NP | UV (254) | 50 |
| Amphetamines | | RP | UV (220,254) | 18 |
| Amphetamine | 1-[(4-Nitrophenyl)sulfonyl]prolyl chloride | NP,RP<br>NP,RP | UV (262)<br>UV (254) | 51<br>17 |
| Oxaprotiline | *N*-Trifluoroacetyl-*S*-(-)-prolyl chloride | NP | UV (260) | 52 |
| Amphetamines | Benoxaprofen | NP | FL (312/365) | 47 |
| Amines | (+)-10-Camphorsulfonyl | NP | UV (254) | 55 |
| Amino acids | (+)-10-Camphorsulfonyl chloride + *p*-nitrobenzyl chloride | NP | UV (254) | 31 |
| Propranolol | Symmetrical anhydride of *tert*-butoxycarbonyl-L-leucine + trifluoroacetic acid | RP | FL (210/340) | 8 |
| Alprenolol, metroprolol | | RP | FL (193/---) | 9 |
| Amino acids | *N*-carboxy-L-phenylalanine (or L-leucine) anhydride | RP | UV (210) | 21 |
| 2,5-Dimethoxy-4-methylamphetamine | | NP | UV (280) | 49 |
| Amino acids | *tert*-Butoxy-L-leucine-*N*-hydroxysuccinimidester | IE<br>RP | <br>UV (208) | 22<br>25 |
| Thyronines | | RP | UV (230) | 54 |
| Penicillamine | | RP | UV (215) | 55 |
| γ-Vinyl-γ-amino butyric acid | | RP | UV (210) | 43 |
| Amino acid methyl esters | (-)-1,7-Dimethyl-7-norbornyl isothiocyanate or (+)-neomenthyl isothiocyanate | NP | UV (250) | 38 |

Table 5.1 (continued)

| Compounds to be separated | Reagent(s) | Stationary phase[a] | Detection[b] | Ref. |
|---|---|---|---|---|
| Amino acid *tert*-butyldimethyl-silyl esters | | NP | UV (250) | 38 |
| Amino acids | 2,3,4-tri-O-acetyl-α-D-arbinopyranosyl iso-thiocyanate (AITC) | RP | UV (250) | 24 |
| Epinephrine, nor-epinephrine | | RP<br>RP | UV (250)<br>UV (250) | 27<br>58 |
| Amphetamines | | RP | UV (220,254) | 18 |
| Amino acids | 2,3,4,6-tetra-O-acetyl-β-D-glucopyranosyl iso-thiocyanate (GTIC) | RP<br>RP | UV (250)<br>UV (250) | 24<br>27 |
| Beta-blocking agents | | RP | UV (254) | 4 |
| 4-Hydroxypropranolol sulfate | | RP | UV (313) | 3 |
| Prenalterol sulfate | | RP | UV (254) | 59 |
| Epinephrine | | RP | UV (254) | 60 |
| Epinephrine, nor-epinephrine | | RP | UV (250) | 58 |
| Epinephrine | | RP | UV (254) | 20 |
| Ephedrines | | RP | UV (254) | 61 |
| Mexiletine | | RP | UV (254) | 11 |
| Amphetamines | | RP | UV (220,254) | 18 |
| Amino acids | N-fluorenylmethoxycarbo-nyl-L-*p*-isothiocyanate phenylalanine methylester | RP | UV | 26 |
| Propranolol | 1-Phenylethyl isocyanate | RP | FL (220/340) | 5 |
| Propranolol, 4-hydroxypro-pranolol | | RP | UV (313) | 6 |
| Acebutolol, di-acetolol | | RP | FL (238,450) | 10 |
| Amphetamines | | RP | UV (220,254) | 18 |
| Beta-blocking agents | 1-Naphthylethyl isocy-anate | RP | UV (290)<br>FL (285,330) | 7<br>7 |
| Betaxolol | | RP | FL (285,330) | 57 |
| Oxprenolol | | NP | UV (275) | 56 |
| Amino acids | | RP | UV (340) | 29 |

Table 5.1 (continued)

| Compounds to be separated | Reagent(s) | Stationary phase[a] | Detection[b] | Ref. |
|---|---|---|---|---|
| Amino acids | *N-tert*-Butyloxycarbonyl-L-cysteine + ortho-phthalaldehyde | RP | FL (360/405) | 30 |
| | | RP | | 32 |
| | | RP | FL (344,433) | 33 |

[a]Stationary phase: NP, normal phase; RP, reversed phase; IE, ion exchange.

[b]Detection: UV, ultraviolet-visible absorption, wavelength in nm; FL, fluorescence, excitation and emission wavelengths in nm.

Table 5.2 Chiral Derivatization of Primary and Secondary Hydroxyl Groups

| Compound(s) to be separated | Reagent(s) | Stationary phase[a] | Detection[b] | Ref. |
|---|---|---|---|---|
| Warfarin | Carbobenzyloxy-L-proline + dicyclohexylcarbodiimide + imidazole | NP | UV (254) | 15 |
| Warfarin | 1-Menthylchloroformate | NP | UV (310) | 62 |
| Proxyphylline | (-)-Camphanoyl chloride | RP | UV (254) | 16 |
| Beta-blocking agents | (*R,R*)-*O,O*-diacetyl (dibenzoyl, di-*p*-toluoyldimethyl, diethyl, dibenzyl)tartaric acid anhydride | RP | UV (254) | 45 |
| Alcohols | (+)/(-)-2-Methyl-1-1'-binaphthalene-2-carbonyl nitrile | NP | FL (342,420) | 65 |
| Zeaxanthine, lutein | 1-Naphthylethyl isocyanate | NP | UV (443) | 63 |

[a]Stationary phase: NP, normal phase; RP, reversed phase; IE, ion-exchange.

[b]Detection: UV, ultraviolet-visible absorption, wavelength in nm; FL, fluorescence, excitation and emission wavelengths in nm.

Table 5.3 Chiral Derivatization of Carboxylic Acids

| Compound(s) to be separated | Reagent(s) | Stationary phase[a] | Detection[b] | Ref. |
|---|---|---|---|---|
| Naproxen | (*S*)-(+)-2-Octanol | NP | FL (342/420) | 66 |
| Ibuprofen | | NP | UV (220) | 14 |
| Carprofen | 1-(-)-α-Methylbenzylamine + carbodiimidazole | NP | TLC radiom. | 71 |
| 2-Arylpropionates | | NP | UV (225) | 12 |
| Loxoprofen | 1-1-(4-Dimethylamino-1-naphthyl)ethylamine | NP | FL (313/420) | 68 |
| Naproxen | | NP | FL (320/410) | 67 |
| N-Protected amino acids | D-/L-*O*-(4-nitrobenzyl)-tyrosine methylester + carbodiimides | NP | UV (270) | 44 |
| Pyrethroid insecticides | (-)-1-(1-Phenyl)ethylamine + thionyl chloride | NP | UV (240) | 69 |
| 2-Phenylpropionic acid | | NP | UV (254,272) | 70 |
| | | RP | UV (254,272) | 70 |
| Ketoprofen | | NP | UV (254,272) | 70 |
| | | RP | UV (254,272) | 70 |
| 2-[3-(2-Chlorophenoxyphenyl)]-propionic acid | 2-Aminobutanol + oxalyl chloride | NP | UV (280) | 73 |
| Isoprenoid acids | *R*-(-)/*S*-(-)-α-Methyl-*p*-nitrobenzylamine + oxalyl chloride | NP | UV (254) | 39 |
| Citronellic acid | | NP | UV (254) | 72 |
| Mandelic acid | 1-2-Bromooctane + 18-crown-6 | NP | UV (254) | 74 |

[a]Stationary phase: NP, normal phase; RP, reversed phase; IE, ion exchange.

[b]Detection: UV, ultraviolet-visible absorption, wavelength in nm; FL, fluorescence, excitation and emission wavelengths in nm.

contains more or less bulky ring systems in the vicinity of the chiral center. The exceptionally high separation factors obtained with, for instance, 2,3,4,6-tetra-*O*-acetyl-β-D-glucopyranosyl isothiocyanate (GITC) and the naphthalene derivatives is also considered to be due to the bulkiness of the ring systems [11]. In the former and some other cases another feature positively influencing the resolution is that the chiral center is part of the ring system.

The situation is more complicated with respect to the bulkiness of remote groups. For example, Nambara et al. [38] claim that neomenthyl thiourea derivatives of amino acids can only be separated if their carboxyl groups are esterified with the bulky *tert*-butyldimethylsilyl group; the methyl esters are not separable. At the same time the *tert*-butyloxycarbonyl (BOC) protecting group of the BOC-L-Leu-*O*-Su, etc., reagent has to be removed after the derivatization reaction of, for instance, amino acids in order to obtain good separation in RP systems [8,9,22,25,43]. It is interesting to note, however, that in the NP systems of Görög et al. [44] several amino acid enantiomers with bulky protecting groups at the amino moieties were separated successfully.

As for the importance of the possibility of the formation of intramolecular hydrogen bonds, the case of propranolol and other α-hydroxylamine-type β-receptor blocking agents will be taken as an example. Here the success of the enantiomeric separation after derivatization of their amino groups with an appropriate isocyanate or isothiocyanate to form urea or thiourea derivatives is believed to be based on the formation of hydrogen bonds between the free hydroxyl group and the carbonyl group of the urea moiety. Etherification of the hydroxyl group dramatically reduces the separability of the enantiomers [5]. The review of Lindner and Petterson [36] contains several other interesting examples to demonstrate the importance of intramolecular interactions.

(2) The derivatives should have favorable chromatographic properties. In addition to the resolution of the enantiomers as the diastereomeric derivatives, discussed in the previous section, the chromatographic properties of the diastereomers should be at least as good as those of the parent compounds enabling high column efficiencies. This is a very important point in the case of the biological application of the technique, namely, here traces of the diastereomeric derivatives have to be determined in the presence of several other components.

The majority of reagents in Tables 5.1-5.3 fulfills these prerequisites. Of course, careful optimization of the chromatographic conditions (e.g., selection of RP or NP system, search for optimal solvent composition, additives, pH) is necessary to achieve the highest possible column efficiency. Some of the papers cited in Tables 5.1-5.3 contain useful information on this optimization procedure [8, 9,45].

(3) In this subsection some general features of the reagents and reactions are summarized without going into detail because these are not special requirements in chiral derivatizations but apply to achiral derivatizations as well. Furthermore, they are discussed extensively in Chap. 1. These features are as follows: Both the reagent and the reaction product should be stable under the reaction and chromatographic conditions; the reaction conditions should be as simple and mild as possible; and the reaction should yield a single product from one enantiomer or a pair of diastereomers from a racemic mixture.

(4) In the case of chiral derivatization the reagent and reaction product should not only be chemically but also chirally stable. This means that the chiral reagent should be optically pure (containing no detectable quantity of enantiomeric impurities) and this optical purity should not change during storage, and in the course of the derivatization reaction no racemization of either the reagent or the chiral analyte or the reaction product should occur. If both enantiomers of the analyte are available in an optically pure form, the extent of racemization or the enantiomeric impurity of the reagent can be estimated by checking the relative peak area of the diastereomeric impurity. In most cases the investigators claim this impurity to be less than 1% and normally this is sufficient whenever biological samples are investigated. A contamination of less than 1% is achieved with almost all the reagents listed in Tables 5.1 to 5.3. However, there are some exceptions, e.g., *N*-trifluoroacetyl-(*S*)-(-)-prolyl chloride may racemize during storage, meaning that the commercially available reagent may contain several percents of the (*R*)-(+) antipode [46]. If it is possible, the optical purity of the reagent should be checked by an independent method, e.g., chiral phase GC [47].

If the aim of the investigation is the determination of traces of enantiomeric impurities, then of course the above-mentioned 1% maximum level of enantiomeric impurity in the reagent is not sufficient. If the reagent is optically pure, the separation factor of the diastereomers is sufficiently high, and the detectability of the derivatives is good enough to enable traces of the diastereomeric impurity to be determined, then the lowest measurable quantity of the contaminating optical antipode can be as little as 0.01% as is demonstrated for amino acids by Takaya et al. [21].

In addition to the sources of error discussed so far, insufficient optical purity or optical stability of the chiral reagent or racemization during the derivatization reaction, there is another danger to be taken into consideration. This is the possibility of enantioselectivity of the rate of the derivatization reaction; if there is a difference between the rate constants of the reaction of the two enantiomers and the reaction is not quantitative, then the ratio of the peak areas of the diastereomers will, of course, not reflect the real enantiomeric ratio. In the majority of cases this danger can be avoided by careful optimization of the reaction conditions which can be checked easily by measuring the ratio of the peak area of the racemate.

From the point of view of decreasing the detection level of an enantiomeric impurity, it is advantageous to be sure that the impurity peak elutes before the main peak. This can be easily achieved if both antipodes of the reagent are available [44].

(5) The presence of favorable UV-Vis or fluorimetric properties. Although the primary aim of the chiral derivatization is to transform the enantiomers to well-separable diastereomeric derivatives with the aid of chiral reagents, a really good reagent will furnish at the same time good detectability for the separated derivatives. This is especially important in the case of the investigation of biological samples where the concentration of the separated enantiomers is usually low and also whenever traces of enantiomeric impurities in relatively pure enantiomers must be analyzed.

Looking at the fourth column of Tables 5.1-5.3 it can be seen that normally the chromatograms are scanned with an UV-Vis detector. Some of the reagents do not possess UV-Vis active groups or contain only weak chromophores (e.g., isolated phenyl groups). In these cases either the compound to be derivatized is sufficiently UV-Vis-active or a low wavelength (below 210 nm) is selected to scan the chromatogram. In many cases, however, the detection limits can be improved remarkedly as a consequence of the derivatization reaction with a highly UV-Vis active chiral reagent. With the aid of reagents with a naphthalene group or the thiourea chromophore, formed by the reaction of amines with isothiocyanates, the lowest detectable quantity is in the 10-ng region. The detection level can be decreased easily to 1 ng (absolute) or lower if fluorimetric detection is applied as is demonstrated with some examples in Tables 5.1 and 5.2. For example, the detection limit in the method using *o*-phthalaldehyde (OPA) and *N*-acetyl-L-cysteine as the reagents is claimed to be 5 pmol amino acid per injection [30]. For the details of derivatizations aiming at increasing the sensitivity of UV-Vis and fluorescence detection, Chaps. 7 and 9 will give additional information.

## CHIRAL DERIVATIZING AGENTS AND CORRESPONDING REACTIONS

### Derivatization of Amino Acids

The majority of chiral drugs to be separated by chromatographic methods are primary and secondary amino derivatives and the greatest variety of derivatizing reagents are therefore also available for these compounds. Table 5.1 contains a selection of the reagents and reactions for the chiral derivatization of primary and secondary amines. Among the described reagents several carboxylic acid derivatives are found. However, the free carboxylic acid function reacts only with amines if coupling agents such as dicyclohexylcarbodiimide (DCC) or related compounds are added to the reaction mixture. More widespread is the use of acid chlorides, acid anhydrides, or active ester derivatives which directly transform the primary and secondary amines to the corresponding diastereomeric carboxamide or sulfonamide derivatives.

As a typical example the procedure of Barksdale and Clark [17] is presented for the determination of the enantiomeric composition of amphetamine and related drugs using 4-nitrophenylsulfonyl-(*S*)-prolyl chloride as the reagent (Eq. 5).

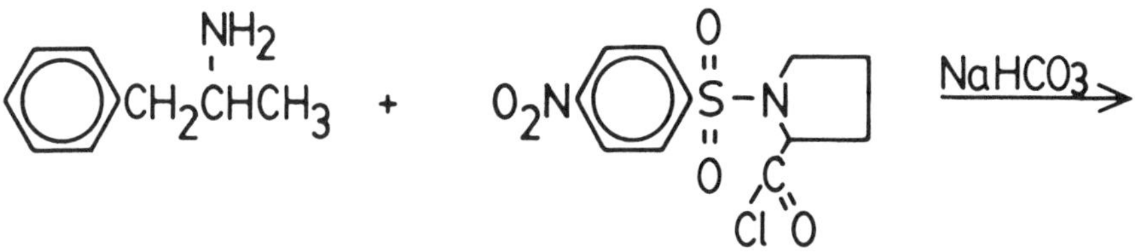


(S,R)-AMPHETAMINE (S)-REAGENT

(5)

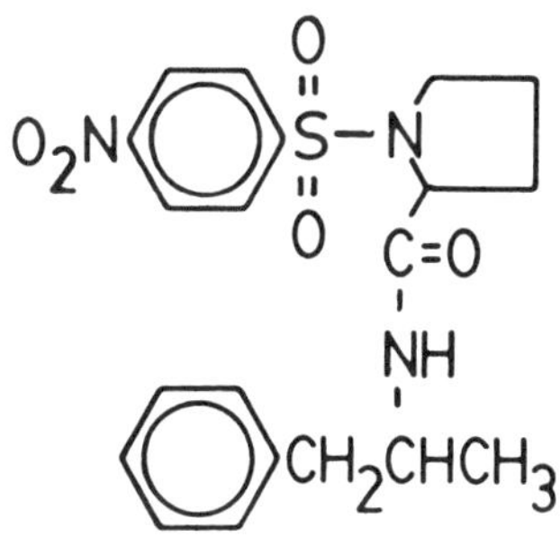

(SR) (SS)

To a reaction vial containing 2 ml tetrahydrofuran and 1 ml freshly prepared solution of 4-nitrophenylsulfonyl-(*S*)-prolyl chloride (0.335 M in tetrahydrofuran), 1 ml of an aqueous solution containing approximately 1 mg of the amphetamine is added, followed by 0.07 ml of a 10% aqueous solution of sodium bicarbonate. The vial is sealed with a Teflon-lined cap and allowed to react at 338 K for 1 hr. The resulting solution is extracted with chloroform (3 times 10 ml). The combined organic phases are washed with 10 ml water, dried over magnesium sulfate, and evaporated to dryness. The residue is dissolved in 2.5 ml of the mobile phase (chloroform/heptane 4:1) and a 5-μl aliquot is injected into the chromatographic system. The 15 cm × 4.6 mm I.D. stainless steel column is packed with 5-μm porous Supelcosil LC-Si and the chromatogram is monitored at 254 nm.

Another important group of reagents are the isocyanates and isothiocyanates which form urea and thiourea derivatives, respectively, with primary and secondary amines. This reaction can be made selective for amines: in the case of amino alcohols only the amino groups react under the chosen conditions. Equation 6 shows the reaction between amino acid ethyl esters and the GITC reagent as described by Nimura et al. [24,27].

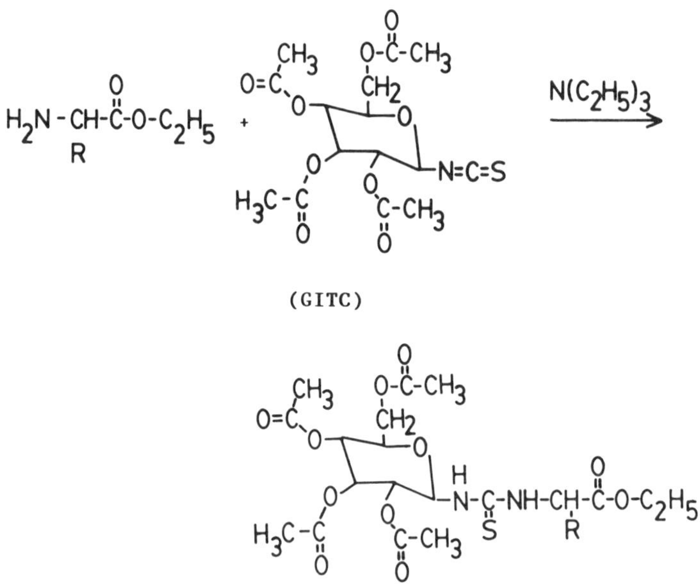


(GITC) (6)

The derivatization procedure is extremely simple. The sample containing approximately 50 μg amino acid ethyl ester is dissolved in 50 μl acetonitrile containing 0.2% triethylamine. Subsequently, 50 μl of 0.5% GITC in acetonitrile is added and after 60 min at room temperature a 5 μl aliquot is injected directly into the LC system (column 25 cm × 4.0 nm I.D. packed with LiChrosorb RP-18; eluent 60% aqueous methanol; detection at 250 nm).

Another method from the same investigators [30] using OPA and *N*-acetyl-L-cysteine as the derivatizing agent expels with its extremely high resolution power and sensitivity and also with an interesting reaction mechanism (see Eq. 7 and Fig. 5.1). It is worth mentioning that in this case, unlike in the preceding example, unesterified amino acids are derivatized and separated in a very simple way.

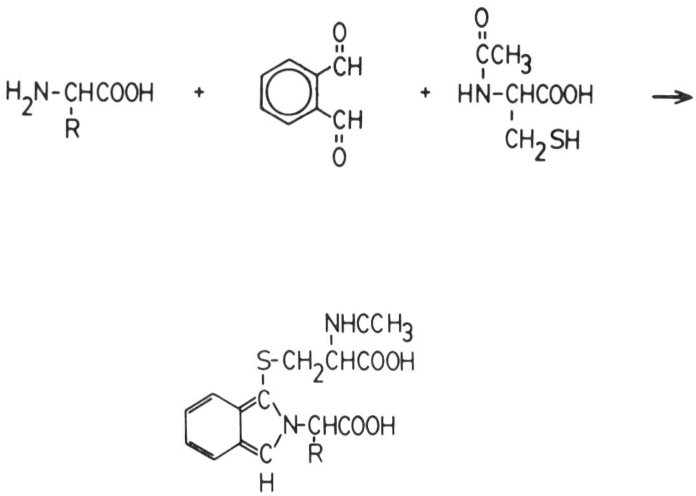


(7)

To 10 μl of a solution of the amino acids in 0.1% aqueous hydrocloric acid 30 μl 0.1 M sodium borate and 20 μl of the reagent solution (8 mg OPA and 20 mg *N*-acetyl-L-cysteine in 1 ml methanol) are added. After a reaction time of 2 min at room temperature an aliquot is injected directly into the LC equipment. The conditions of the chromatographic system are presented in the legend of Fig. 5.1. The fluorimetric detection (excitation 360 nm, emission beyond 405 nm) enables subnanogram quantities of the amino acids to be determined.

## Derivatization of Hydroxyl Groups

A few methods have been described for the chiral derivatization of primary and secondary alcohols. The reagents are carboxylic acid derivatives which form diastereomeric esters (among them free

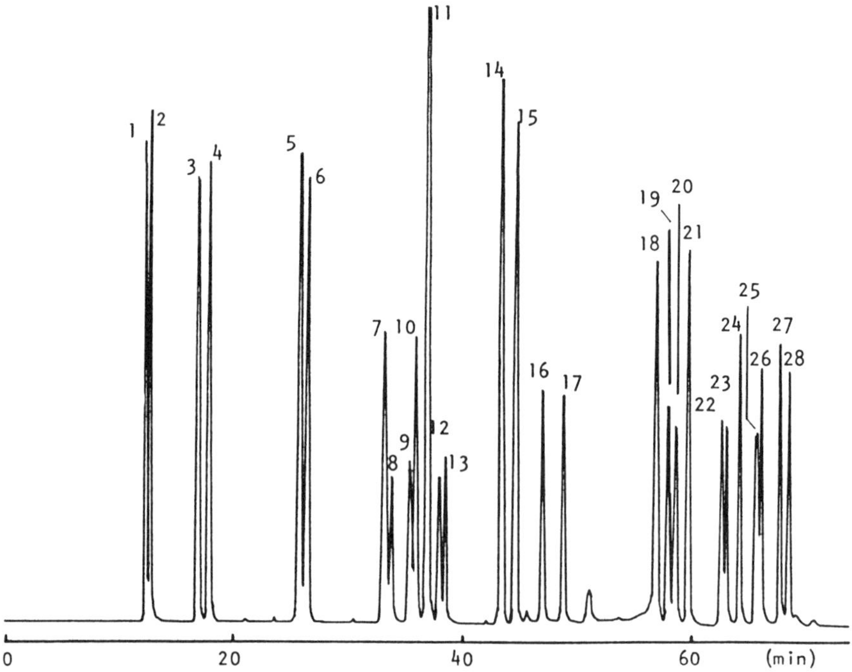


Figure 5.1 Chromatogram of *o*-phthalaldehyde-*N*-acetyl-L-cysteine derivatives of amino acid enantiomers. Column 20 cm × 6 mm packed with Develosil ODS 5 μm. Mobile phase: A, 50 mM sodium acetate; B, methanol. Gradient: 0-16 min 0-20% B, 16-22 min 20% B, 22-40 min 20-40% B, 40-46 min 40% B, 46-65 min 40-60% B. Peaks: 1. D-Asp; 2. L-Asp; 3. L-Glu; 4. D-Glu; 5. D-Ser; 6. L-Ser; 7. D-Thr; 8. L-His; 9. D-His; 10. L-Thr; 11. Gly; 12. L-Arg; 13. D-Arg; 14. D-Ala; 15. L-Ala; 16. L-Tyr; 17. D-Tyr; 18. L-Val; 19. D-Met; 20. L-Met; 21. D-Val; 22. D-Phe; 23. L-Phe; 24. L-Ile; 25. D,L-Lys; 26. D-Ile; 27. D-Leu; 28. L-leu. [Reprinted with permission from *J. Chromatogr.*, *352*:172 (1986).]

carboxylic acids with DCC as the coupling agent, carboxylic anhydrides, chlorides, chloroformates) with the hydroxylic function. Some data on these reagents and the reactions are collected in Table 5.2.

It is interesting to note that one of the reagents in Table 5.2, 1-(1-naphthyl)ethylisocyanate, which forms diastereomeric carbamates with the secondary hydroxyl groups of zeaxanthin and lutein [63], is also included in Table 5.1 as a selective reagent of amino groups in amino alcohols to form urea derivatives. In the latter case the reaction time is 20 min [7] while in the former case a reaction time of 20 hr was adopted.

The recently published paper of Gal [64] is worth mentioning because the resolution of chiral epoxides is described. In a two-step derivatization procedure the epoxide first reacts with achiral primary amines to form diastereomeric amino alcohols which in turn react in the next step with GITC (see Table 5.1) to form the diastereomeric thioureas.

## Derivatization of Carboxyl Groups

Several methods have been published for the chiral derivatization of carboxylic acids with LC separation of the resulting ester or carboxamide diastereomers. Table 5.3 presents an overview of these methods.

The majority of the reagents are chiral alcohols or primary amines. Although at elevated temperature the direct esterification of, e.g., ibuprofen with (*S*)-(+)-2-octanol is described [14], normally coupling agents (carbodiimides) or catalysts (crown ethers) are used or the carboxylic acid is transformed to the active chloride derivative by oxalyl chloride or thionyl chloride prior to the reaction with the chiral amine reagent.

From the methods listed in Table 5.3 the reaction (Eq. 8) and procedure of the derivatization of flurbiprofen using (*S*)-(-)-1-phenylethylamine as the reagent and 1,1'-carbonyldiimidazole (CDI) as the coupling agent are shown as characteristic examples [12].

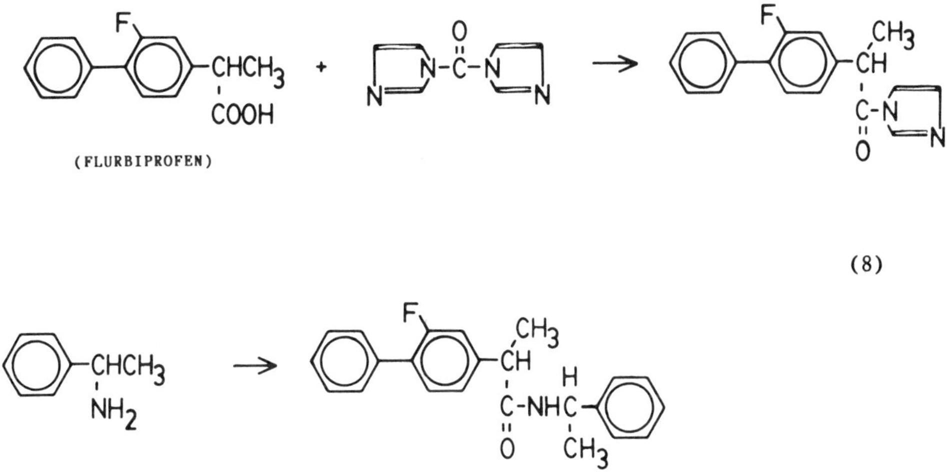


(8)

The sample containing 0.05-0.5 μmol flurbiprofen (or another arylpropionic acid) is dissolved in 1.5 ml chloroform, the internal standards, (+)-suprofen, and 0.8 ml of a freshly prepared solution of

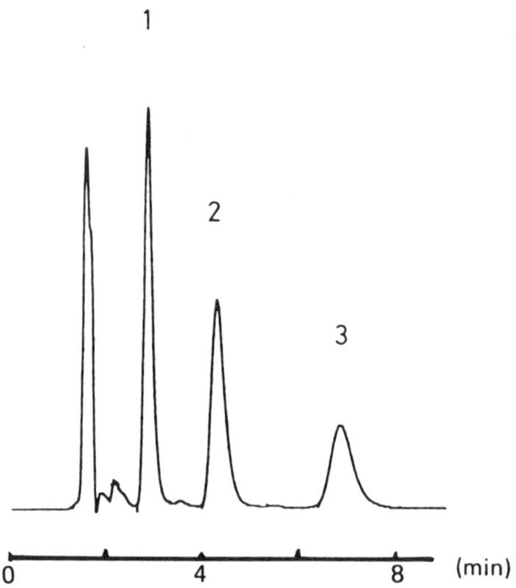


Figure 5.2 Chromatogram of the extract from rat liver containing flurbiprofen enantiomers after derivatization with 1-phenylethylamine. For the chromatographic conditions see text. Peaks: 1. (*R*)-flurbiprofen; 2. (*S*)-flurbiprofen; 3. (*S*)-suprofen (internal standard). [Reprinted with permission from *J. Chromatogr.*, *299*:402 (1984).]

CDI (65 mg/ml in chloroform) are added and the mixture is agitated mechanically for 10 min. After addition of 10 μl acetic acid and 10 min agitation, 80 μl (*S*)-(-)-1-phenylethylamine is added. The mixture is shaken for 25 min and centrifuged. After a period of 30 min to ensure complete reaction, 3 ml of a 0.2 M ammonia solution and 5 ml *n*-hexane are added, the mixture is shaken for 15 min and centrifuged. A 4-ml portion of the organic layer is washed with 3 ml of 0.3 M hydrochloric acid. After centrifugation, 3 ml of the organic layer is evaporated in vacuo and the residue is dissolved in 1 ml of the mobile phase (cyclohexane/2-propanol 93:7); column 25 cm × 3.0 mm I.D. packed with LiChrosorb Si-60 10 μm; detection at 245 nm. The separation of the diastereomeric amides can be seen in Fig. 5.2.

Investigating the examples in Tables 5.1 and 5.3 one can see that the chiral derivatization of the bifunctional amino acids takes place almost exclusively at their amino groups. The only exception is the method of Görög et al. [44] whereby the enantiomeric purity of N-protected amino acids was checked after derivatization with *O*-(4-nitrobenzyl)tyrosine methyl ester as the chiral reagent.

## ACHIRAL DERIVATIZATIONS TO IMPROVE CHIRAL SEPARATIONS

This chapter, dealing mainly with covalent derivatizations with chiral reagents, would not be complete without mentioning briefly the possibilities of derivatization reactions with achiral reagents with the aim of improving the separation efficiency of enantiomers on chiral stationary phases. A typical example is the necessary derivatization of amino acids before separation of their enantiomers on a chemically bonded β-cyclodextrin column.

As described by Hinze et al. [75], no separation of the underivatized enantiomers can be achieved. The reason for this is that free amino acids are too small to be tightly bound to the cavity of β-cyclodextrin which would be the prerequisite of any enantioselective separation. The same applies to N-protected amino acids where the protecting group is not bulky enough (e.g., *tert*-butyloxycarbonyl, benzyloxycarbonyl). If, however, bulkier derivatizing groups such as the dansyl group are employed for the derivatization of the amino groups or when the carboxyl groups are transferred to the β-naphthylamide or β-naphthol ester derivative, separation factors above 1.12 can be obtained. Another advantage of these derivatives is their high UV-Vis and fluorescence activity.

Another example is the separation of enantiomeric α-amino alcohols, among them the pharmaceutically important β-phenylethylamines and catecholamines, by ligand exchange chromatography using chemically bonded chiral stationary phases such as *N*-ω-undecenyl-L-proline *tert*-butyl ester-modified silica columns. The derivatizing agent in the study of Gelber et al. [76] is salicylaldehyde which converts the selectant to the Schiff base. The latter coordinates to copper(II) around its asymmetrical center even at the acidic pH of the eluent thus enabling effective resolution of the antipodes and monitoring of the chromatogram at 350 nm.

The third example is the method of Wainer et al. [2] for the determination of propranolol antipodes in human serum using a Pirkle-type chiral stationary phase [77]. In this case the derivatization reaction is the ring closure to the oxazolidone derivative using phosgene as the reagent. The rigid structure of the ring is more favorable from the point of view of chromatographic behavior. Other separations described by the same group using Pirkle-type columns are the resolution of the enantiomers of amphetamine after the introduction of the bulky 2-naphthoyl group to the primary amino moiety [19] and the separation of ibuprofen enantiomers after transformation of their carboxyl groups to the corresponding 1-naphthalene-methylamide derivatives [13].

## CONCLUSIONS AND FUTURE TRENDS

From the data in Tables 5.1-5.3 it can be concluded that practically all problems associated with the enantiomeric separation of drugs, metabolites, and other organic compounds containing suitable functional groups (primary or secondary amino, hydroxyl or carboxyl groups) meeting some structural requirements can be solved after diastereomeric covalent derivatization and separation on achiral stationary phases. This permits solution of the overwhelming majority of the enantiomeric separation problems in pharmaceutical and biomedical analysis. As was described earlier, the same problems also can be solved using dynamic diastereomeric complex formation or chiral stationary phases. The possibilities of the latter methods are even greater because in such a way even resolution can be achieved of those enantiomeric pairs which lack suitable functional groups for the resolution via covalently bonded diastereomeric derivatives.

Generally speaking, the present situation can be characterized as the "confusion of abundance." This means that at least three or four different methods are available for the solution of a given problem: in addition to the above-listed three variants of LC, GC, and to an increasing extent, TLC are also available.

In the present situation one can say, without greatly exaggerating, that in the majority of cases choice of method depends only on the working style of the laboratory, the available stationary phases, expertise, and to some extent the personal taste of the analyst.

As discussed earlier in this chapter, the importance of the chromatographic separation of enantiomers is increasing and a further increase can be expected because in the near future there will be certainly an increasing tendency to use the experiences obtained in analytical chromatography on the semipreparative or preparative level.

As far as the role of covalently bonded diastereomeric derivatization is concerned, this is by no means the most up-to-date method in this field as reflected by the much greater number of publications in recent years from the other two branches of LC separation of enantiomers. However, it is the authors' opinion that the proportion of papers published in a certain period of time does not necessarily reflect the proportion of the techniques in practical use in the same period. In this special case this is motivated by the fact that many of the papers published from the field of enantiomeric separations on chiral stationary phases reflect their authors' interest in the theoretical and technical problems of the preparation of chiral stationary phases rather than in their routine scale applications. The not declining importance of the separation methods based on diastereomeric chiral derivatization up to the present time is reflected by the fact that the average year of publication of the papers listed in Tables 5.1-5.3 (which contain some of the "classical" publications also) is 1982. Even the most up-to-date volume on enantiomeric separation by LC [34] contains two chapters dealing with covalent diastereomeric derivatization. The author of this chapter strongly believes that this method will continue for many years to be an important method, especially in those laboratories where enantiomeric separation is only a part of the everyday routine work, and for this reason it would be very advantageous if this could be done using the same stationary and mobile phases as for the other, achiral separations.

## REFERENCES

1. Ariens, A., W. Soudijn, and P. Timmermans (eds.), *Stereochemistry and Biological Activity of Drugs*, Blackwell Scientific, Oxford, 1983.
2. Wainer, I. W., T. D. Doyle, K. H. Donn, and J. R. Powell, *J. Chromatogr.*, *306*:405 (1984).
3. Walle, T., D. D. Christ, U. K. Walle, and M. J. Wilson, *J. Chromatogr.*, *341*:213 (1985).
4. Sedman, A. J., and J. Gal, *J. Chromatogr.*, *278*:199 (1983).
5. Thompson, J. A., J. L. Holtzman, M. Tsuru, C. L. Lerman, and J. L. Holtzman, *J. Chromatogr.*, *238*:470 (1982).
6. Wilson, M. J., and T. Walle, *J. Chromatogr.*, *310*:424 (1984).
7. Gübitz, G., and S. Mihellyes, *J. Chromatogr.*, *314*:462 (1984).
8. Hermansson, J., *Acta Pharm. Suec.*, *19*:11 (1982).
9. Hermansson, J., and C. von Bahr, *J. Chromatogr.*, *227*:113 (1982).
10. Gulaid, A. A., G. W. Houghton, and A. R. Boobis, *J. Chromatogr.*, *318*:393 (1985).
11. Grech-Belânger, O., and J. Turgeon, *J. Chromatogr.*, *337*:172 (1985).
12. Maître, J.-M., G. Boss, and B. Testa, *J. Chromatogr.*, *299*:397 (1984).
13. Wainer, I. M., and T. D. Doyle, *Liq. Chromatogr.*, *2*:88 (1984).
14. Lee, E. J. D., K. M. Williams, G. G. Graham, R. O. Day, and G. D. Champion, *J. Pharm. Sci.*, *73*:1542 (1984).
15. Banfield, C., and M. Rowland, *J. Pharm. Sci.*, *72*:921 (1983).
16. Ruud-Christensen, M., and B. Salvesen, *J. Chromatogr.*, *303*:433 (1984).
17. Barksdale, J. M., and C. R. Clark, *J. Chromatogr. Sci.*, *23*:176 (1985).
18. Miller, K. J., J. Gal, and M. M. Ames, *J. Chromatogr.*, *307*:335 (1984).

19. Wainer, I. M., T. D. Doyle, and W. M. Adams, *J. Pharm. Sci., 73*:1162 (1984).
20. Kirchhoefer, R. D., G. M. Sullivan, and J. F. Allgire, *J. Assoc. Off. Anal. Chem., 68*:163 (1985).
21. Takaya, T., Y. Kishida, and S. Sakakibara, *J. Chromatogr., 215*:279 (1981).
22. Mitchell, A. R., S. B. H. Kent, I. C. Chu, and R. B. Merrifield, *Anal. Chem., 50*:637 (1978).
23. Weinstein, S., *Trends Anal. Chem., 3*:16 (1984).
24. Nimura, H., H. Ogura, and T. Kinoshita, *J. Chromatogr., 202*:375 (1980).
25. Cahill, W. R., E. P. Kroeff, and D. J. Pietrzyk, *J. Liq. Chromatogr., 3*:1319 (1980).
26. Davies, J. S., and A. K. A. Mohammed, *J. Chem. Soc., Perkin Trans., 11*:1723 (1984).
27. Kinoshita, T., Y. Kasahara, and N. Nimura, *J. Chromatogr., 210*:77 (1981).
28. Goto, J., M. Hasegawa, S. Nakamura, K. Shimada, and T. Nambara, *J. Chromatogr., 152*:413 (1978).
29. Marfey, P., *Carlsberg Res. Commun., 49*:591 (1984).
30. Nimura, N., and T. Kinoshita, *J. Chromatogr., 352*:161 (1986).
31. Furukawa, H., Y. Mori, Y. Takeuchi, and K. Ito, *J. Chromatogr., 136*:428 (1977).
32. Aswad, D. A., *Anal. Biochem., 137*:428 (1977).
33. Buck, R. H., and K. Krummen, *J. Chromatogr., 315*:279 (1984).
34. Hara, S., and J. Cazes (eds.), Optical resolution by liquid chromatography, *J. Liq. Chromatogr., 9*:241 (1986).
35. Lindner, W. F., Resolution of optical isomers by gas and liquid chromatography, in *Chemical Derivatization in Analytical Chemistry* (R. W. Frei and J. F. Lawrence, eds.), Plenum Press, New York, 1982, p. 145.
36. Lindner, W., and C. Petterson, Resolution of optical isomers by liquid chromatographic techniques, in *Chromatography in Pharmaceutical Development* (I. M. Wainer, ed.), Aster, Oregon, 1985, p. 63.
37. Tamegai, T., M. Ohmae, K. Kawabe, and M. Tomoeda, *J. Liq. Chromatogr., 2*:1229 (1979).
38. Nambara, T., S. Ikegawa, M. Hasegawa, and J. Goto, *Anal. Chim. Acta, 101*:111 (1978).
39. Scott, G. G., M. J. Petrin, and R. McCorkle, *J. Chromatogr., 125*:157 (1976).
40. Helmchen, G., G. Nil, D. Flockerzi, W. Schuele, and M. S. K. Youssef, *Angew. Chem., Int. Ed., 18*:62 (1979).
41. Helmchen, G., H. Voelter, and W. Schuele, *Tetrahedron Lett., 16*:1417 (1977).

42. Rose, H. C., R. L. Stern, and B. L. Karger, *Anal. Chem.*, *38*:469 (1966).
43. Chen, T.-M., and J. J. Contario, *J. Chromatogr.*, *314*:495 (1984).
44. Görög, S., B. Herényi, and M. Löw, *J. Chromatogr.*, *353*:417 (1986).
45. Lindner, W., Ch. Leitner, and G. Uray, *J. Chromatogr.*, *316*:605 (1984).
46. Silber, B., and S. Riegelman, *J. Pharmacol. Exp. Ther.*, *215*:643 (1980).
47. Weber, H., H. Spahn, E. Mutschler, and W. Möhrke, *J. Chromatogr.*, *307*:145 (1984).
48. Helmchen, G., and W. Strubert, *Chromatographia*, *7*:713 (1974).
49. Goto, J., N. Goto, A. Hikichi, and T. Nambara, *J. Liq. Chromatogr.*, *2*:1179 (1979).
50. Coleman, M. W., *Chromatographia*, *17*:23 (1983).
51. Clark, C. R., and J. M. Barksdale, *Anal. Chem.*, *56*:958 (1984).
52. Dieterle, W., and J. W. Faigle, *J. Chromatogr.*, *242*:289 (1982).
53. Souter, R. W., *Chromatographia*, *9*:635 (1976).
54. Lankmayr, E. P., W. Budna, and F. Nachtmann, *J. Chromatogr.*, *198*:471 (1980).
55. Nachtmann, F., *Int. J. Pharm.*, *4*:337 (1980).
56. Dieterle, W., and J. W. Faigle, *J. Chromatogr.*, *259*:311 (1983).
57. Darmon, A., and J. P. Thenot, *J. Chromatogr.*, *374*:321 (1986).
58. Nimura, N., Y. Kasahara, and T. Kinoshita, *J. Chromatogr.*, *213*:327 (1981).
59. Hoffmann, K.-J., A. Arfwidsson, and K. O. Borg, *Drug Metab. Dispos.*, *10*:173 (1982).
60. Allgire, J. F., E. C. Juenge, C. P. Damo, G. M. Sullivan, and R. D. Kirchhoefer, *J. Chromatogr.*, *325*:249 (1985).
61. Gal, J., *J. Chromatogr.*, *307*:220 (1984).
62. Jeyaraj, G. L., and W. R. Porter, *J. Chromatogr.*, *315*:378 (1984).
63. Ruettimann, A., K. Schiedt, and M. Vecchi, *J. High Resolut. Chromatogr. Chromatogr. Commun.*, *6*:612 (1983).
64. Gal, J., *J. Chromatogr.*, *331*:349 (1985).
65. Goto, J., N. Goto, and T. Nambara, *Chem. Parm. Bull.*, *30*:4597 (1982).
66. Johnson, D. M., A. Reuter, J. M. Collins, and G. F. Thompson, *J. Pharm. Sci.*, *68*:112 (1979).
67. Goto, J., N. Goto, and T. Nambara, *J. Chromatogr.*, *239*:559 (1982).
68. Nagashima, H., Y. Tanaka, H. Watanabe, R. Hayashi, and K. Kawada, *Chem. Pharm. Bull.*, *32*:251 (1984).
69. Jiang, M., and D. M. Soederlund, *J. Chromatogr.*, *248*:143 (1982).

70. Sallustio, B. C., A. Abas, Y. J. Purdie, and P. J. Meffin, *J. Chromatogr.*, *374*:329 (1986).
71. Kemmerer, J. M., F. A. Rubio, R. M. McClain, and B. A. Koechlin, *J. Pharm. Sci.*, *68*:1274 (1979).
72. Valentine, D., K. K. Chan, C. G. Scott, K. K. Johnson, K. Toth, and G. Saucy, *J. Org. Chem.*, *41*:62 (1976).
73. Tamegai, T., T. Tanaka, T. Kaneko, S. Ozaki, and M. Ohmae, *J. Liq. Chromatogr.*, *2*:551 (1979).
74. Wainer, I. M., *J. Chromatogr.*, *202*:478 (1980).
75. Hinze, W. L., T. E. Riehl, D. W. Armstrong, W. DeMond, A. Alak, and T. Ward, *Anal. Chem.*, *57*:237 (1985).
76. Gelber, L. R., B. L. Karger, J. L. Neumeyer, and B. Feibush, *J. Am. Chem. Soc.*, *106*:7729 (1984).
77. Pirkle, W. H., J. M. Finn, J. L. Schreiner, and B. C. Hamper, *J. Am. Chem. Soc.*, *103*:3964 (1981).

# 6
# Enzymatic Derivatization

PETRA C. VAN KRIMPEN* and MARIAN A. J. VAN OPSTAL
*State University of Utrecht, Utrecht, The Netherlands*

## INTRODUCTION

In biological systems most reactions do not occur at a perceptible rate in the absence of a biological catalyst, an enzyme [1-3]. Enzymes are proteins with a specific geometric three-dimensional structure. This so-called active site facilitates the reaction and is therefore essential for the catalytic activity and specificity of the enzyme.

Enzymes react very specifically, not only with respect to the type of catalyzed reaction but also in the choice of reactants, the substrates. This specificity can be attributed to a stereochemical complementarity of the active site toward the substrate (lock-and-key fit). Due to the binding of the substrate, a structural rearrangement of the enzyme may occur as a result of which another part of the enzyme becomes available and can interact with a different part of the substrate (induced fit).

The huge number of enzymes has introduced the necessity of a systematic nomenclature. The systematic names and numbers, which were introduced by the International Commission of Enzymes in 1956 and have been extended by the Enzyme Commission of the International Union of Biochemistry (1961 and 1975), are based on the catalyzed chemical reaction. Six different types of reactions can be distinguished: (1) oxidation-reduction; (2) functional group transfer;

**Present affiliation*: Unilever Research Laboratorium, Vlaardingen, The Netherlands

(3) hydrolysis; (4) lysis; (5) isomerization; and (6) ligation. Lysis reactions involve a nonhydrolytic removal of a group resulting in the occurrence of double bonds. Ligation is the combination of two species with concomitant consumption of energy, provided by the breakdown of adenosine-5'-triphosphate (ATP). Further differentiation can be achieved by considering the class of the involved substrate.

A number of enzymes also possess a non-protein-like function, the prosthetic group (if bound to the enzyme) or coenzyme (cofactor). These enzymes are called holoenzymes and the protein part is the apoenzyme. Coenzymes act as donor or acceptor of certain reactants, such as protons or electrons, and they can be regenerated also. Coenzymes can be very useful for the analytical chemist due to their fixed stoichiometric relation to the substrate or product. Several detection methods are based on the coenzyme assay. A well-known sample is the spectrophotometric detection of the dihydronicotinamide adenine dinucleotide and its oxidized form system, NADH/$NAD^+$.

Furthermore, enzymes can be selective for a group of substrates (e.g., β-lactamases, which hydrolyze the β-lactam ring of penicillins and cephalosporins) or for one compound in a group of derivatives (e.g., urease, which catalyzes only the conversion of urea to ammonia). Although an enzyme can be the catalyst of a number of reactions, involving a common functional group, it should be mentioned that the rates of the reactions of each of the substrates can be very different.

In living organisms a large number of different enzymes are present and the concentration and types vary strongly in different parts of the organism. This feature explains why the analysis of enzyme activity has become an important diagnostic tool in clinical chemistry. The concentration of an enzyme in the organism or the ratio of two or more enzymes can be altered enormously due to a large number and variety of events. For example, some nerve gases or agricultural poisons inhibit cholinesterase, resulting in an excess of the neural stimulator acetylcholine. Heart and liver diseases in particular alter the enzyme pattern in the blood in a striking way.

Enzymes are used commercially in the food and drug industry. The hydrolysis of starch, for example, is well known. Other products which depend on the use of enzymes are beer, milk, meat, and cheese.

### Why Enzymes?

The use of enzymes in analytical chemistry suffers from several disadvantages. In comparison to other analytical techniques, the main disadvantages are the limited availability and purity of the enzymes in combination with the instability of the compounds and the difficulties in handling of enzymes. During the last decade, the importance of enzymes as analytical tools has increased dramatically. This is due to a combination of factors: increased knowledge of protein synthesis and purification and the development of immobilization techniques for enzymes.

Furthermore, immobilization of enzymes permits their reuse, thus creating the possibility to perform more experiments with the same batch of enzymes and thus saving the costs of the analysis. Reactors with immobilized enzymes have proved to be very suitable for combinations with continuous flow processes. This includes the use of enzymes as analytical reagents for selective degradation or derivatization procedures and for various analytical purposes.

The analytical use of enzymes has one major advantage over analytical techniques: the selectivity of the enzyme for its substrate. In combination with the sensitivity which can be achieved nowadays, enzymatic analysis is a very useful tool for the modern analytical chemist. For example, an enzyme reactor can be used for the conversion of the substrate to a product which can be separated from or detected specifically in the sample.

## THEORETICAL DESCRIPTION OF ENZYMATIC CATALYSIS

As mentioned before, enzymes are active because of their influence on the reaction kinetics [1,2]. Michaelis and Menten developed the basic enzyme kinetics for the homogeneous catalysis of enzymes in batch experiments. One molecule of the enzyme (E) combines with one molecule of the substrate (S) to form an intermediate (ES), which subsequently decomposes, resulting in the liberation of the reaction product (P) and the enzyme molecule:

$$E + S \underset{k_{-1}}{\overset{k_1}{\rightleftharpoons}} ES \xrightarrow{k_2} E + P \qquad (1)$$

$k_1$, $k_{-1}$, and $k_1$ are the reaction constants of the formation and dissociation of the intermediate and the formation of the product, respectively. The effect of an enzyme is expressed by the initial rate of the formation of P:

$$\frac{d[P]}{dt} = \frac{k_2^{*}[E]_t^{*}[S]}{(k_{-1} + k_2)/(k_1) + [S]} \qquad (2)$$

where $[E]_t$ is the total concentration of the enzyme. Equation 2 can be rearranged into the widely used Michaelis-Menten equation:

$$v = \frac{v^*_{max}[S]}{K_m + [S]} \tag{3}$$

in which v is the initial reaction velocity, $v_{max}$ the maximum reaction velocity for a given enzyme concentration ($[E]_t$), and $K_m$ the Michaelis constant $[(k_{-1} + k_2)/k_1]$. The relationship between v and [S] is half-hyperbolic and for experimental data a linear transformed plot, 1/v versus 1/[S], the Lineweaver-Burk plot, is widely used to obtain the enzymatic characteristics. The slope of the line is $K_m/v_{max}$; the intercept on the 1/v axis is 1/v and on the 1/[S] axis is $-1/K_m$.

The Michaelis-Menten mechanism is a simplification, but it is widely used to describe simple enzyme systems.

### Inhibition or Activation of Enzyme Activity

Inhibition of activity is likely to occur to some extent when enzymes are used analytically. The inhibition process can be reversible or irreversible. A reversible inhibitor forms complexes with the enzyme, but, since these complexes are not stable, the enzyme can be released. The maximum velocity at a given enzyme concentration $v_{max}$ is unchanged in the presence of a reversible inhibitor, but will only be obtained at a higher substrate concentration. The Michaelis constant, however, increases.

An irreversible inhibitor also complexes with the enzyme, but this complex subsequently reacts with the substrate resulting in a second complex. This second complex does not degrade into the enzyme and a product, and therefore the overall reaction is a blockade of the enzyme. In case of irreversible inhibition, the maximum velocity at a given enzyme concentration and excess substrate will be less than the maximum velocity of the uninhibited reaction.

It should be mentioned that in many cases the inhibiting agent has a mixed reversible-irreversible action. For analytical purposes the exact mechanism of the enzyme inhibition is of limited importance. Both reversible and irreversible inhibitors affect the activity of the enzyme solution or enzyme reactor. Limiting the inhibition is therefore essential for achieving more sensitivity and longer use of the enzyme. Many drugs and toxic agents form more or less stable complexes with active sites of enzymes. Examples of inhibitors of the enzyme activity, which can be important in analytical chemistry, are heavy-metal ions (especially mercury ions) and structural substrate analogs.

Activation of an enzyme can also occur. An activator has been defined as a substance which is required for an enzyme to be an an active catalyst or one which increases the efficiency of an active enzyme [1]. An activator does not change during the reaction. Examples of activators are electrolytes and metal ions, but low concentrations of organic solvents have also been reported to increase the activity of an enzyme [4].

### Effect of Environmental Parameters on Enzymes

Changes of the activity of an enzyme are mostly caused by changes in accessibility of the active site (denaturation). These changes may result from physical (e.g., temperature) or chemical (e.g., pH, ionic strength, specific reagent effects) parameters of the analytical system.

One of the major parameters is the temperature of the system. Temperature influences both the activity and the stability of the enzyme. The effect of the temperature on reaction kinetics can be quantified by the temperature coefficient $Q_{10}$ [1]. For most enzymes, $Q_{10}$ is about 1-2, which means that the reaction velocity is doubled when the temperature increases 10°. However, possibilities to increase the temperature are limited because of the denaturation of the enzymes at higher temperature. The combination of these two phenomena explains why many enzymes have an optimum in their activity with respect to temperature. For analytical purposes the temperature of an enzyme system should be kept constant.

Another important parameter is the acidity of the system. The relation between the activity of the enzyme and the pH often shows an optimum. This can be explained by the fact that various prototropic forms of the enzyme have different activities or affinities to the substrate. Comparatively, the enzyme-substrate complexes can also be (de)protonated, resulting in a different stability of the complexes. Buffers are regularly used in enzymatic analysis in order to overcome this problem. However, these buffers can have other effects on the activity of the enzyme apart from controlling the pH. The increase of the ionic strength in the solution can change the tertiary structure of the enzyme. The active site of the enzyme may be blocked or can be more accessible for the substrate. Other buffer effects, caused by changing of the surface tension of the enzyme by the buffer species, have also been reported.

## COMBINATION OF LIQUID CHROMATOGRAPHY AND ENZYMATIC ANALYSIS

### Effect on Kinetic Properties of Enzyme

The most important effect of a flow system on the activity of enzymes is the kinetic change. In the case of a reaction in an unstirred bulk solution, the reaction rate of the enzyme with the substrate is totally diffusion-controlled. In flow systems this diffusion pattern is disturbed by convection, especially when immobilized enzymes are used. This phenomenon often results in a totally different physicochemical environment at the enzyme surface compared to the bulk solution. The exact composition of the microenvironment of the enzyme in flow systems is difficult to determine. The Michaelis-Menten mechanism is often applied to describe the enzymatic reaction in flow systems. The overall effect on the reaction rate is usually expressed by substituting $K_m$ (Eq. 3) by the apparent reaction constant $K_m^{app}$ and $v_{max}$ by $v_{max}^{app}$. The effectivity factor of an enzyme detector $\eta$ indicates to what extent the enzyme is utilized [5]:

$$\eta = \frac{K_m^{app}}{K_m} \qquad (4)$$

It should be noted that Eq. 4 is valid for a first-order reaction. Generally, an excess of enzyme is used in flow systems, resulting in an increase of $K_m$, indicated by $K_m^{app}$.

The combination of enzymatic derivatization with liquid chromatography increases the use of immobilized enzymes. However, enzymes in solution can be used for postchromatographic derivatization as well, but immobilization is preferable because

Immobilization permits reuse of the enzyme, consequently saving enzyme and having the ability of using the same batch for a set of analytical runs.

Immobilized enzymes are in general more stable and less sensitive to inhibition than enzymes in solution.

Immobilized enzymes do not interfere with the final detection method, e.g., by pollution of cuvettes.

Immobilization of an enzyme has specific effects on its kinetic behavior. Not only the differences due to the flow conditions and the heterogeneous catalysis should be considered; the effect of the binding to a rigid matrix, resulting in a less flexible enzyme, is also important.

Another important feature is the change in the microenvironment of the enzyme due to the matrix. For instance, an acidic matrix can (apparently) change the optimum in the pH-activity relationship. The ideal matrix does not interfere with the active site of the enzyme and should be inert toward other components of the system. Unfortunately, no matrix exists which fulfills both requirements.

An important parameter of a reactor loaded with enzyme is the time (t) required to obtain a certain percentage conversion of the substrate:

$$t = \frac{K_m^{app}}{V_{max}^{app}} \left[ \frac{X}{B} + \ln \frac{1}{1 - X} \right] \tag{5}$$

where B is $K_m^{app}/[S]_o$ and X is the relative conversion ($([S]_o - [S]_t)/[S]_o$ (from Eq. 3; $S_o$ and $S_t$ are the initial substrate concentration and substrate concentration at time t, respectively). The maximal available time for a reaction is controlled by the reactor geometry and the flow rate.

## Types of Enzymatic Systems

Enzymes can be used in several ways in chromatography to improve selectivity or to enhance the detector response. Although the emphasis in the subsequent discussion will be on the use of enzymes combined with LC, the list of references has been extended by several applications of enzymes in flow analysis systems: continuous flow analysis (CFA) or flow injection analysis (FIA). The application of enzymes involves either enzymes with a broad class specificity to quantify a family of related compounds with LC or enzymes with a high specificity to selectively quantify one compound in a mixture with LC or flow analysis. The applications can be divided into three different types.

In the first type the enzymatic reaction occurs in a separate reactor. The enzyme catalyzes the conversion of the substrate to the corresponding product. The increase of a specific product or, alternatively, the increase of a coproduct or the decrease of coreactant is measured downstream by means of a suitable detector. The detector signal is proportional to the amount of analyte (substrate) to be measured. The variety of detectors is only limited by the requirement that it be provided with a flow-through detection cell. Both enzyme solutions and immobilized enzymes can be employed.

In the second type the enzyme is directly attached to the sensor or transducer resulting in enzyme probes. Such a measuring system can be seen as a reactor linked to the detector. The most obvious members of this class are the enzyme electrodes, where potentiometric or amperometric detectors are used. Only immobilizable enzymes can be employed. In this approach again the formation of product or coproduct or the consumption of coreactant is measured. General aspects of the use of immobilized enzymes in reactors and in sensors have been extensively discussed by Bowers [6,7]. One of the key advantages of immobilized enzyme reactors (IMER) compared with immobilized enzyme probes (IEP) is their potential to achieve a high yield due to the relatively high enzyme concentration. Almost 100% conversion of substrate can be obtained on a single pass through the reactor resulting in a relative high response. A second advantage of the IMER over IEP is the great variety of available detectors.

In the third type the enzyme is used as a label in immunoassays. Such enzyme immunoassays have been combined with flow analysis, either off-line or on-line, to contribute to the type of applications of enzymes discussed here and have been included in the subsequent list of applications.

### Immobilization Procedures

Enzymes have been immobilized on several support materials using several procedures. The choice of matrix and procedure depends mainly on the application of the method. The only limitation on the choice of the matrix and the immobilization procedure is that the enzyme is not totally inhibited. In addition, some restrictions have to be made due to the combination with LC.

The basic procedures for immobilization are as follows [1,2]:

(1) Adsorption. This is the simplest method for immobilization and is performed by immersing the matrix in a solution of the enzyme. The major disadvantage of this method is desorption; the matrix "bleeds" enzyme. This bleeding can be influenced by changes in the chemical environment.

(2) Physical entrapment. This includes the trapping of the enzyme molecule in the pores of an inert matrix. Matrices that can be used are starch or polymer gels. The advantage of this method over chemical binding is that the activity of the enzyme is unchanged by the chemical procedure or the binding. A disadvantage is that the substrates must diffuse into the pores, which can be impossible for large molecules.

This method is suitable for preparing enzyme electrodes, but due to the limited rigidity of the available polymers is less suitable for enzyme reactors.

(3) Encapsulation. The enzyme is encapsulated into semipermeable microspheres. The transport of substrate and products is governed by diffusion into and out of the microsphere. This technique is not widely used because the formation of the microspheres often decreases the activity of the enzyme.

To avoid this problem the enzyme can be brought in hollow fibers, which are sealed thereafter. This method shows several advantages over the chemical binding: the filling procedure is relatively simple, and the method allows recovery of the enzyme, which is generally impossible when the enzyme is chemically bonded. The major disadvantage is that substrate and product have to diffuse into the fiber, which is restricted by the permeability of the fiber wall.

(4) Chemical crosslinking to the matrix. The enzyme is chemically tied to a matrix by means of a bifunctional agent. A well-known agent is glutaraldehyde, which links the primary amino group of the enzyme to a primary amino group of the matrix. Other examples of agents are disulfonic acids (e.g., bisdiazobenzidine-2,2'-disulfonic acid and diphenyl-4,4'-dithiocyanate-2,2'-disulfonic acid), isocyanates (e.g., toluene-2-isocyanate-4-isothiocyanate and hexamethylenediisocyanate) and succinimides (e.g., maleimidobenzoyl-hydroxysuccinimide ester). Chemical crosslinking is relatively simple and the binding to the matrix is stable, but it is possible that the enzyme is partly deactivated during the crosslinking process.

(5) Direct linking to the matrix. This is the most widely used method for the immobilization of enzymes. The popularity of this method can be explained by the simplicity and the efficiency of the coupling procedure and the increased stability of the enzyme. After activation of the matrix the solution of enzyme is brought into contact with the matrix and the nonbound enzyme is washed off the matrix. The yield of immobilization and the activity of the bonded enzyme is governed by the activation step. Multiple bonding of the enzyme leads to decreased activity, while insufficient activation results in a poor yield. The binding of the enzyme should take place at amino acid residues which are not essential for the activity.

It is essential for all matrices that there be the possibility of binding to the enzyme. A general description of several groups of matrices with examples of each group is given in Table 6.1 [2,3,8].

### Solvent Effects

Another parameter in the combination of enzymes with LC is the presence of organic solvent(s) in the system. There is no general description of the effects of organic solvents on the enzyme activity but changes in the three-dimensional structure, specific binding, solvation, competition at the active site, and shifting equilibria are obvious. Increased release (bleeding) of the enzyme from the matrix may occur due to reaction of the organic cosolvent with the enzyme-matrix bonds. Some cosolvents can act as an alternative substrate for the enzyme. In most cases enzymes are denatured when large concentrations of organic solvents are used. However, the presence of small amounts of organic solvents may even increase the activity, e.g., up to 10% methanol with immobilized β-glucuronidase [2]. This can be explained by comparing the analytical system to the natural environment of enzymes, the cell. It is known that enzymes interact with other nonpolar cellular components, such as membranes. Therefore, the addition of some organic solvent can result in a better resemblance to the natural environment, resulting in an

Table 6.1 Matrices Used for Immobilization of Enzymes[a]

| Matrix | Remarks |
|---|---|
| *Inorganics* | |
| Controlled pore glass (CPG) beads | Good mechanical stability; different sizes available; insensitive to microbiological degradation |
| Silica gel | Good chemical stability; unstable in alkali |
| *Natural Polymers* | |
| Agarose | Bad mechanical stability; wide pore gel; very suitable for large enzymes or substrates, sensitive to microbiological degradation |
| Cellulose | High permeability; bad mechanical stability; sensitive to microbiological degradation |
| Dextran | Semisynthetic product; bad mechanical stability; within-batch variation of pore width large; sensitive to microbiological degradation |
| Collagen | Polymer on protein basis; bad mechanical stability; sensitive to microbiological degradation; suitable for enzyne membranes |
| *Synthetic Polymers* | |
| Polyacrylic | Bad mechanical stability; many different derivatives available; sensitive to microbiological degradation |
| Polystyrenes | Many different structures available; insensitive to microbiological degradation; bad mechanical stability |
| Polyamides | Suitable for covalent binding of the enzyme; better mechanical stability than the other polymers |

[a]These matrices are suitable for crosslinking and covalent binding procedures [2,3,8].

increased activity. Examples of stabilizing organic cosolvents are glycerol and other short-chain polyalcohols and ketones [4].

Organic cosolvents are used in enzymatic analysis for purposes such as antifreeze, shifting equilibria, or solubilization of enzyme. In the combination LC enzymes the organic cosolvent is used to achieve a better or faster chromatographic result. The amount of organic cosolvent is in general five to 10 times higher than in enzymatic analysis. For the combination of LC with enzymes the cosolvent concentration should be adapted, possibly resulting in poorer chromatographic results or, alternatively, a less effective or less stable enzyme reactor.

The specific effect of the inclusion of an enzyme in a chromatographic system is difficult to predict. In general, the chromatographic conditions have a negative influence on the effectivity of the enzyme detector. The effect of organic solvents on a number of enzymes have been reviewed [4]. Other constituents of the mobile phase, such as buffers and metal ions in the chromatographic system, may also have dualistic effects on the enzyme. Optimization of the chromatographic parameters is recommended for each specific system.

## APPLICATIONS

The applications of enzyme reactors, sensors, and labels in LC or flow analysis will be discussed successively in this section. Then some miscellaneous applications will be referred.

### Enzyme Reactors

Although the advantages of immobilization are numerous, as mentioned before, some applications using enzymes in the soluble state have also been described in the literature [9-11] and have been included in the subsequent discussion. However, the major part of the applications handle with IMERs. Three prerequisites should be fulfilled during the development of a useful reactor: a high conversion of substrate to product and/or measurable compound, a minimum mass dispersion in the flow stream in order to obtain good sensitivity, and a detectability property which is proportional to the concentration of the analyte. Recently, two review papers were published handling IMERs in LC [12,13].

#### *On-Line Enzyme Reactors in Liquid Chromatography*

3-α-HYDROXYSTEROID DEHYDROGENASE (HSD) HSD can be applied for the detection of bile acids. The total concentration of bile acids and especially their biological profile provide an important diagnostic tool for several diseases. The compounds can be separated in a reversed phase (RP) LC system but are poorly detectable. An enzyme reactor with immobilized 3-α-HSD has been successfully applied as postchromatographic IMER for the enzymatic modification of the separated bile acids [14-20]. The enzyme catalyzes the oxidation of bile acids in the presence of a coreactant, NAD, to the corresponding ketone. The 3α position is oxidized and NADH, the reduced form of NAD, is formed as a coproduct as shown in Fig. 6.1.

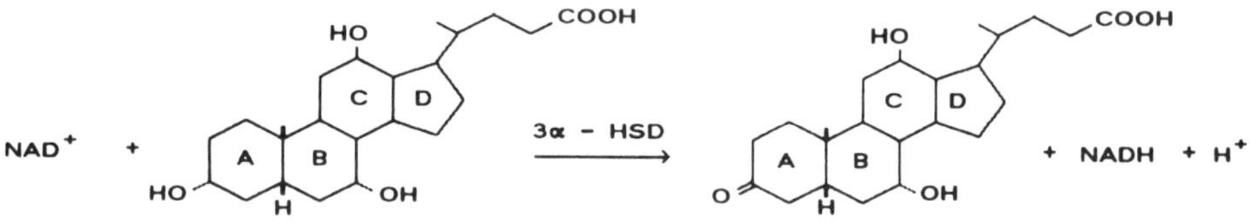


Figure 6.1 Reaction scheme of the enzymatic oxidation of the 3-α position in cholic acid by 3-α-hydroxysteroid dehydrogenase in the presence of NAD.

The produced NADH can be measured in two different ways: NADH has an ultraviolet (UV) absorbance at 365 nm, which is absent in NAD, and a fluorescence emission at 470 nm after excitation at 365 nm. In most cases the fluorescence is applied for detection [14-19]. The reoxidation of NADH to NAD provides the possibility for electrochemical detection [20]. The earlier work deals with the immobilization of the enzyme on aminosilanized glass using the glutaraldehyde coupling [14]. A mixture of 14 bile acids was separated on a RP column within 40 min. The detection limit in serum samples was about 10 ng. Later on cellulose beads were used as support material [15,16]. A mixture of 15 bile acids was separated using gradient elution in combination with RP chromatography at pH 9.7, corresponding to the pH optimum of 3-α-HDH. The detection limits varied from 0.01 to 0.04 μmol. One reactor could be used for about 300 analyses. A microbore LC system was developed and equipped with an IMER of 3-α-HDH immobilized on controlled pore glass (CPG/200-400 mesh) [17]. Again, a mixture of 15 bile acids was separated (120 min) and detection limits of 0.09 to 0.7 ng were achieved. The same system was further optimized [18] resulting in an improved separation within 100 min and detection limits in the pmol range. Another LC system combined with an immobilized 3-α-HDH reactor was introduced, especially intended for the analysis of rat bile [19]. The detection limits varied from 0.8 to 1.5 ng and the reactor could be used for at least 200 analyses. Kamada et al. applied a pre-separation of the bile acids into free bile acids, blycine and taurine conjugates. The three fractions were analyzed separately using a suitable analytical column combined with a 3-α-HDH IMER followed by electrochemical detection at pmol level [20].

β-GLUCURONIDASE Exogenous and endogenous lipophilic compounds, in humans are generally metabolized to more hydrophilic compounds, such as the glucuronide conjugates, which can be excreted in the

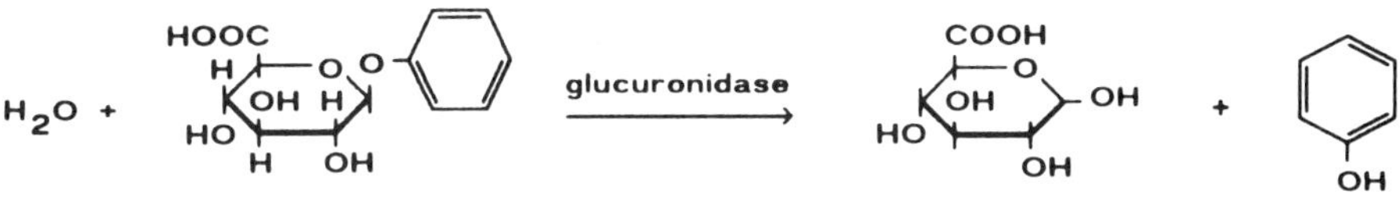


Figure 6.2 Reaction scheme of the enzymatic hydrolysis of phenyl-β-D-glucuronide by β-glucuronidase.

urine. The enzyme β-glucuronidase can be applied for analysis of the glucuronides in urine. The glucuronides are cleaved to glucuronic acid and the aglycone, e.g., phenol, according to the scheme in Fig. 6.2. The enzyme has been immobilized and the IMER was inserted into a chromatographic system, either prior to [21-24] or after [25,26] the analytical column. After the enzymatic hydrolysis either a specific detection of the parent aglycone can be applied or the produced glucuronic acid can be measured. The latter method results in a more general approach for the detection of glucuronides and is based on reduction of glucuronic acid in the presence of lucigenin leading to a chemiluminescent product. Most frequently, the enzyme is immobilized by the glutaraldehyde crosslinking method on aminopropyl CPG beads. Estriol and estradiol glucuronides in urine were determined by using a column-switching technique [21,22]. Both glucuronides were hydrolyzed by the immobilized β-glucuronidase at the optimal pH 7; the steroid mixture was subsequently separated by gradient elution on a RP system and finally detected by measuring the UV absorbance at 280 nm. The enzyme retained its 100% activity after being used continuously during 150 hr and the exposure to 10% methanol. A separate RP precolumn was inserted into the same system resulting in extra purification and shorter analysis time [23]. Another β-glucuronidase prechromatographic method has been described by Klopf et al. [24], who used the analytical column for the separation of the enzymatically produced parent aglycone(s) from glucuronic acid. Lucigenin was added postchromatographically and the produced chemiluminescence quantified. Such procedures determine the total amount of glucuronides and do not distinguish between the different aglycones. Detection limits varying from 5 to 10 μM were measured.

A postchromatographic β-glucuronidase IMER was used for quantitation of phenolic glucuronides, such as the drug trimethoprim [25]. This drug is metabolized in mammals to phenol and subsequently the phenolic glucuronide is produced. The enzymatic analysis of this glucuronide provided the regeneration of phenol, which was measured by UV detection and by electrochemical detection as well.

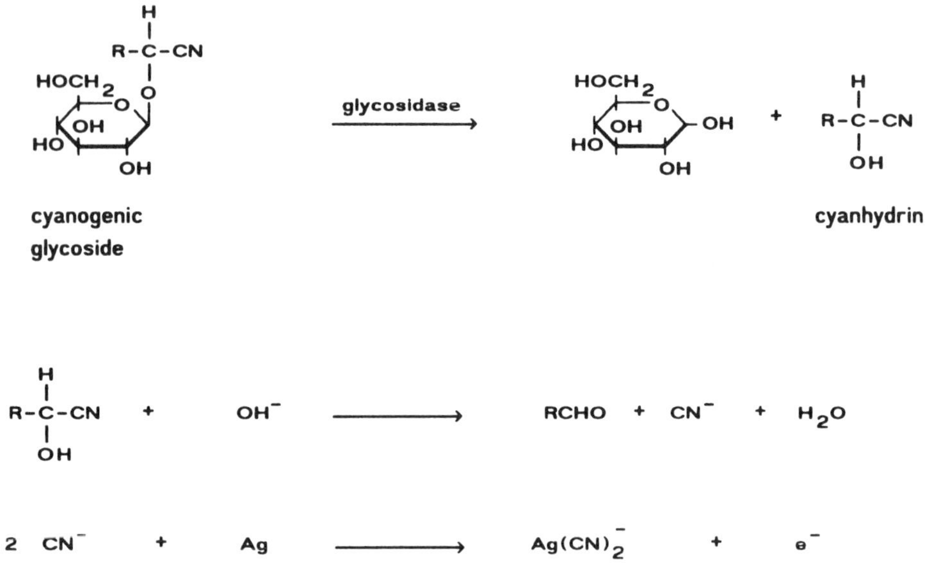


Figure 6.3 Reaction scheme of the enzymatic hydrolysis of cyanogenic glycosides followed by the alkaline hydrolysis of cyanhydrin and the electrochemical oxidation of silver by cyanide.

The detection limit was 10 pmol for both methods of detection but the selectivity of the electrochemical detector was better. Another postchromatographic application has been described for the assay of diastereomeric glucuronides of fenoldopam in plasma and urine [26]. The electrochemically active parent aglycone, fenoldopam, was detected at +0.6 mV versus Ag/AgCl. The IMER retained its activity in a mobile phase containing 0-25% methanol and also retained its stability during continuous usage (over 4 months) and during storage (over 6 months).

GLYCOSIDASE The enzyme glycosidase is related to β-glucuronidase. While glucuronides are common metabolites in mammals, glycosides are widely distributed in the plant kingdom. Dalgaard et al. [27] immobilized glycosidase and determined phenolic glycosides in crude plant material applying the chromatographic system similar to that described above by the same authors [25]. Nondetectable metabolites could be quantitated in this way. The majority of the glycosides, however, belong to the cyanogenic glycosides, which can be hydrolyzed by glycosidase resulting in the formation of cyanhydrin. Cyanhydrin is rather unstable and in basic solution it is cleaved to an aldehyde or ketone and cyanide, which can be detected electrochemically using a silver electrode (Fig. 6.3). Brimer et al. [28] analyzed cyanogenic glycosides using this detection method. The method had proven to be specific for the cyanogenic glycosides.

OXIDASES Four different oxidases have been immobilized and are used as postchromatographic IMER in LC: amino acid oxidase [29], choline oxidase [10,30-32], cholesterol oxidase [33], and xanthine oxidase [34-38].

Amino acid oxidase was coimmobilized with peroxidase on aminoaryl porous glass [29]. The immobilization was achieved by diazo coupling. The enzymatic oxidation leads to the formation of oxidized amino acid, while two coproducts, ammonia and hydrogen peroxide, are formed. Homovanillic acid, which reacts with hydrogen peroxide, was added and the fluorescence at 425 nm after excitation at 315 nm measured. The reactor has to be used at elevated temperatures and therefore its activity decreased rapidly (2 weeks at 313 K to 3 days at 323 K).

The use of choline oxidase provides a sensitive detection method for acetylcholine, which possesses neither chromophoric nor electroactive nor derivatizable groups. Enzymatic conversion, however, was satisfactory and acetylcholine could be separated from interfering substances in tissue extracts by RP chromatography. Potter et al. [10] used a mixed-enzyme solution of cholinesterase and choline oxidase which was combined with the column effluent. Acetylcholine was deacetylated by the former and thereafter converted to betaine and peroxide by the latter. Hydrogen peroxide was measured using a platinum electrode in an amperometric detector. Later on the same enzymatic procedure was applied using immobilized enzymes. The enzymes were adsorbed at an ion exchange resin and used as a postchromatographic IMER [30,31]. The method enabled the detection of 5 pmol choline [31]. The enzyme reactor had to be reloaded every 5-10 days [30]. Damsma et al. [32] described the same method but applied reactors in which the enzymes were covalently bound to cyanobromide-activated Sepharose 4B. If the reactor was used continuously for 2 weeks, the sensitivity decreased by about 25%. Choline-containing phospholipids have been selectively detected by enzymatic reaction in a postchromatographic effluent by using commercial kits [11]. The kit contained phospholipase D, choline oxidase, and peroxidase. An enzymatic solution was prepared by dissolving one vial of each enzyme in a suitable buffer and the solution was added to the column effluent. The enzymatic reaction was carried out at 313 K in a $20.0 \times 0.5$ mm Teflon reactor tube. After passage through the reactor tube the products were detected at 500 nm.

Cholesterol oxidase catalyzes the conversion of cholesterol to a more conjugated product, cholest-4-ene-3-one, with coproduction of hydrogen peroxide. The highly conjugated oxidized cholesterol shows strong UV absorption at 202 nm ($\varepsilon$ = 15,000). The enzyme has been immobilized and used as a postchromatographic reactor [33]. The relatively strong organic column effluent (75% ethanol) was diluted with an aqueous buffer to adjust the pH and to lower the ethanol concentration.

Xanthine oxidase converts hypoxanthine and xanthine, both metabolites of adenine, to uric acid with the simultaneous formation of hydrogen peroxide. Uric acid can be detected at 290 nm, meaning that both metabolites can be determined in one run by separating them chromatographically (RP), followed by enzymatic conversion. The enzyme was immobilized by glutaraldehyde crosslinking and inserted as IMER after the chromatographic column [34,35]. However, relatively high levels of endogenous uric acid disturb these detection methods. Therefore a second enzyme reactor was placed prior to the xantine oxidase reactor [36]. In that enzyme reactor uricase and catalase (10:1) were coimmobilized. Uricase converts uric acid to a non-UV-active compound with the concomitant formation of hydrogen peroxide which is degraded by the coimmobilized enzyme catalase. The use of a xanthine oxidase reactor provided with such a preenzyme reactor leads to reactor lifetimes of at least 7 weeks. Detection limits are in the nanogram range. To improve the detection limit to picogram levels, the xanthine oxidase was combined with fluorescence detection [37,38]. Therefore, xanthine oxidase was coimmobilized with peroxidase which oxidizes coproduced peroxide. This reaction was performed in the presence of a fluorophore, *p*-hydroxyphenylacetic acid, and the fluorescence was measured at 407 nm after excitation at 320 nm.

ALKALINE PHOSPHATASE Alkaline phosphatase hydrolyzes phosphate esters and has been applied for the detection of inositol bi- and triphosphates [39]. These compounds do not possess chromophores or fluorophores or electroactive groups. However, enzymatic conversion to detectable products appeared to be possible. The phosphates were separated by anion exchange chromatography. Subsequently the column effluent was led through a column with immobilized alkaline phosphatase to hydrolyze the phosphate esters. Subsequently, the stream was continuously mixed with a molybdate solution to allow detection of the inorganic phosphates formed. The enzyme was immobilized simply by adsorbing it to a moderately hydrophobic support, phenoxyacetylcellulose. The limit of detection was less than 1 nmol [39].

UREASE Urease is a substrate-specific enzyme which only catalyzes the hydrolysis of urea to carbon dioxide and ammonia. It has been applied in a postchromatographic enzyme detector for the determination

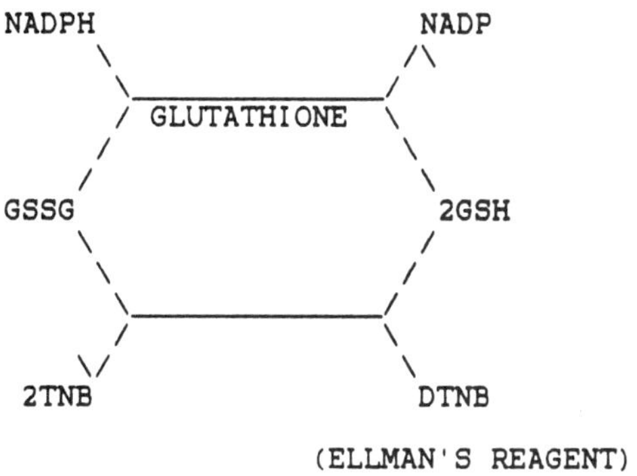


Figure 6.4 Enzymatic recycling reaction of glutathione by glutathione reductase in the presence of coenzyme NADPH and color reagent 5,5'-dithiobis-2-nitrobenzoic acid.

of urea and ammonia in samples from a urea plant and in wastewater samples of a fertilizer plant [40]. The enzymatically produced ammonia was mixed with *o*-phthaldehyde (OPA) to form a fluorescent compound, which was measured by excitation at 340 nm and emission at 455 nm. Urea was detectable in the $10^{-7}$ M range. Another application of the urease reactor in conjunction with ion pair chromatography has been described for the simultaneous determination of the inorganic ions $K^+$, $Na^+$, and $NH_4^+$, and neutral urea in human urine and in samples from fertilizer plants [41].

GLUTATHIONE REDUCTASE Glutathione reductase reduces the oxidized glutathione (GSSG) to the reduced glutathione (GSH) in the presence of nicotinamide adenine dinucleotide phosphate, reduced form (NADPH). The latter is oxidized to nicotinamide adenine dinucleotide phosphate (NADP) during the enzymatic reaction. The formed GSH reoxidizes spontaneously to GSSG. When additionally Ellman's reagent (DTNB: 5,5'-dithiobis-2-nitrobenzoic acid) is added, the recycling redox reaction can be detected spectrophotometrically. DTNB is reduced to 5-thio-2-nitrobenzoate (TNB), which absorbs strongly at 412 nm ($\varepsilon$ = 13,600). A schematic presentation of the recycling reaction is shown in Fig. 6.4. The method is sensitive to both GSH and GSSG, but the compounds can be separated chromatographically. The enzymatic recycling reaction has been performed with a postchromatographic reactor using the enzyme in the soluble state [9]. The sensitivity of such an approach is enhanced by chemical amplification due to the recycling reaction. GSSG levels of 2 pmol could be determined in biological samples [9].

### *Off-Line Enzymatic Reactions in Liquid Chromatography*

An enzymatic reaction can be carried out prior to the chromatographic separation, as a prechromatographic derivatization procedure.

PENICILLINASE Penicillinase can be applied to hydrolyze the β-lactam ring in penicillins, resulting in the formation of the corresponding penicilloates. The procedure has been carried out by a multiresidue method for the determination of eight penicillins in milk [42]. The respective penicilloates were subsequently labeled with the fluorescence label dansylhydrazine and finally the derivatives were injected onto the chromatographic system.

CHONDROITINASE For the analysis of chondroitin sulfate and dermatan isomers (both glycosaminoglycans), a method using two different enzymatic reactions followed by chromatographic separation, quantitation, and identification has been described [43]. Chondroitinase generates sulfated disaccharides from chondroitin sulfates or from dermatan sulfate isomers. The second enzyme, chondrosulfatase, has been applied for desulfatation prior to the chromatographic analysis.

PROTEINASES Sublitisin A and proteinase K catalyze the protein hydrolysis. The enzymes can be applied to release drugs from blood proteins prior to the chromatographic analysis of the blood samples. This approach has been described for the determination of the antineoplastic compounds etoposide and teniposide in blood [44]. The treatment has been proven to prevent clogging of the chromatographic system.

3-α-HYDROXYSTEROID DEHYDROGENASE A reactor containing immobilized HSD has been applied off-line in the prechromatographic made by Kawasaki et al. [45]. The enzymatically generated bile acids were labeled with dansylhydrazine. The derivatives were separated using a RP system and fluorescence detection at 520 nm after excitation at 365 nm. The detection limit was 0.5 pmol and the linear range 3-40 pmol.

### Enzyme Sensors

The research on enzyme sensors, whereby an enzyme layer is fixed at the surface of an electrode or other sensor, is expanding rapidly, especially in the area of biosensor development. The enzyme-based electrodes belong to the electrochemical sensors, which are most frequently applied for this purpose [46]. The concept is based on the use of a multilayer membrane containing one or more immobilized enzymes to isolate and transform the desired analyte into a species that can be monitored potentiometrically or amperometrically (Fig. 6.5).

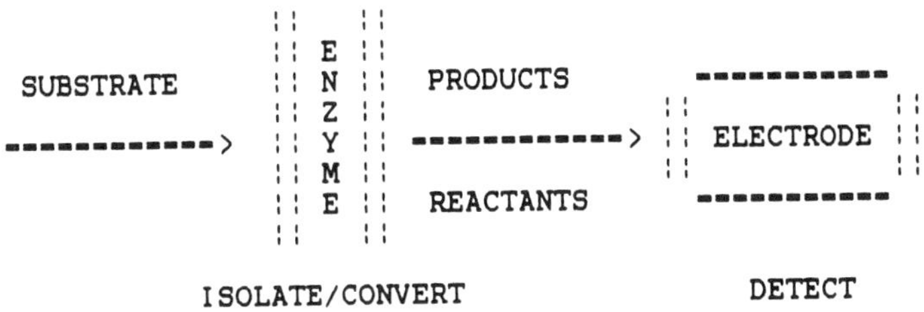


Figure 6.5 Enzyme based electrode conversion of the substrate (analyte) to an electroactive species by the membrane-bonded enzyme(s).

Generally, the membranes are multilayered composites containing one or more immobilized enzymes. If the membrane is brought into contact with the substrate (analyte), the membrane-bound enzyme catalyzes the breakdown of the substrate to form electroactive products (e.g., hydrogen peroxide) and/or to consume electroactive reactants (e.g., oxygen). The goal of the research in this area is the development of a probe which can be applied directly in the matrix without the need of another analysis technique, e.g., chromatography, to enhance selectivity. Applications of such enzyme electrodes are given in two textbooks concerning the analytical use of immobilized enzymes [1,47]. The enzyme sensors are rarely applied as detector devices in flow systems because of their low effectivity in these systems. The enzymes have to be attached to the electrode surface in a flow-through cell resulting in a limited enzyme concentration and a limited yield of the enzymatic reaction due to a relatively short residence time at the electrode. Despite these restrictions some enzyme electrode flow-through cells have been applied [48,49]. Yao et al. [48] used an enzyme electrode as a detector device in LC. The authors describe a simple and rapid method for the determination of acetylcholine and choline by using LC and enzyme-based electrochemical detection. Acetylcholinesterase (AChE) and choline oxidase (ChO) or AChE and ChO and peroxidase (POD) were coimmobilized in a membrane layer on a platinum electrode by glutaraldehyde coupling. The AChE-ChO assay was based on the amperometric detection of the enzymatically generated hydrogen peroxide while at the AChO-ChO-POD electrode the hydrogen peroxide improved the oxidation of hexacyanoferrate(II) by the peroxidase, according to the reaction shown in Fig. 6.6. The method was highly sensitive and specific for the determination of acetylcholine and choline. The detection limit was 1.5 pmol for choline and 2.9 pmol for acetylcholine. The enzyme electrodes could be used continuously for more than 20 days. Another enzyme electrode was employed for the analysis of a champagne sample by measuring the effluent of a chromatographic column.

$$H_2O_2 + 2Fe(CN)_6^{4-} + 2H^+ \xrightarrow{\text{peroxidase}} 2Fe(CN)_6^{3-} + 2H_2O$$

Figure 6.6 Reaction scheme of the oxidation of hexacyanoferrate(II) by hydrogen peroxide, catalyzed by peroxidase.

The detector response has been compared to that of a refractometric detector [49].

The number of applications of enzyme electrodes in LC is rather limited while some more applications are described for the use of such electrodes in continuous flow systems. The determination of glucose has been carried out in a FIA system using a glucose sensor [50]. The enzymatic oxidation of glucose is catalyzed by glucose oxidase; the reaction consumes oxygen and produces glucuronic acid and hydrogen peroxide. The formation of hydrogen peroxide was detected amperometrically. The response was linear from 2 μM up to 2 mM β-D-glucose. The enzyme was immobilized by direct adsorption at a carbon electrode. The adsorbed enzyme layer was estimated to convert 12% of the glucose. New electrodes were allowed to stabilize for 30 min and were useful for 1 day with typically less than 1% decrease in response during 3 hr due to enzyme desorption and deactivation.

A POD electrode has been applied in FIA [51]. POD was coupled by glutaraldehyde to a silanized platinum sheet. The amperometric POD electrode was included in the system to measure the current caused by the oxidation of hexacyanoferrate(II) (Fig. 6.6). The hydrogen peroxide was generated by injection of the cholesterol-containing sample (human serum) onto a packed bed reactor of immobilized cholesterol esterase and cholesterol oxidase. Both the reactor and the POD electrode retained more than 90% of their original activities, even after repetitive use of 4 and 2 months, respectively.

## Enzyme Labels

Immunoassay is based on the use of an antibody as a selective chemical reagent for the determination of the antigen or a hapten analyte. Unfortunately, the fundamental immunochemical reaction often does not result in chemical or physical changes which can be readily applied for quantitation of the analyte. The most commonly used solution for this problem is to label the analyte, and enzymes are successfully applied as labels in immunoassays. The enzyme catalyzes the formation of a detectable product and the rate at which it is formed is related to the unknown analyte concentration in the sample. The general reaction can be described as follows:

$$\text{Ab} + \text{Ag} + \text{Ag*} \begin{cases} \rightarrow \text{Ab-Ag} + \text{Ag*} \\ \rightarrow \text{Ab-Ag*} + \text{Ag} \end{cases} \tag{6}$$

where Ab is the antibody. The enzyme-labeled (Ag*) and unlabeled (Ag) antigens compete for the available antibody sites.

First the immunological reaction is carried out and then the suitable substrate is added. The enzymatic reaction can take place and the corresponding product is formed, which finally has to be detected [52]. In a typical off-line combination of enzyme immunoassay (EIA) and LC, the EIA resulting final product is analyzed by LC. During this analysis the enzyme-generated product is separated from solvent and other interfering components and detected by a suitable technique. An EIA-LC analysis of this type has been published for the assay of digoxin [53,54]. Alkaline phosphatase catalyzing the hydrolysis of phenyl phosphate ester to phenol and phosphate is used as the label. The enzyme-generated phenol is detected amperometrically after separation on a RP column. The procedure allows the determination of digoxin in its therapeutic range with a detection limit of 50 pg/ml. The same combination of EIA and LC has been applied for the assay of $\alpha_1$-acid glycoprotein (5 ng/ml) [55], immunoglobulin (10 pg/ml) [56], and phenytoin [57].

Immunoassays can also be based on high-performance immunoaffinity chromatography (HPIC) with a suitable detection of the column effluent [58]. The HPIC system is schematically shown in Fig. 6.7. In this technique the antibody (Ab) is immobilized on a chromatographic column. Antigen labeled with enzyme (Ag*) is mixed with the sample Ag and then injected into the immunoaffinity column. Competitive binding occurs with the immobilized Ab. The appropriate substrate is then injected and the product generated and measured. The amount of product is proportional to the amount of Ag* bound to the column, which is inversely proportional to the amount of Ag in the sample. After the analysis a switching valve is used to pass an acidic buffer through the column to displace bound Ag and Ag* and the column is prepared for the next analysis. The Ab-Ag reaction is sufficiently fast. Reaction yields over 90% in 8-10 sec [59]. The method enables carefully controlled conditioning of the immunosorbent (e.g., washing), which is essential to achieve a good precision. The immunosorbent columns are stable for at least 3 months (500-600 assays). It has been demonstrated that for a HPIC-based assay of IgG detection limits of 1 fmol could be achieved [60]. Glucose oxidase, L-amino acid oxidase, amino oxidase, catalase, glucose-6-phosphate dehydrogenase, and urease have all been successfully applied in EIAs with electrochemical detection [58].

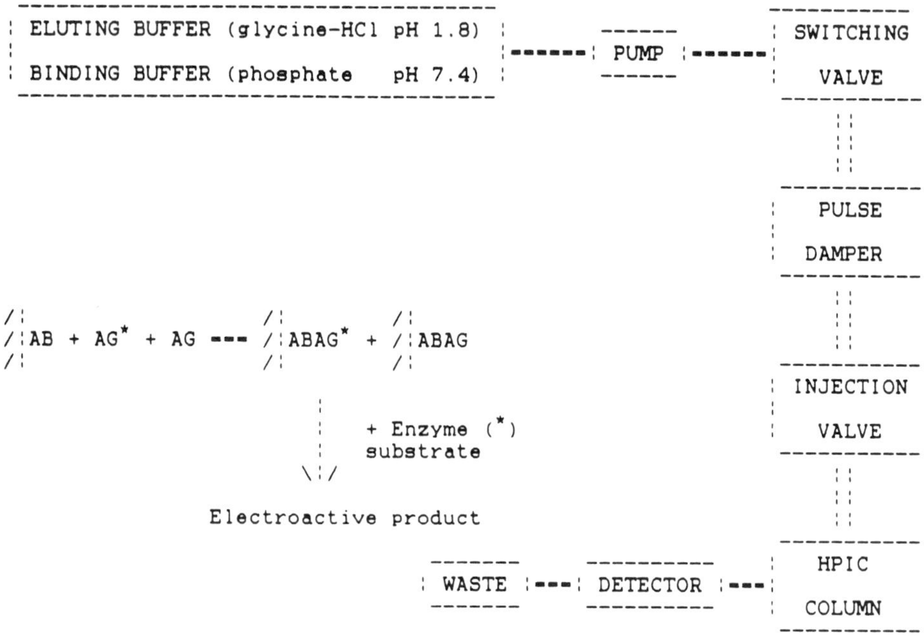


Figure 6.7 Generalized setup for high-performance immuno affinity chromatography.

## Miscellaneous

In this section some alternative applications of the use of enzymes in flow systems will be discussed.

An immobilized alcohol dehydrogenase reactor has been exploited as chromatographic column in affinity chromatography [61]. Adenine nucleotides, adenine nucleosides, and triazine dyes were separated by the immobilized enzyme affinity column (IMEAC). The column was eluted by 0.25 M phosphate buffer (pH 7.5) containing zinc sulfate. The compounds were detected at 260 nm. The resolution and retention times increase with increasing enzyme loading. The separation is based on the different values of the dissociation constants of the enzyme-substrate complex.

An immobilized carboxypeptidase reactor has been inserted in a LC system prior to the analytical column for the detection of traces of zinc based on enzyme activity measurements [62]. Zinc ions activate the enzyme and the activity of the enzyme is related to the zinc concentration in the sample. First, the enzyme reactor is switched off from the chromatographic system and the zinc-containing sample is injected. Then the reactor is switched into the system and the substrate (hippuryl-L-phenylalanine) is injected. Two

$NADH + MB^+$ --- $NADH\ MB^+$ (1)

$NADH\ MB^+$ --> $NAD^+ + MBH$ (2)

$MBD$ --- $MB^+ + 2e^- + H^+$ (3)

$NADH$ --> $NAD^+ + 2e^- + H^+$ (4)

Figure 6.8 A proposed reaction scheme of the Meldola blue-mediated oxidation of NADH.

products are formed (L-phenylalanine and hippuryl acid), which can be separated chromatographically and detected at 228 nm. The detection limit was 0.1 nmol. The same enzyme has been employed in a totally different manner for the sequencing of amino acids in peptides [63]. The enzyme reactor has been combined with a LC mass spectrometric method. The peptides are separated and hydrolyzed on-line by immobilized carboxypeptidase. Both the generated amino acids and the residual peptides are transported to the mass spectrometer and detected as molecular ions.

In another kind of analysis enzymatic activity in biological samples is determined using LC combined with an enzymatic reaction. In such applications the sample which contains the enzyme is injected and the enzyme is separated. The column effluent is mixed with the appropriate substrate and the enzymatically formed product is detected. Peptidase and esterase activity in rat urinary kallikrein have been measured by such a procedure [64,65]. Peptidase activity was measured by adding propylphenylalanylarginine-4-methylcoumaryl-7-amide. The enzymatic reaction was carried out in a 20-m helically coiled Teflon tube (0.25 mm ID) at 313 K. The generated 7-amino-4-methylcoumarin was monitored fluorometrically at 460 nm with excitation at 380 nm. Esterase activity was detected with *N*-α-tosyl-L-arginine methyl ester and the resulting fluorescence was detected at 510 nm with excitation at 410 nm.

Since all enzyme reactions cause a change of heat, the detection system may be based on the measurement of the heat of the reaction. Especially when there is neither an electroactive nor an UV-Vis absorbing species present, such a detection method may be useful. Detection of generated heat in a flow system by using a thermistor-like probe has been applied, e.g., for the measurement of penicillin [66] with a detection limit of 1 μg.

Modified electrodes can be exploited in enzymatic determination. The surface of the electrode is modified by introducing mediating functionalities. A modified electrode has been applied, e.g., for the determination of glucose in a flow system by catalytic oxidation of the nicotinamide coenzyme [67]. Graphite rods were modified by

Table 6.2 Analytical Applications of Enzymes in Combination with Liquid Chromatography or Flow Injection Analysis

| Enzymes | Analytes | Type | Method | Detection | Ref. |
|---|---|---|---|---|---|
| Alcohol dehydrogenase | Nucleotides, triazine dyes | IMEAC | HPAC | UV | 61 |
| Alkaline phosphatase | Inositol bi- and triphosphates | IMER | LC | TUR | 39 |
| Alkaline phosphatase | Digoxin | Label | EIA-LC[1] | EC | 53,54 |
| Alkaline phosphatase | $\alpha_1$-Acid glycoprotein | Label | EIA-LC[1] | EC | 55 |
| Alkaline phosphatase | Immunoglobulin G | Label | EIA-LC[1] | EC | 56 |
| Alkaline phosphatase | Phenytoin | Label | EIA-LC[1] | EC | 57 |
| Amino acid oxidase (peroxidase[3]) | Amino acids | IMER | LC | F | 29 |
| Carboxypeptidase | Zinc (traces) | IMER[2] | LC | UV | 62 |
| Carboxypeptidase | Amino acids (sequencing) | IMER | LC | MS | 63 |
| Cholesterol oxidase | Cholesterol | IMER | LC | UV | 33 |
| Choline oxidase (choline-esterase) | Cholines | Solution | LC | EC | 10 |
| Choline oxidase (phospho-lipase | (Choline-containing) phospholipids | Solution | LC | VIS | 11 |
| Choline oxidase (choline-esterase[3]) | Cholines | IMER | LC | EC | 30-32 |
| Choline oxidase (choline-esterase[3]) | Cholines | Sensor | LC | EC | 48 |
| Chondroitinase (chondro-sulfatase) | Chondroitin sulfate or dermatan sulfate isomers | Solution | LC[1] | UV | 43 |
| Esterase | Esterase activity | Sample | LC | F | 64,65 |

| | | | | | |
|---|---|---|---|---|---|
| Glucose oxidase | Glucose | Sensor | FIA | EC | 50 |
| Glucose oxidase | Immunoglobulin G | Label | HPIC | EC | 60 |
| Glucose dehydrogenase | Glucose | IMER | FIA | EC | 67 |
| Glucuronidase | Estriol/estradiol | IMER[2] | LC | UV | 21,22 |
| Glucuronidase | Estriol/estradiol | IMER[2] | LC | F | 49 |
| Glucuronidase | Several glucuronides | IMER[2] | LC | LUM | 24 |
| Glucuronidase | Trimethoprim | IMER | LC | UV/EC | 25 |
| Glucuronidase | Fenoldopam | IMER | LC | EC | 26 |
| Glycosidase | Phenolic glycosides | IMER | LC | EC | 27 |
| Glycosidase | Cyanogenic glycosides | IMER | LC | EC | 28 |
| Glutathione reductase | Glutathione | Solution | LC | VIS | 9 |
| Hydroxysteroid dehydrogenase | Bile acids | IMER | LC | F | 14–19 |
| Hydroxysteroid dehydrogenase | Bile acids | IMER | LC | EC | 20 |
| Hydroxysteroid dehydrogenase | Bile acids | IMER | LC[1] | F | 45 |
| Penicillinase | Penicillins | Sensor | LC | CAL | 66 |
| Penicillinase | Penicillins (traces) | Solution | LC[1] | F | 42 |
| Peptidase | Peptidase activity | Sample | LC | F | 64,65 |
| Peroxidase | Cholesterol | Sensor | FIA | EC | 51 |
| Proteinase K | Plasma | Solution | LC[1] | F | 44 |
| Urease | Urea | IMER | LC | F | 40 |
| Urease | Urea | IMER | LC | CON | 41 |
| Xanthine oxidase | Uric acid and (hypo)xanthine | IMER | LC | UV | 34,35 |

Table 6.2 (continued)

| Enzymes | Analytes | Type | Method | Detection | Ref. |
|---|---|---|---|---|---|
| Xanthine oxidase (combined with second IMER with uricase/catalase) | (Hypo)xanthine | IMER | LC | UV | 36 |
| Xanthine oxidase (peroxidase[3]) | (Hypo)xanthine | IMER | LC | F | 37,38 |
| PEIQ-$NH_2$ enzyme (artificial decarboxylase) | Oxalacetate | Sensor | Batch | EC | 68 |

Abbreviations: IMEAC, immobilized enzyme affinity column; IMER, immobilized enzyme reactor; HPAC, high-performance affinity chromatography; LC, high-performance liquid chromatography; EIA, enzyme immunoassay; FIA, flow injection analysis; HPIC, high-performance immunoaffinity chromatography; CAL, calorimetric; CON, conductivity; EC, electrochemical; F, fluorescence; LUM, chemiluminescence; MS, mass spectrometric; TUR, turbidimetric; UV, ultraviolet absorbance; VIS, visible absorption; [1], off-line; [2], prechromatographic; [3], coimmobilized.

dipping them into solutions of 7-dimethylamino-1,2-benzophenoxazinium salt (Meldola blue). A proposed reaction scheme for the mediated NADH oxidation has been given in Fig. 6.8. Glucose dehydrogenase was immobilized and the reactor inserted prior to the modified electrode and the production of NADH was detected. Straight calibration graphs were obtained up to 1 mM β-D-glucose. The detection limit was 0.25 μM. The electrode was kept at -50 mV versus a saturated calomel electrode (SCE), while measurement of the oxidation of NADH at SCE in an unmodified electrode has to be carried out at 750 mV or higher versus SCE. The use of this voltage implies that interferences are likely.

Table 6.2 summarizes several applications of enzymes.

## FUTURE TRENDS

Enzymes have been successfully adapted to LC to provide specific and sensitive detection methods. The number of published applications, especially those dealing with IMER, increases steadily. The wide variety of enzymes available and the advantages of immobilization promise to make the LC-IMER combination a potent analytical tool. However, rather laborious immobilization procedures in which predominantly the method of trial and error is used and incomplete characterization of most immobilized enzyme systems put obstacles in the way of extensive use. Further study of immobilization procedures and more unified ways of measuring activity and stability would be highly useful. Furthermore, the effect of diffusion on the reaction kinetics is crucial and has to be recognized in evaluating and optimizing reactor performance. It is clear that a theoretical understanding of enzyme kinetics in a reactor is necessary. Better understanding and theoretical formulations will lead to commercial availability of immobilized enzymes or even of complete IMERs, and consequently to more ready use of them.

Practical limitations of enzymes, e.g., insufficient lifetimes or cofactor requirements, have led to the use of whole cells or tissue slices or live bacteria as alternate biocatalysts. These recent developments have already been successfully applied in the area of biosensors, but in the future such an approach will be applied in a much broader sense.

Another important limitation accompanied by the use of biological material in analytical systems is that the reaction has to take place under nearly biological conditions. In LC it means, e.g., that only a low content of organic cosolvent can be tolerated and that a compromise has to be found between the optimum pH for the separation and the optimum pH for enzyme activity. To overcome such limitations the synthesis and application of artificial enzymes have been reported recently [68]. Artificial enzymes are synthetic polymer chains having functional groups that mimic the biocatalytic activity of natural enzymes. Such synthetic or artificial enzymes may provide long lifetimes and simple operating conditions.

## REFERENCES

1. Guilbault, G. G., *Analytical Uses of Immobilized Enzymes.* Modern Monographs in Analytical Chemistry, Vol. 2, Marcel Dekker, New York, 1984.
2. Bowers, L. D., and W. D. Bostick, in *Chemical Derivatization in Analytical Chemistry* (R. W. Frei and J. F. Lawrence, eds.), Plenum Press, New York, 1981.
3. van der Velde, E., The use and applications of immobilized enzymes in liquid chromatography: Post- and pre-column enzyme reactors. Thesis, Free Unversity, Amsterdam, 1985.
4. Butler, L. G., *Enzyme Microb. Technol.*, *1*:253 (1979).
5. Johansson, G., B. Olsson, and L. Ogren, Enzyme reactors in analytical flow systems. Symposium on electrochemical detection in flow analysis, Matrafüred, 1982.
6. Bowers, L. D., *Trends Anal. Chem.*, *1*:191 (1982).
7. Bowers, L. D., *Anal. Chem.*, *58*:513A (1986).
8. Schellenberger, A., and R. Ulbrich, *Z. Chem.*, *25*:349 (1985).
9. Alpert, A. J., and H. F. Gilbert, *Anal. Biochem.*, *144*:553 (1985).
10. Potter, P. E., J. L. Meek, and N. H. Neff, *J. Neurochem.*, *41*:188 (1983).
11. Okazaki, M., N. Hagiwara, and I. Hara, *J. Chromatogr.*, *231*: 13 (1982).
12. Dalgaard, L., *Trends Anal. Chem.*, *5*:185 (1986).
13. Bowers, L. D., *Chromatogr. Sci.*, *34*:195 (1986).
14. Okuyama, S., N. Kokubun, S. Higashidate, D. Kemura, and Y. Hirata, *Chem. Lett.*, *12*:1443 (1979).
15. Hasegawa, S., R. Uenoyama, F. Takeda, J. Chuma, and S. Baba, *J. Chromatogr.*, *278*:25 (1983).
16. Hasegawa, S., R. Uenoyama, F. Takeda, J. Chuma, K. Suzuki, F. Kamiyama, K. Yamazaki, and S. Baba, *J. Liq. Chromatogr.*, *7*:2267 (1984).
17. Takeuchi, T., S. Saito, and D. Ishii, *J. Chromatogr.*, *258*:125 (1983).
18. Ishii, D., S. Murata, and T. Takeuchi, *J. Chromatogr.*, *282*: 569 (1983).
19. Hayashi, M., Y. Imai, Y. Minami, S. Kawata, Y. Matsuzawa, and S. Tarui, *J. Chromatogr.*, *338*:195 (1985).

20. Kamada, S., M. Maeda, A. Tsuji, Y. Umezawa, and T. Kurahashi, *J. Chromatogr.*, *239*:773 (1982).
21. Bowers, L. D., and P. R. Johnson, *Anal. Biochem.*, *116*:111 (1981).
22. Bowers, L. D., and P. R. Johnson, *Clin. Chem.*, *27*:1554 (1981).
23. Johnson, P. R., and L. D. Bowers, *Anal. Chem.*, *54*:2247 (1983).
24. Klopf, L. L., and T. A. Nieman, *Anal. Chem.*, *57*:46 (1985).
25. Dalgaard, L., and L. Nordholm, *J. Chromatogr.*, *265*:183 (1983).
26. Boppana, V. K., K. L. Fong, J. A. Ziemniak, and R. K. Lynn, *J. Chromatogr.*, *353*:231 (1986).
27. Dalgaard, L., and L. Brimer, *J. Chromatogr.*, *303*:67 (1984).
28. Brimer, L., and L. Dalgaard, *J. Chromatogr.*, *303*:77 (1984).
29. Kiba, N., and M. Kaneko, *J. Chromatogr.*, *303*:396 (1984).
30. Meek, J. L., and C. Eva, *J. Chromatogr.*, *317*:343 (1984).
31. Eva, C., M. Hadjiconstantinou, N. H. Neff, and J. L. Meek, *Anal. Biochem.*, *143*:320 (1984).
32. Damsma, G., B. H. C. Westerink, and A. S. Horn, *J. Neurochem.*, *45*:1649 (1985).
33. Ogren, L., I. Csiky, L. Risinger, L. G. Nilsson, and G. Johansson, *Anal. Chim. Acta*, *117*:71 (1980).
34. Tawa, R., M. Kito, and S. Hirose, *Chem. Lett.*, 745 (1981).
35. Tawa, R., M. Kito, S. Hirose, and K. Adachi, *Chem. Pharm. Bull.*, *30*:615 (1982).
36. Kito, M., R. Tawa, S. Takeshima, and S. Hirose, *J. Chromatogr.*, *231*:183 (1982).
37. Kito, M., R. Tawa, S. Takeshima, and S. Hirose, *J. Chromatogr.*, *278*:35 (1983).
38. Kito, M., R. Tawa, S. Takeshima, and S. Hirose, *Anal. Lett.*, *18*:323 (1985).
39. Meek, J. L., and F. Nicoletti, *J. Chromatogr.*, *351*:303 (1986).
40. Jansen, H., R. W. Frei, U. A. Th. Brinkman, R. S. Deelder, and R. P. J. Stellings, *J. Chromatogr.*, *325*:255 (1985).
41. Uchiyama, S., Y. Tohfuku, S. Suzuki, and G. Muto, *Anal. Chim. Acta*, *174*:313 (1985).
42. Munns, R. K., W. Shimoda, J. E. Roybal, C. Vieira, *J. Assoc. Off. Anal. Chem.*, *68*:968 (1985).
43. Murata, K., and Y. Yokoyama, *Anal. Biochem.*, *149*:261 (1985).
44. Werkhoven-Goewie, C. E., C. de Ruiter, U. A. Th. Brinkman, R. W. Frei, G. J. de Jong, C. Little, and O. Stahel, *J. Chromatogr.*, *255*:79 (1983).
45. Kawasaki, T., M. Maeda, and A. Tsuji, *J. Chromatogr.*, *272*:261 (1983).
46. Czaban, J. D., *Anal. Chem.*, *57*:345A (1985).

47. Carr, P. W., and L. D. Bowers, *Immobilized Enzymes in Analytical and Clinical Chemistry: Fundamentals and Applications.* Wiley-Interscience, New York, 1980.
48. Yao, T., M. Sato, and T. Wasa, *Nippon Kagaku Kaishi, 7*: 1501 (1985).
49. Johnson, J. M., *Pat. Appl. Ep., 104*:935 (1984).
50. Gorton, L., F. Scheller, and G. Johansson, *Studia Biophysica, 109*:199 (1985).
51. Yao, T., M. Sato, Y. Kobayashi, and T. Wasa, *Anal. Biochem., 149*:387 (1985).
52. Heineman, W. R., and H. B. Halsall, *Anal. Chem., 57*:1321A (1985).
53. Wehmeyer, K. R., M. J. Doyle, D. S. Wright, H. M. Eggers, H. B. Halsall, and W. R. Heinman, *J. Liq. Chromatogr., 6*: 2141 (1983).
54. Wehmeyer, K. R., H. B. Halsall, W. R. Heineman, C. P. Volle, and I. Chen, *Anal. Chem., 58*:135 (1986).
55. Doyle, M. J., H. B. Halsall, and W. R. Heineman, *Anal. Chem., 56*:2355 (1984).
56. Wehmeyer, K. R., H. B. Halsall, and W. R. Heineman, *Clin. Chem., 31*:1546 (1985).
57. Eggers, H. M., H. B. Halsall, and W. R. Heineman, *Clin. Chem., 28*:1848 (1982).
58. Sittampalam, G. S., and G. S. Wilson, *Trends Anal. Chem., 3*:96 (1984).
59. Sportsman, J. R., J. D. Jiddil, and G. S. Wilson, *Anal. Chem., 55*:771 (1983).
60. de Alwis, W. U., and G. S. Wilson, *Anal. Chem., 57*:2754 (1985).
61. Nilsson, K., and P. O. Larsson, *Anal. Biochem., 134*:60 (1983).
62. Risinger, L., L. Ogren, and G. Johansson, *Anal. Chim. Acta, 154*:251 (1983).
63. Kim, H. Y., D. Pilosof, D. F. Dyckes, and M. L. Vestal, *J. Anal. Chem. Soc., 106*:7304 (1984).
64. Funae, Y., H. Akiyama, S. Imaoka, M. Takaoka, and S. Morimoto, *J. Chromatogr., 264*:249 (1983).
65. Morimoto, S., H. Akiyama, M. Takaoka, H. Okamura, S. Imaoka, and Y. Funae, *J. Chromatogr., 295*:226 (1984).
66. Decristoforo, G., and B. Danielsson, *Anal. Chem., 56*:263 (1984).
67. Appelqvist, R., G. Marko-Varga, L. Gorton, A. Torstensson, and G. Johansson, *Anal. Chim. Acta, 169*:237 (1985).
68. Ho, M. Y. K., and G. A. Rechnitz, *Anal. Chem., 59*:536 (1987).

# 7

# Ultraviolet-Visible Derivatization

JAN A. P. MEULENDIJK* and WILLY J. M. UNDERBERG
*State University of Utrecht, Utrecht, The Netherlands*

## INTRODUCTION

Ultraviolet-visible (UV-Vis) detection in liquid chromatography (LC) is often hampered by the poor spectral properties of the analytes at the applied analytical wavelengths. Several attempts have been made to improve the general detectability of the analytes in this respect, such as indirect UV-Vis [1,2] detection and the derivatization of functional groups [3-8]. Although indirect UV detection in some cases is advantageous, it will not be discussed in this chapter as it lies beyond the scope of this book.

Derivatization in LC to improve the detectability of compounds is performed using various detection-sensitive properties (e.g., electrochemical, absorbance, fluorescence). The advantage of UV-Vis detection over other methods, however, is its extended applicability combined with a good sensitivity. Although the selectivity is limited in comparison with other methods, a good LC separation compensates for this disadvantage. UV-Vis detection, after derivatization, offers the advantage of a relatively free choice of the label enabling the introduction of a large absorptivity of the derivative. Moreover, contrarily to other detection systems, the UV-Vis detectors are very reliable instruments.

Modern UV-Vis LC detectors present the possibility of variable-wavelength detection, usually in the range of 190-600 nm, thus offering the potential to optimize selectivity as well as the sensitivity of detection. Detection at short wavelengths (190-220 nm) increases

**Present affiliation*: Organon International B. V., Den Bosch, The Netherlands

Table 7.1 Derivatization Reagents for Ultraviolet-Visible Derivatization[a]

| Reagent | Abbreviation | Functional group | Ref. |
|---|---|---|---|
| Acetic anhydride | | Amine (prim.) | 122 |
| *p*-(9-Anthroyloxy)phenacyl bromide | | See NBr | 27 |
| Benzoyl chloride | BCI | Amine (prim., sec.), hydroxyl, phenol | 89,103 |
| 1-Benzyl-3-*p*-tolyltriazene | BTT | Carboxyl (reactive hydrogen) | 44 |
| *p*-Bromophenacyl bromide | | See NBr | 30 |
| *d*-10-Camphorsulfonyl chloride | | See Dns-Cl | 64 |
| 1-Chloromethylisatine | ClMIS | Carboxyl (acidic hydrogen, amine) | 33 |
| *N*-Chloromethyl-4-nitrophthalimide | ClMNPI | See ClMIS | 33 |
| *N*-Chloromethyl-4-phthalimide | ClMPI | See ClMIS | 33 |
| Diethylchlorophosphite | | Anhydride | 46 |
| Sodium diethyldithiocarbamate/TNBS | | Carbonyl | 114.115 |
| 5-Dimethylamino-1-naphthalenesulfonyl chloride | Dns-Cl | Amine (prim., sec.) (hydroxyl), imidazole, phenol, thiol | 63,74 |
| 4-*N*,*N*′-Dimethylaminoazobenzenesulfonyl chloride | DABS-Cl | See Dns-Cl | 57-60 |
| 4-*N*,*N*′-Dimethylaminoazobenzene-4′-isothiocyanate | | See PITC | 100 |
| 2,4-Dinitrobenzenesulfonic acid | DNBS | Amine | 123 |
| 3,5-Dinitrobenzoyl chloride | DNBCl | See BCl | 107 |
| 2,4-Dinitrophenylhydrazine | DNPH | Carbonyl | 124 |
| 6,6′-Dithiodinicotinic acid | DTNA | Thiol | 118 |

| | | | |
|---|---|---|---|
| 5,5'-Dithio-2-nitrobenzoic acid | DTNB | Thiol | 118 |
| 1-Fluoro-2,4-dinitrobenzene | FDNB | Amine (prim., sec.), phenol | 75-83 |
| 4-Fluoro-3-nitrotrifluoromethylbenzene | | See FDNB | 84 |
| p-Methoxyaniline/triphenylphosphine | | Carboxyl | 34 |
| p-Methoxyphenacyl bromide | | See NBr | 23 |
| p-Methoxybenzoyl chloride | | See BCl | 89 |
| 2-Naphthacyl bromide | NBr | Carboxyl (reactive hydrogen) | 9,10 |
| 1,2-Naphthoquinone-4-sulfonate | | Amine | 125 |
| 1-Naphthylamine/oxalyl chloride | | Carboxyl | 37 |
| 1-(2-Naphthyl)diazomethane | | Reactive hydrogen | 126 |
| Naphthylisocyanate | NIC | Amine (prim., sec.), hydroxyl | 94 |
| Naphthylisothiocyanate | NITC | See PITC | 92 |
| α-Naphthylselenyl chloride | | Amine | 116 |
| Ninhydrin | | Amine (prim.) | 70-72 |
| *p*-Nitrobenzenediazoniumborate | | Phenol | 127 |
| *p*-Nitrobenzoyl chloride | pNBB | See BCl | 89,104 |
| 1-*p*-Nitrobenzyl-*N*,*N'*-diisopropylisourea | PNBDI | See BTT | 32 |
| *p*-Nitrobenzylhydroxylamine | PNBA | See DNPH | 111 |
| *N*-*p*-Nitrobenzyl-*N*-*n*-propylamine | | Isocyanate | 128 |
| 1-*p*-Nitrobenzyl-3-*p*-tolyltriazine | PNBTT | See BTT | 44 |
| 2-Nitrophenylhydrazine/dicyclohexyl-carbodiimide | | Carboxyl | 38,39 |

Table 7.1 (continued)

| Reagent | Abbreviation | Functional group | Ref. |
|---|---|---|---|
| 2-Nitrophenylhydrazine/*N*-ethyl-*N*′-(3-dimethylaminopropyl)carbodiimide | | Carboxyl | 45 |
| Pentafluorobenzylhydroxylamine | | Carbonyl | 129 |
| Phenacyl bromide | PBr | See NBr | 9,10 |
| *p*-Phenylbenzoyl-isothiocyanate | | See PITC | 93 |
| Phenyldimethylsilyl chloride | | Hydroxyl | 109 |
| Phenylisothiocyanate | PITC | Amine (prim., sec.) (hydroxyl, phenol) | 73,98, 100,101 |
| *o*-Phthaldialdehyde | OPA | Amine (prim.) | 68,69 |
| Quinoline-8-sulfonyl chloride | | Amine | 130 |
| Salicylaldehyde | | Hydrazine | 131 |
| *p*-Toluenesulfonyl chloride | | Amine | 132 |
| *m*-Toluol chloride | | See BCl | 89 |
| 2,4,6-Trinitrobenzenesulfonic acid | TNBS | See FDNB | 85,86 |

[a]Reactive hydrogen: carboxyl, hydroxyl, phenol, thiol, imide, amine.

the sensitivity combined with a loss in selectivity, whereas long-wavelength detection, with its higher selectivity, usually requires higher concentration (less sensitivity) of the analyte due to the lower molar absorptivities ($\varepsilon$) at these wavelengths.

Derivatization reactions to improve the UV-Vis detectability in LC can be performed pre- and postchromatographically. The pros and cons of both systems mentioned earlier (see Chap. 1) apply in this case.

The majority of the analytes, subjected to LC analysis and UV-Vis absorption derivatization, belong to one of the following classes, with respect to the appropriate functional group:

Carboxylic acids and related compounds
Amino acids
Amines, mostly biogenic in nature
Carbohydrates and other hydroxyl-bearing compounds

Furthermore, some compounds of different nature have been studied and will also be discussed. An overview of the applied derivatization reagents for UV-Vis derivatization is presented in Table 7.1. The same table lists the abbreviations used in this chapter for the labeling reagents.

## CARBOXYLIC ACIDS AND RELATED COMPOUNDS

The most widely used method for the derivatization of carboxylic acids is the formation of esters using UV-Vis absorbing alcohols or halides as labels. The most commonly performed derivatizations use phenacyl bromide (PBr) or naphthacyl bromide (NBr) to form the corresponding phenacyl- or naphthacylcarboxylates [7,9-31]. Benzyl esters [32] and phthalimido-*n*-methyl esters [33] are also employed.

The formation of amides is also described, mostly after activation of the carboxylic acid function [34-37] and coupling with aromatic amines. Derivatization of carboxylic acids with phenylhydrazines [38-41,45,46] is still under investigation, while aryltolyltriazines [44] failed as UV-Vis absorbance introducing labels in the analysis of fatty acids, probably due to the formation of a variety of reaction products and the toxicity of the labels.

### Prechromatographic Derivatizations

#### *Phenacyl Bromide, Naphthacyl Bromide, and Related Compounds*

The studies of Durst et al. [9] and Grushka et al. [10] introduced a very popular method for the derivatization of carboxylic acids.

These investigators used PBr and NBr and analogs as alkylating agents. These halides react with carboxylic acids to form phenacyl or naphthacyl esters but the derivatization yield is far from complete. Moreover, the reaction is nonselective, since many other compounds undergo alkylation with these reagents. Selectivity can be introduced using catalysts, activating the carboxylic function. Crown ethers are widely used for this purpose. The crown ether complexes the cation of the neutralized acid, thus activating the anion to react in a nucleophilic substitution reaction with PBr or NBr (see Chap. 2). The average yield under these conditions is over 97%.

A typical example for this kind of procedure is the following [9]: A sample of a carboxylic acid (total acid concentration 0.001-0.05 mM) dissolved in methanol is neutralized with a potassium hydroxide solution in methanol to a phenolphthalein end point. The solvent is then removed at reduced pressure and the residue transferred to a 3-ml reaction vial which is fitted with a stirring bar. An excess of a solution of *p*-bromophenacyl bromide, the label, and 18-crown-6 in a molar ratio of 10:1 and a sufficient amount of acetonitrile are added to bring the total volume to 0.5-1.5 ml. The reaction vial is capped and heated at 353 K for 15 min while stirring. After cooling the solution is subjected to reversed phase (RP) LC.

The derivatization reaction can be performed in any aprotic solvent; small amounts of water do not influence the reaction yield significantly [9].

LC analysis of the resulting derivatives can be performed either in the normal phase (NP) or in the RP mode. Since in a number of cases gradient elution is preferable, especially in the analysis of fatty acids [9], RP-LC has some advantages, e.g., a better control of solvent polarity. Not only monocarboxylic acids can be analyzed; the technique is also suitable for the analysis of dicarboxylic acids [10]. Generally, the detection limits are in the nanogram region, ranging from 1 ng for acetic acid to 50 ng for a $C_{20}$ carboxylic acid, using 254 nm as the analytical wavelength [10].

The alkylating method has been used in the analysis of various compounds containing carboxylic acid functions, e.g., low molecular weight carboxylic acids [11,12], biotin [13], and penicillins [14]. In some cases [2] extractions from aqueous solutions precede the derivatization procedure in order to concentrate the sample and to remove excessive water.

Since the vast majority of organic solvents are suitable for derivatization, the assay of carboxylic acids in biological matrices is not expected to present major problems [9]. This is clearly demonstrated with studies on the analysis of valproic acid in serum and plasma samples [15-19]. Concentration and removal of excess water can be achieved by extraction of the acids from the biological fluid

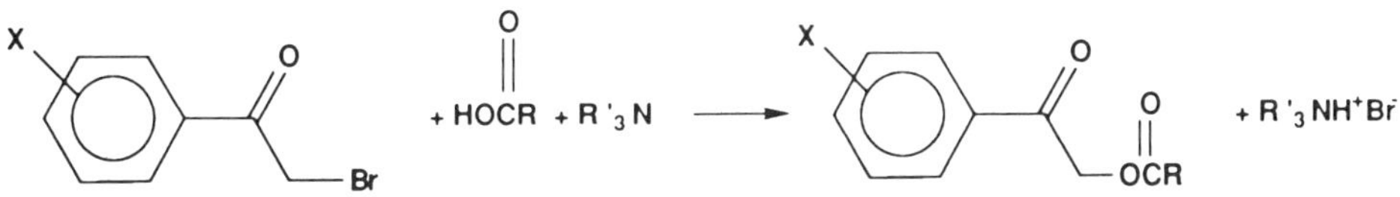

Figure 7.1 Reaction of a carboxylic acid with phenacyl bromide using a tertiary amine as hydrobromide scavenger.

with organic solvents [15,16,18] or by using a solvent-demixing technique [17]. In this case, after addition of acetonitrile to the serum sample the solution is saturated with potassium chloride. As a result the water and acetonitrile phases separate and derivatization with subsequent LC analysis is performed in the acetonitrile layer. Extraction can even be omitted, e.g., after addition of an excess acetonitrile to the serum sample derivatization is carried out in this mixture followed by LC separation [19]. Recoveries are good, detection limits are around 60 nmol/ml, and analysis times are short (6 min). Another method for the removal of water is described in a study concerning the RP-LC analysis of fatty acids in tooth plaque [20]. After extracting the acids with diluted alkali, neutralizing and freeze-drying the extract, derivatization is performed in acetonitrile with NBr. In the procedure Kryptofix (a polyoxadiazamacrobicyclic compound) is used as a catalyst instead of a crown ether, since the former compound forms more stable complexes with the cations of the salts. Detection is carried out at 254 nm; fluorescence detection also can be employed.

In general, activation of the carboxylic acid anions with crown ethers or other cation-complexing agents requires neutralization of the solutions prior to derivatization, e.g., with potassium hydroxide [9-11], potassium carbonate [17], or sodium hydrogen carbonate [18]. Activation of the carboxylic acid can also be achieved with solely a tertiary amine acting as hydrobromide scavenger [2], as depicted in Fig. 7.1. Frequently used tertiary amines are *N,N'*-diisopropylethylamine [21-26] and triethylamine [27,29]. Acetonitrile [25,26,28] and dimethylformamide [21-24] are mainly used as solvents in the derivatization reactions and RP-LC is performed using gradient or isocratic elution systems [27]. NP-LC systems are used to determine several prostaglandins (Alpostradil, $PGA_1$, and $PGB_1$) after derivatization with NBr and trimethylamine [25,26]. With this procedure impurities of other prostaglandins as low as 0.1% could be determined.

Fatty acid derivatization is performed using substituted phenacyl and naphthacyl halides as labeling compounds and *N,N'*-diisopropylethylamine as base [22-24]. The best results are obtained with methoxyphenacyl bromide. The method is also suitable for fatty acids containing hydroxyl groups, unsaturated bonds, and cyclopropane

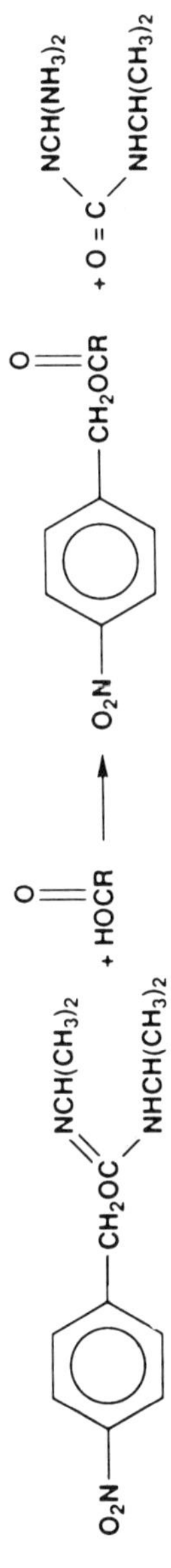

Figure 7.2 Alkylation of carboxylic acids using *p*-nitrobenzyl-*N*,*N'*-diisopropylisourea (PNBDI).

structures [23]. For the derivatization of eicosanoides (arachidonic acid metabolites and prostaglandins), *p*-(9-anthroyloxy)phenacyl bromide is used and trimethylamine as bromide scavenger [27]. Using RP-LC the detection limit is 280 pg at an analytical wavelength of 254 nm. Bile acids can be derivatized with NBr; the LC analysis is performed on octadecyl-modified silica columns [28]. Employing this system no tailing of the derivatives occurs, unlike a system with a nonmodified silica column.

Another catalyst in carboxylic acid derivatization is potassium fluoride [29]. Quinic acid, after isolation from plant material, is derivatized with *p*-bromophenacyl bromide in dimethylformamide [30]. This method does not require preneutralization of the acid and hence the use of (expensive) crown ethers can be avoided.

Alkylation with the described procedures is not restricted to carboxylic acids. Other acidic compounds, like barbiturates, are also subject to this type of derivatization. Barbiturates, when protonated, show no appreciable absorbance at 254 nm to enable LC detection. Hulshoff et al. reported the derivatization of phenobarbital with NBr in acetone with cesium carbonate as catalyst [31]. The derivative is subjected to RP-LC and at 254 nm sufficient sensitivity can be obtained.

### *Benzyl Derivatives*

A derivatization method for bile acids has been developed using *p*-nitrobenzyl-*N*,*N'*-diisopropylisourea (PNBDI) as alkylating agent without the need of a catalyst (Fig. 7.2) [32]. Separation of the resulting esters is achieved on NP-LC with isopropanol-isooctane mixtures as mobile phases and detection at 254 nm. The detection limit is about 100 ng.

### *N-Methylphthalimide Derivatives*

*N*-Chloromethyl-4-phthalimide (ClMPI), *N*-chloromethyl-4-nitrophthalimide (ClMNPI), and *N*-chloromethylisatine (ClMIS) are used as labeling compounds for several solvents such as methanol, diethyl ether, and acetonitrile, with crown ethers or triethylamine as catalysts. The reactivity of these labels toward prototropic functions is caused by the high mobility of the chlorine atom. Consequently, the selectivity is limited: carboxylic acids and other acidic functions as well as primary and secondary amines react promptly [33]. Separation of the derivatives is achieved using RP-LC.

### *Derivatization After Conversion into Acyl Chlorides*

So-called reversed labeling methods are based on the principles used in the derivatization of the amino function, which reacts with carboxylic acid chlorides. In turn, carboxylic acids can be converted

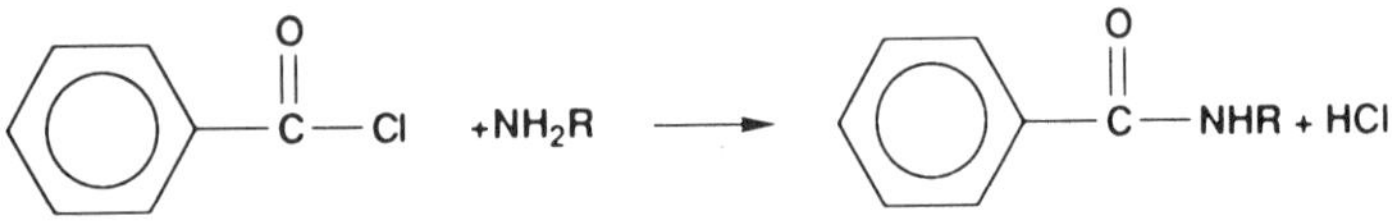


Figure 7.3 Reaction of acyl chlorides with primary amines.

to their corresponding acyl chlorides and these intermediates can react with a suitable amine to form an UV-Vis absorbing derivative (Fig. 7.3).

Hoffman and Liao use mixtures of triphenylphosphine and carbon tetrachloride to convert carboxylic acids to acyl chlorides [34]. Subsequently, these compounds are converted to amides with *p*-methoxyaniline and the derivatives are subjected to RP-LC. The system produces good results but the method is time consuming because the gradient elution with acetonitrile-water and methanol-water mixtures results in long retention times. To overcome the disadvantage of pollution of the chromatographic system with organophosphines and other interfering reaction products, a polymer-bound diphenylphosphine-carbon tetrachloride mixture has also been applied.

Other methods use a similar activation of the carboxylic acid. Bissinger et al. [35] were able to introduce an isocratic LC system, which is an advantageous simplification of the method described just above [34]. Initially, thionyl chloride [36] is applied to convert the carboxylic function of fatty acids but later on this method is modified by the introduction of oxalyl chloride [37] to prepare the acyl chloride after isolation of free fatty acids from serum by ion exchange (IE) chromatography on a solid phase cartridge.

The isolated fatty acids are converted into their corresponding acid chlorides after dissolving of the residue in benzene. To 0.6 ml of this solution containing 2-1000 nmol of the free fatty acids a known amount of margaric acid, in benzene, as internal standard and 0.6 ml of a 2% solution of oxalyl chloride in benzene is added. After homogenization the mixture is kept at 343 K for 30 min. After the reaction is completed the solvent is evaporated under reduced pressure, 0.1 ml of a solution of 400 mmol of 1-naphthylamine in 10 ml of benzene and 0.01 ml of a solution of 4 mmol triethylamine in 10 ml benzene is added to the residue. The mixture is kept at 303 K for 15 min in order to form the 1-naphthylamide derivatives. 0.002 ml of the reaction mixture is subjected to RP-LC with detection at 280 nm. The sensitivity of the method is satisfactory: detection limits are in the range of 3-5 ng.

Although several investigators report satisfactory results with this method, the general disadvantages of the reversed labeling

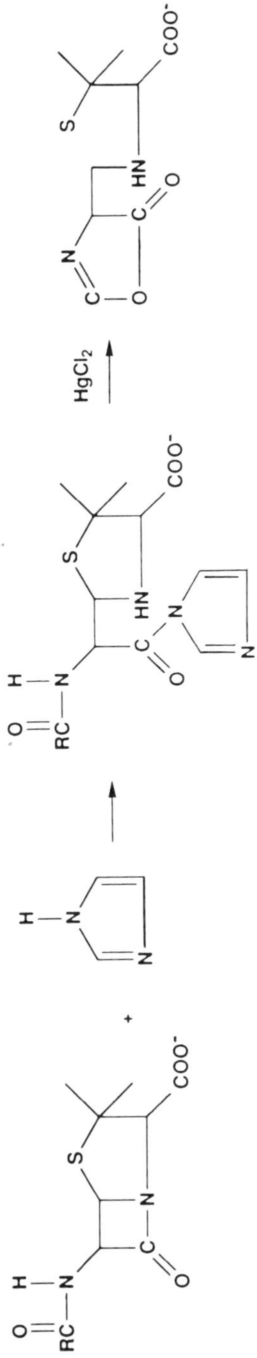


Figure 7.4 Derivatization of penicillins by means of imidazole and mercuric chloride.

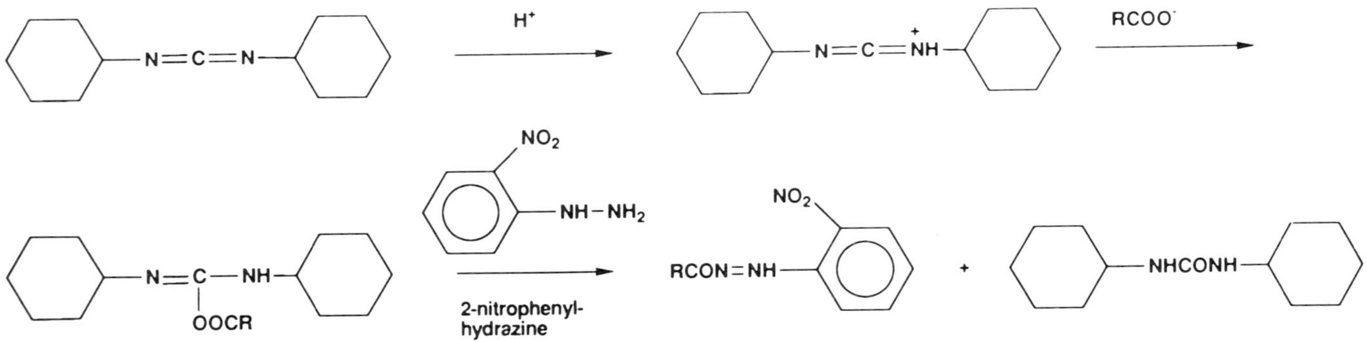


Figure 7.5 Derivatization of carboxylic acids with 2-nitrophenylhydrazine using *N,N'*-dicyclohexylcarbodiimide (DCC) as activator.

method using acyl chlorides are the high reactivity of the compounds, implying the easy occurrence of side reactions, the two-step derivatization procedure with a high degree of complexicity, and the toxicity of the reagents.

### Postchromatographic Derivatizations

The number of postchromatographic derivatization methods for carboxylic acids in the literature is very limited. The most prominent one is a method for penicillins developed by Westerlund et al. [42]. They isolate the compounds by extraction from serum or urine, achieve separation using RP-LC, and accomplish derivatization with a reagent containing imidazole and mercuric chloride. In this reaction an initial β-lactam ring opening occurs, catalyzed by imidazole. The resulting intermediate after rearrangement reacts with mercuric chloride to form a mercuric mercaptide (Fig. 7.4). Detection of this derivative is possible at 310 nm. The reaction is performed at 313 K and the detection limit is about 100 ng/ml. The method is applicable in the analysis of amoxicillin [43].

### Miscellaneous Derivatizations

A number of publications describe the derivatization procedures of carboxylic acids and related compounds which may be useful for UV-Vis detection prior to or after LC analysis of these compounds but which have not yet been thoroughly tested. Most of these procedures involve derivatization with 2-nitrophenylhydrazine after activation of the carboxylic function.

Munson reports the reaction of carboanhydrides or acid chlorides with 2-nitrophenylhydrazines [38]. In a subsequent publication, Munson and Bilous report the reaction also to occur with a carboxylic acid after activation of this function with *N,N'*-dicyclohexylcarbodiimide (DCC) [39]. Both alcohols and amino acids, including hydrazines, can act as labels (Fig. 7.5). Similar reactions are described in a number of other studies [40,41,45]. Besides DCC [40,41], 1-ethyl-3-(dimethylaminopropyl)carbodiimide (EDCI) [45] is used for the activation of the carboxylic acid function. The latter compound, being more water-soluble than DCC, can be applied in aqueous media. Moreover, the activator exhibits no UV-Vis absorption, thus diminishing the possible interference in the detection of the derivatives at 550 nm. However, due to the occurrence of side reactions, the reagent blanks usually also show significant absorbances at the analytical wavelength, limiting the applicability of the reaction as a postchromatographic detection method. LC separation systems for the derivatives and the excess reagent have not yet been tested, but the possibilities seem limited. In general, derivatization yields are good (over 90%) [45].

Görög et al. [46] describe a modified version of the reaction. A mixed anhydride of the carboxylic acid with diethylchlorophosphite reacts with 2-nitrophenylhydrazine without considerable side reactions. According to the authors, this reaction should be suitable in a prechromatographic derivatization procedure. After LC, the derivative must be detected at 550 nm following a postchromatographic reaction with alkali, since the currently available columns usually cannot withstand the pH over 12 required for this detection.

## AMINO ACIDS

A great variety of methods have been published for the modification of amino acids to make them suitable for UV-Vis detection in LC. Both the amino function and the carboxylic group offer possibilities for labeling. Most methods, however, concentrate on labeling of the amino function.

Prechromatographic derivatization predominates, using various reagents, such as phenylisothiocyanate (PITC) [47-56], 4-*N,N'*-dimethylaminoazobenzenesulfonyl chloride (DABS-Cl) [57-60], or nitrobenzene derivatives [61,62]. 5-Dimethylamino-1-naphthalenesulfonyl chloride (Dns-Cl) is also widely applied, but only in some occasions is UV-Vis detection applied [63]. For the separation of diastereomeric amino acids several chiral labels have been described [64-67], sometimes involving derivatization of both reactive functions in the compound [64]. These labels mostly show appreciable UV absorbance, enabling a sensitive LC detection. For more detailed information see Chap. 5.

Postchromatographic methods to improve the detection of amino acids also are available [68-72], although these methods do not always involve simple labeling reactions.

### Prechromatographic Derivatizations

#### *Phenylisothiocyanate Derivatives*

PITC reacts with the amino function of amino acids to form a phenylthiohydantoin (PTH) amino acid (Fig. 7.6) [73]. This reaction is performed in aqueous solutions and several investigators use it prior to LC analysis of peptide hydrolysates. Heinrikson and Meredith [47], after hydrolysis, derivatize the amino acids with PITC for 5 min at room temperature in an acetonitrile-pyridine-triethylamine-water (10:5:3:2) mixture. Subsequent RP-LC separation requires a complicated gradient elution system. The method of Edman [72] is also applied for the labeling of amino acids [48].

To achieve this 1.0 ml of a solution of 500 mg/liter L-norleucine in a 1 M sodium bicarbonate solution as internal standard and 0.05 ml of PITC as label are added to 5.0 ml urine or 1.0 ml serum, and

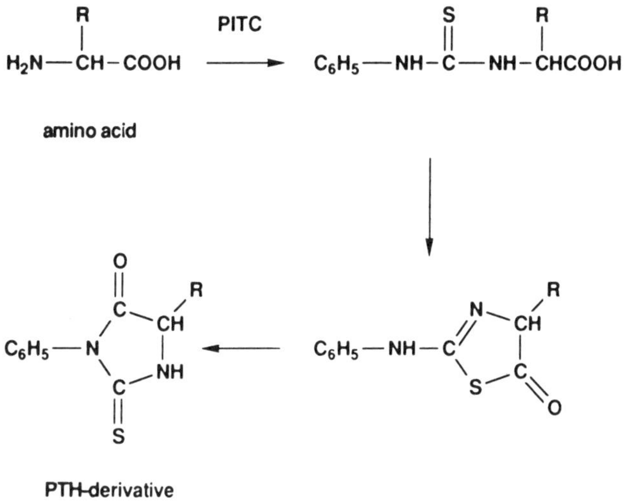


Figure 7.6 Derivatization of primary amines with phenylisothiocyanate (PITC).

the total volume of the mixture is adjusted to 6 ml with water if necessary. The mixture is heated at 333 K for 40 min. Excess of PITC is removed by extraction with 4 ml of diethyl ether. The diethyl ether layer is discarded and to the aqueous phase 0.25 ml of 12 M hydrochloric acid is added. After cooling the PTH derivatives are extracted with two portions of 4 ml of diethyl ether. After evaporation of the combined diethyl ether extracts at room temperature, the residue is dissolved in methanol-acetonitrile (4:1, v/v) and 0.01 ml of this solution is subjected to a RP-LC procedure with gradient elution. Quantitation of the amino acids seems to be satisfactory with detection limits varying from 5 to 20 ng/ml.

Various investigators have attempted to simplify the LC procedure. For instance, after considering the fact that the available LC systems suffer from long retention times and/or high column temperatures [50-55], a binary gradient elution system is developed enabling complete separation in 14 min at 308 K, with good reproducibility [49]. Another gradient elution also accomplished complete separation of 20 PTH-amino acids in a short time (20 min), but the column temperature was high (343 K) [56]. Using an isocratic elution system the analysis time could be reduced to 15 min with the disadvantage of coelution of PTH-serine and PTH-arginine.

*Dimethylaminoazobenzenesulfonyl Derivatives*

DABS-Cl reacts with the amino function of amino acids to form a sulfonamide. This reaction is the basis for a prechromatographic labeling procedure developed by Lammens and Verzele [57]. The advantage of the method is the long analytical wavelength (450 nm), which excludes interferences from endogenous substances to a large extent. The reaction is performed at 353 K at pH 8.9 with subsequent isocratic RP-LC of the mixture and detection at 464 nm. This system enables separation of 17 amino acids. Due to the low pH, however, no separation is obtained between histidine and arginine. The reaction time is short and yields are quantitative but the chromatography requires a long analysis time. The method is also suitable for amino acid assays in urine [58]. The derivatives have to be isolated from the matrix in case amines are present in the urine samples. Since DAB-sulfonamide, resulting from the reaction of ammonia with DABS-Cl, coelutes with DABS-methionine, the urine samples should be lyophilized prior to the derivatization to avoid this interference. Chang et al. [59,60] use this method to determine several amino acids in brain tissue. After suspension in dilute perchloric acid and centrifugation, derivatization is performed in the supernatant at pH 9 with subsequent RP-LC analysis with a gradient solvent system and detection at 436 nm. The detection limit is as low as 0.3 pmol.

*Nitrobenzene Derivatives*

1-Fluoro-2,4-dinitrobenzene (FDNB) reacts with amino functions to form substituted aromatic amines (Fig. 7.7). Hoppel et al. [61] apply this reaction, after isolation of 6-*N*-trimethyllysine from urine samples with IE chromatography, to label the compound in alkaline solution at 310 K in 1 hr. RP ion pair (IP) chromatography is performed using heptanesulfonate as the ion-pairing agent.

2,4,6-Trinitrobenzenesulfonic acid (TNBS) is also used for amino acid labeling in alkaline medium (Fig. 7.8) [62]. After isolation by means of an extraction, RP-LC or RP-IE analysis of the derivatives is performed. Although trinitrophenyl derivatives exhibit maximum

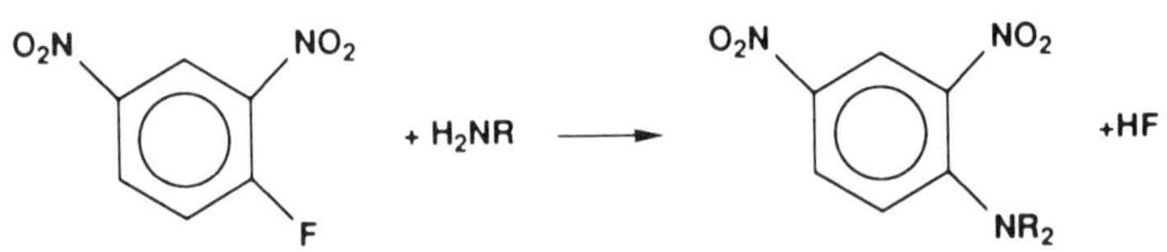


Figure 7.7 Derivatization of primary amines with 1-fluoro-2,4-dinitrobenzene (FNDB).

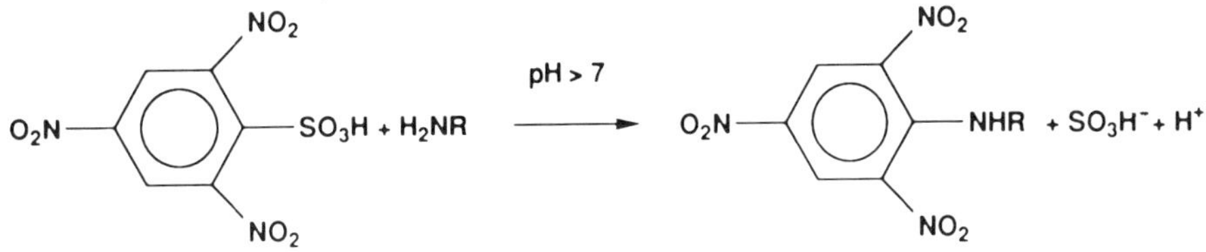

Figure 7.8 Derivatization of primary amines with 2,4,6-trinitrobenzenesulfonic acid (TNBS).

absorptivity at 350 nm, the detection wavelength used is 254 nm. The limit for detection is 25 pmol.

Both FDNB and TNBS are widely used for the derivatization of amines other than amino acids and are dealt with extensively in one of the following sections of this chapter.

*Dansyl Derivatives*

Dansyl chloride (Dns-Cl) is widely applied for acylating amino groups in amino acids, but only occasionally is detection of the derivatives performed using their UV-Vis absorbances. Wilkinson [63], using a method first described by Gray [74], acylates amino acids and subsequently subjects the compounds to RP-LC analysis with phosphate buffer (pH 7.7)-acetonitrile mixtures in a gradient system. Fluorescence detection is troublesome in this case due to the quenching effect of the phosphate ions. Thus, UV detection at 250 nm is applied. Complete resolution is achieved within 30 min; the detection limit is 0.5 nmol. Extended information about the use of Dns-Cl can be found in Chap. 9.

### Postchromatographic Derivatizations

A few postchromatographic derivatization methods for amino acids have been described in the literature. The reaction between *o*-phthaldialdehyde (OPA) and primary amines, often used for fluorimetric detection, is employed by Radjai and Hatch to detect amino acid derivatives using their UV-Vis absorbances at 340 nm [68]. The reaction is performed in a reducing medium obtained after addition of 2-mercaptoethanol (see Chap. 9). The detection limit is around 50 ng. For the analysis of proline and hydroxyproline pretreatment with sodium hypochlorite is necessary [69].

The reaction with ninhydrin is applied by Lou and Siena for the assay of methylated amino acids in urine [70]. Using an amino acid analyzer the compounds can be separated on an IE-LC system using postchromatographic derivatization with ninhydrin in dimethylsulfoxide, reacting with the amino acids to form a purple anion (Fig. 7.9) [71].

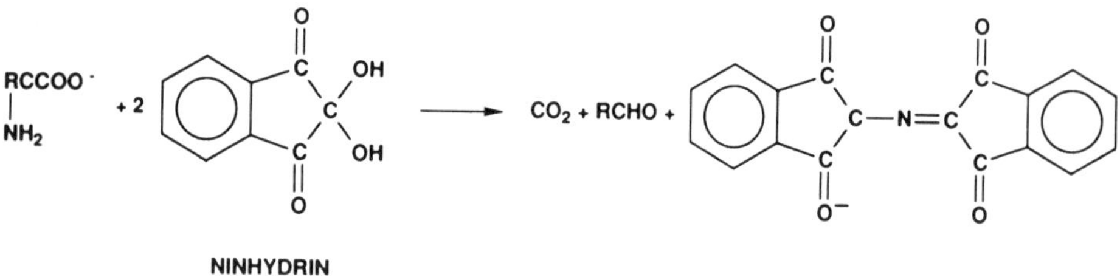


Figure 7.9 Derivatization of primary amines with ninhydrin.

An original combination of pre- and postchromatographic techniques is developed by LePage and Rocha [72]. Separation of the amino acids is achieved by IP chromatography using dodecyl sulfonate as the ion-pairing agent and octyl-modified silica as the stationary phase. Ninhydrin or 1,2,3-perinaphthindantrione (Peri) in dimethylsulfoxide is applied as the derivatization reagent. The column effluent, during transportation through a 2.5-m-long Teflon tube, is heated to 413 or 373 K, respectively, in which the reaction takes place and subsequent detection can be achieved. Derivatization with Peri is advantageous, since contrarily to ninhydrin this reaction is not influenced by the low pH of the effluent, which is necessary for separation of the amino acids. However, the $\varepsilon$ of the ninhydrin-amino acid reaction products, at the wavelength of maximum absorbance ($\varepsilon$ is 20,000 at 570 nm) allows the detection of lower concentrations of amino acids in comparison to the Peri reaction products ($\varepsilon$ is 1,800 at 450 nm).

## AMINES

Primary and secondary amino functions are most commonly labeled by means of a nucleophilic substitution reaction, introducing a suitable chromophore. Commonly used labels, introducing nitrophenyl chromophores, are FDNB [75-83], 4-fluoro-3-nitrotrifluoromethylbenzene [84], and TNBS [85,86]. These reagents are also used to label amino acids. As already mentioned in the section on carboxylic acids, acid chlorides react readily with amino groups and this reaction is also used to label amino functions [87-90]. Isothiocyanates [92,93] and isocyanates [94] are sometimes applied. Methods for the detection of tertiary amines involve quaternization [95] and postchromatographic condensation [96].

## Prechromatographic Derivatizations

### *Nitrobenzene Derivatives*

FDNB is nowadays commonly used for the derivatization of aminoglycosides (Fig. 7.7). The kinetics of the labeling reaction are quite complicated as shown in a recent study about tobramycin derivatization [76]. The labeling of all amino groups present in the compound, the hydrolytic degradation of the derivative, and the hydroxyl derivatization were thoroughly investigated. Prevention of labeling the latter function combined with the formation of a totally N-derivatized product requires accurately standardized reaction conditions, especially where concerning the pH, due to a sharp optimum in proton activity in this reaction. FDNB is used as a label for neomycin [75, 77], tobramycin [76,81-83], fortimicin A [78], amikacin [79,80], gentamycin and sisomicin [81-83]. The derivatization requires a fairly long reaction time (30-60 min) at high temperature (333-373 K) in alkaline solution (pH 9). Barends et al. [80] succeeded in derivatizing amikacin within 5 min in 0.05 M sodium hydroxide-methanol in serum.

The procedure for the analysis of amikacin starts with the addition of 0.02 ml of a solution of 250 mg/liter kanamycin sulfate in water as internal standard to 0.2 ml of serum containing amikacin. After mixing thoroughly, the solution is dispersed on top of an appropriately prepared CM-Sephadex C-25 column. The system is eluted with 2 ml of a solution containing 1 mM hydrochloric acid and 0.2 M sodium sulfate and the eluate is discarded. Subsequently, the column is eluted with 0.25 ml of 0.05 M sodium hydroxide solution. Again this eluate is discarded, being the void volume of the column. The column is then eluted with 1 ml of 0.05 M sodium hydroxide solution and the eluate collected in a 5-ml ampule. After addition of 2.5 ml of a solution of 30 mg/ml FDNB in methanol the ampule is heat-sealed and kept at 373 K for 5 min. After cooling 0.15 ml of the mixture is subjected to RP-LC analysis. Prior to the analysis in plasma the proteins can also be removed by precipitation with methanol [79] or acetonitrile [81-83].

In some cases extraction of the labeled compounds with a nonpolar solvent is necessary, especially when NP-LC is applied [75,77, 79]. Application of RP-LC analysis does not require such an isolation procedure, indicating the suitability of these systems in allowing the analysis of complex mixtures without extended clean-up procedures. When NP-LC analysis is applied the water content of the samples must be thoroughly standardized in order to obtain constant capacity factors [79]. The detection wavelength is normally about 350 nm.

4-Fluoro-3-nitrotrifluoromethylbenzene is used to label putrescine and some related compounds [84]. This reagent does not attack molecules containing polar groups, such as amino acids [84]. Secondary amines do not react either. Derivatization is accomplished at 333 K and completed within 20 min, whereafter the excess reagent is removed with histidine (in contradiction of the fact that amino acids are not susceptible to the reagent). The derivatives are separated in a RP-LC procedure at 313 K. Separation is accomplished within 10 min and detection is performed at 242 or 410 nm. The detection limits range from 5 to 25 pmol.

TNBS, like FDNB, is used for the determination of aminoglycosides in biological fluids. The resulting *N*-trinitrophenyl derivatives (Fig. 7.8), e.g., the derivatives of tobramycin and kanamycin [85, 86], are analyzed by RP-LC at 323 K. Apparently the resolution is not sufficient, since prechromatographic isolation of the derivatives is necessary. This isolation is accomplished with a $C_8$ cartridge. The advantage of TNBS is the fact that hydroxyl groups are unsusceptible to this reagent [91], with an increase in selectivity as a consequence. Recoveries are almost quantitative and detection limits are about 0.1 ng.

### *Acyl Derivatives*

As was mentioned earlier, acyl chlorides react readily with amino functions to form amides (Fig. 7.3). Suitable acyl chlorides are used to allow sensitive UV-Vis detection of derivatives of nonabsorbing amines. Clark and Wells [89] describe a number of acyl chlorides, such as benzoyl, *p*-chloro, *p*-methoxy, *p*-nitrobenzoyl, *p*-tolyl, and *p*-nitrobenzenesulfonyl chloride. Among all derivatives the methoxybenzamides are the most suitable, with high molar absorptivities at a convenient analytical wavelength (254 nm). After derivatization in tetrahydrofuran-sodium hydroxide solutions, the derivatives are extracted with chloroform and subjected to RP-LC analysis. Polyfunctional amines can be analyzed as *m*-tolylamides [88]. Derivatization requires mild conditions (5 min at room temperature). After clean-up via extraction with dichloromethane RP-LC is applied. The detection limit is about 5 ng and neither water nor ammonia interferes to any extent. Another polyfunctional amino compound, gentamicin, has been determined using RP-LC after labeling with benzenesulfonyl chloride [90]. The detection limit is better than 0.2 μg/ml when netilmicin is used as internal standard, but separation between several gentamicin isomers cannot be accomplished. Metformin (a biguanide) can also be labeled with acyl chlorides [87]. Due to its very polar nature and low absorption maximum the parent compound shows poor chromatographic properties. Reaction with *p*-nitrobenzoyl chloride results in a triazine derivative with a more suitable analytical detection wavelength (280 nm) and lower polarity, thus allowing better LC characteristics. Derivatization is carried out in urine samples after alkalinization, saturation with sodium chloride, and addition of acetonitrile. Due to solvent demixing [17], the acetonitrile phase containing the derivative can be isolated and subjected to RP-LC. Using this solvent-demixing effect the *p*-nitrobenzoyl chloride degradation product remains in the water phase, resulting in relatively clean chromatograms.

### *Isothiocyanate Derivatives*

Isothiocyanates are widely used for labeling of the primary amino function [98]. They have replaced the isocyanates which also react with hydroxyl groups, carboxylic groups, and water and hence lack sensitivity [94,98,99]. However, according to Tocksteinova et al., isothiocyanates are also reactive toward secondary amines, alcohols, and phenols [100]. Nevertheless, the affinity of this label for the amino function seems greater than for the other functional groups [73,94]. An advantage of the method is the pH independence of the reaction. The mechanism is depicted in Fig. 7.6. Labeling with PITC or naphthylisothiocyanate (NITC) is applied to $C_3$-$C_{10}$ alkylamines in ethanol [92]. After dilution with water and extraction with a nonpolar solvent, thin-layer chromatographic (TLC) analysis is applied, but the authors consider the separation promising for RP-LC analysis. This opinion is confirmed by Bjorkqvist, who uses PITC for the derivatization of aliphatic and aromatic amines with subsequent RP-LC analysis [94]. The selectivity of the reaction is reported to be limited: alcohol and water interfere but the amino function appears to be the most reactive. Standardization of the conditions therefore allows accurate quantitative determination of amines. Another label, *p*-phenylbenzoylisothiocyanate, has the advantage of a high wavelength of absorption, limiting the interferences due to nonspecific absorbances originating from the matrix [93]. Although labeling with this reagent in acetonitrile is a good starting point for LC analysis, no chromatographic details are given. Bauer et al. report the derivatization of 6-amino-2-methyl-2-heptanol (Heptaminol) with 4-*N*,*N'*-dimethylaminoazobenzene-4'-isothiocyanate, a label with a chromophore identical to that of DABS-Cl [100]. RP-LC of the resulting derivative gives satisfactory results in the determination of the amino compound in pharmaceutical preparations.

### *Quaternization Reactions*

Mitra et al. describe the alkylation of pilocarpine in aqueous solutions and subject the derivative to RP-LC [95]. The label is *p*-nitrobenzyl bromide (pNBB) and the reaction is performed in sealed ampules for 24 hr at 313 K. IP chromatography is applied and detection is executed at 254 nm. Separation of pilocarpine and isopilocarpine is possible with this system and the detection limits are about 50 ng/ml.

### Postchromatographic Derivatizations

Amines, being basic compounds, are preferably separated using RP/IE-LC as shown by Lingeman et al. [96] and by NP-LC as shown by Kudoh et al. [97] for tertiary amines and alkaloid-type compounds. Detection of these compounds is sometimes difficult due to the lack of UV-Vis absorbance. Kudoh et al. used a fairly complicated method for postchromatographic derivatization of the compounds. Tertiary amines selectively develop a violet-red color with citric acid on acetic acid anhydride, a mixture that is mixed with the chromatographic eluent and heated to 393 K [102]. Absorbance is then measured at 550 nm. Separation is good but some amino compounds show strongly tailing peaks. The reproducibility in detection requires the maintenance of a temperature during the derivatization of 393 ± 0.2 K.

## CARBOHYDRATES AND RELATED COMPOUNDS

Hydroxyl-containing compounds are very suitable for transformation into esters, enabling detection improvement when proper labels are introduced. Most procedures have been developed for the analysis of carbohydrates. Generally, benzoyl chlorides with various spectroscopic properties are utilized [103-108], but phenyldimethylsilyl chloride [109] or phenylisocyanate [110] is also applied. Derivatization of digitalis glycosides and their corresponding aglycones has been described [106,107]. The procedures generally require rigorous standardization in order to label all functional groups. Invariably the labeling is performed prior to LC; postchromatographic techniques are not available. This is probably due to the bad chromatographic properties of the parent compounds. Labeling not only improves the detectability but also the LC performance.

### Benzoyl Derivatives

Benzoyl chlorides, like other acyl chlorides, react readily with alcohols to form the corresponding esters (Fig. 7.3). This reaction is used to label sugars and alcohols [103]. The anomers of the compounds can be chromatographed on a NP-LC column using diethyl ether-hexane in a gradient. Prior to derivatization the carbohydrate solution is kept at 335 K to establish equilibrium between the anomers. The procedure is rather time consuming; the total LC requires about 110 min. A modification is described by Nachtmann and Budna [104]. Several oligosaccharides are labeled with 4-nitrobenzoyl chloride in pyridine and subsequently subjected to NP-LC. Quantitation, using an internal standard not suited to form anomers (like sorbitol) is complicated, but an equation for the carbohydrate content has been derived. Using this equation quantitation seems satisfactory, with relative standard deviations varying from 0.6 to 4.5%. A modified procedure can be applied for the analysis of carbohydrates, aminosugars, and glycosides [105]. Benzoyl chloride is used as label and NP-LC is performed after extraction of the derivatives with dichloromethane. Separation between the various anomers is satisfactory and LC analysis times are relatively short (about 25 min). However, derivatization requires 19 hr. The limits for detection are 20 ng at 254 nm and about 2 ng at the wavelength of maximum absorbance (230 nm). These investigators investigated 4-nitrobenzoyl derivatization, but the derivatization was not complete.

Nachtmann et al. applied their previously described procedure [104] to the analysis of digitalis glycosides and their aglycones [106]. Isolation of the resulting derivatives requires precautions, such as low temperatures, to keep the glycosidic derivatives from degrading. After evaporation of the solvent (pyridine) at 323 K under reduced pressure, extraction with chloroform at pH 8 yields a purified derivative mixture suitable for NP-LC analysis. A large excess of reagent is necessary due to unfavorable equilibrium constants of the various reactions, not the rate constants. The aglycones do not react uniformly: mono- and diesters are formed. The $C_{14}$ hydroxyl group is not reactive due to steric hindrance. The same glycosides are labeled with 3,5-dinitrobenzoyl chloride [107]. The reaction is performed at room temperature in pyridine over 2 hr. RP-LC with mixtures containing acetonitrile, methanol, and water yield accurate results with low detection limits of about 0.5 ng. Retention times, however, are up to 55 min.

Ginseng saponins are derivatized after extraction from biological matrices using benzoyl, 4-nitrobenzoyl, 3,5-dinitrobenzoyl, or naphthoyl chloride as labels [108]. For quantitative purposes benzoylation in pyridine is most satisfactory; NP-LC on silica with *n*-hexane-dichloromethane-acetonitrile mixtures yields "fingerprint" chromatograms that are useful for the identification of ginsennoside mixtures from various sources and varieties.

### Phenyldimethylsilyl Derivatives

It can be concluded that the preparation of acyl derivatives of monosaccharides is difficult; therefore, phenyldimethylsilyl chloride has been introduced as a label [109]. After dissolving of the monosaccharides in dimethylformamide and addition of imidazole the solution is heated at 373 K for 1 hr. Subsequently, the label is added and the reaction takes place over 6 hr. NP-LC is applied after isolation of the derivatives from the reaction mixture with *n*-hexane. In spite of the author's opinion this method seems equally complicated.

The accuracy is low due to the low molar absorptivities of the derivatives (ε is 1000-3000).

### Phenylisocyanate Derivatives

Aliphatic alcohols can be derivatized with phenyl isocyanate prior to LC analysis. Suitable solvents for the derivatization reactions are dimethylformamide, acetonitrile, and dioxane. After removal of the excess reagent, RP-LC with detection at 230 nm is performed [110].

## MISCELLANEOUS COMPOUNDS

Labeling of several compounds of different natures, generally not belonging to one of the previously mentioned classes, is covered in the literature. Both pre- and postchromatographic techniques are applied. The prechromatographic procedures involve the introduction of hydrazino and hydroxylamino groups to the keto function [111-113], derivatization of the cytostatic compounds dianhydrogalactitol and copovithane with sodium diethyldithiocarbamate and TNBS [114,115], and labeling of cyclosporine A with α-naphthylselenyl chloride [116]. Postchromatographic reactions include disulfide analysis with a solid phase reagent [117], 6,6'-dithiodinicotinic acid (DTNA) reaction with thiols [118], cannabinoid assay with Fast Blue Salt B [119], and vitamin B determination with 2,6-dibromoquinone-4-chloroimide [120].

### Prechromatographic Derivatizations

Prostaglandins can be determined after extraction from the biological matrix [111]. Subsequently the carboxyl function is alkylated with diazomethane. The keto function of the methylated compound then is derivatized with *p*-nitrobenzylhydroxylamine in pyridine to form the corresponding *p*-nitrobenzyloxim compound (Fig. 7.10). The reproducibility of the reaction is limited. RP-LC with acetonitrile-water and detection at 254 nm allows the analysis of 25 ng. The advantage of the method is that high background absorbances, occurring after labeling of the carboxylic function can be avoided.

In cases where the carboxylic functions of bile acids are conjugated, like in taurine conjugates, the methods of Stellaard et al. [28] and Shaikh et al. [32] cannot be used. In these cases the 3-α and 7-α-hydroxyl groups of the conjugates can be subjected to enzymatic oxidation to form keto functions with subsequent coupling of the latter with 2,4-dinitrophenylhydrazine (Fig. 7.11) [112]. RP-LC with gradient elution is completed within 7 min. Detection limits are about 20 pmol at 340 nm. A disadvantage of the method

Figure 7.10 Derivatization of ketones with *p*-nitrobenzylhydroxylamine (PBNA).

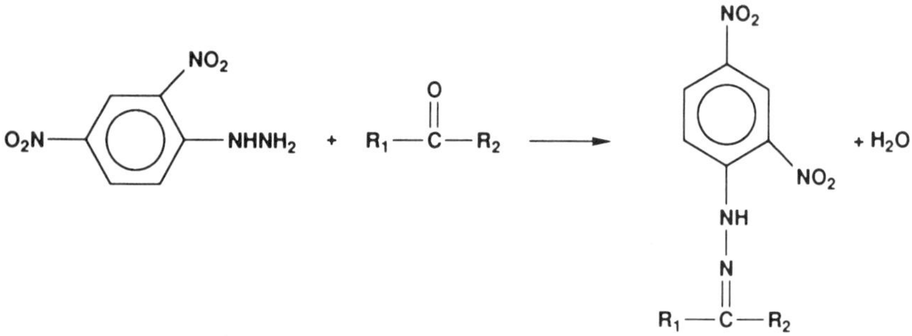

Figure 7.11 Derivatization of ketones with 2,4-dinitrophenylhydrazine (DNPH).

is the occurrence of artifact isomers, causing more than one peak per analyte in the chromatogram.

The reaction developed for keto functions [112], mentioned just before, is employed for the analysis of hydralazine in blood [113]. Using 4-methylhydralazine as the internal standard the analytes are labeled directly in whole blood with *p*-anisaldehyde. The reaction is completed within 10 min at room temperature. After isolation with *n*-hexane the derivatives are subjected to RP-LC on a cyanopropyl column. Since the derivatization can be performed in whole blood, the analysis of suitable solutes with a high clearance rate, like hydralazine, can be performed satisfactorily [121]. The sensitivity is about 1 ng/ml, while the major metabolite of hydralazine, its hydrazone with pyruvic acid, does not interfere.

For copovithane, a polymeric cytostatic drug, an assay is described in plasma and urine samples. After deproteination depolymerization of the analyte is established in the supernatant in 5 M hydrochloric acid at 433 K with subsequent labeling of the monomer with TNBS. After extraction with ethyl acetate RP-LC is applied. The recovery is low, about 40%, and the standard deviation is about 20% with a detection limit 15 ng/μl. An unknown compound originating from the urine samples coelutes with the derivative. All together the method seems to be rather unreliable.

Another cytostatic drug, dianhydrogalactitol, can be determined in plasma after derivatization with sodium diethyldithiocarbamate [115]. The derivatization can be performed directly in plasma samples of pH 7.0; the reaction time is 1 hr. After extraction with chloroform LC on a cyanopropyl column with heptane-chloroform-acetic acid is performed using a detection wavelength of 254 nm. The detection limit is 50 ng.

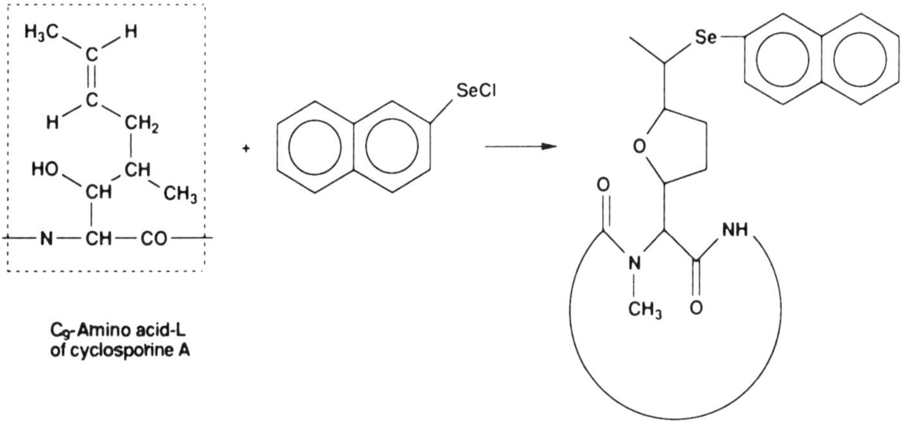


$C_9$-Amino acid-L
of cyclosporine A

Figure 7.12 Derivatization of amines with α-naphthylselenyl chloride.

Cyclosporine A assays generally suffer from insensitivity. The sensitivity increases significantly (3-4 times) after derivatization of the $C_9$-amino acid moiety with α-naphthylselenyl chloride (Fig. 7.12) [116]. The reaction takes place almost instantaneously in carbon tetrachloride. RP-LC analysis is performed and detection at 220 or 230 nm, respectively. The column temperature has to be kept at 346 K. Endogenous peaks from the matrix may disturb the chromatography but may become separable on modification of the conditions.

## Postchromatographic Derivatizations

Thiol analyses are complicated and generally lack sensitivity. After separation on a $C_{18}$ column a reaction with DTNA or 5,5'-dithio-2-nitrobenzoic acid (DTNB) can be performed [118]. The reduced dithio compounds are detectable at 344 or 412 nm, respectively. Chromatography is performed at pH 2.2, whereas the postchromatographic derivatization requires a pH over 6 (DTNA) or over 7.5 (DTNB). The selectivity of the latter reaction is higher due to the higher analytical wavelength but the alkaline medium is disadvantageous in view of the risk of oxidation of the thiols.

For the analysis of disulfides using the same reagents, a reduction to form thiols is necessary prior to the postchromatographic derivatization with DTNA or DTNB. This requires the postchromatographic introduction of a soluble reducing agent prior to the label. In general, procedures like this cause lower signals and peak disturbances. In an attempt to avoid these disturbances a solid phase reagent is applied for the reduction of the analytes [117]. A polymer-bound thiol reduces the disulfides, which after passing the thiopolymer column are directly mixed with DTNB. Peak shapes are satisfactory and the detection limit is better than 0.15 μg.

After a complicated extraction from plasma or brain tissue homogenates, cannabidiol and Δ-tetrahydrocannabinol can be separated completely using RP-LC [119]. Detection can be performed at 490 nm after reaction with Fast Blue Salt B to an unknown product, with a limit of detection of 50 ng. The internal standard, cannabinol, is not completely separated from Δ-tetrahydrocannabinol in this system. Complete isolation of both peaks requires gradient elution and, consequently, extended analysis time.

A postchromatographic derivatization method for vitamin B with 2,6-dibromoquinone-4-chloroimide is described [120]. After RP-IP chromatography, applying hexanesulfonate as the ion-pairing agent, the reagent is mixed with the eluate and subsequently ammonia is added. Pyridoxine, pyridoxal, pyridoxamine, and 4-pyridoxic acid all react causing intense absorbances at 650 nm.

## CONCLUSIONS AND FUTURE TRENDS

The majority of the references deal with derivatization of functional groups with only a few labels, as can be seen from Table 7.1. These reagents are:

Phenacyl and naphthacyl bromide for carboxylic acids
PITC and DABS-Cl for amino acids
Acyl chlorides, FDNB, and TNBS for amines
Acyl chlorides for carbohydrates

The other publications describe pre- and postchromatographic labeling with a wide variety of reagents under very difficult conditions.

Regarding the prerequisites for a manageable derivatization procedure (mild reaction conditions such as a one-step reaction at ambient temperature in a solvent suitable for subsequent LC analysis), the improvements in the labeling procedures for carboxylic acids [9,10] such as the solvent demixing techniques [17] as well as the modification of this technique described by Moody and Allan [19] turn these procedures into simple and reliable methods suitable for routine analysis. The same applies for the PITC labeling of amino acids. However, the LC analysis of the derivatives requires complex and time-consuming gradient elution techniques with high column temperatures, which is unfavorable for a routine method. Perhaps these problems can be reduced by automation of the procedures.

Acyl chlorides as labels for amino functions and carbohydrates have high molar absorptivities at suitable analytical wavelengths and the conditions for labeling are generally mild. These labels are worth considering, although they possess the disadvantage of a fairly high reactivity and the possibility of the occurrence of side products.

The reversed labeling procedures for carboxylic acids after conversion to the corresponding acid chlorides are not to be recommended. These reactions have to be performed with thionyl and oxalyl chloride, which are highly reactive and toxic compounds.

Although the reaction kinetics of the amino labeling with FDNB and TNBS are very complex and the specificity of the labels is limited [76], requiring thorough standardization, this procedure has the advantage of being easily combined with subsequent LC as no isolation of the derivatives is needed.

In bioanalysis sample pretreatment is normally an obligatory manipulation. Therefore, it will be advantageous to combine, whenever possible, clean-up and derivatization into one single step. The major advantage of this combination will be an increased reproducibility of the procedure. An example is the extraction-alkylation of organic acids in aqueous solutions [133]. The hydrophobic XAD-2 resin is impregnated with the alkylation reagent, a pentafluorobenzyl derivative, and the reaction takes place in the aqueous-solid mixture [133]. After the reaction has been completed the acid derivatives are washed off the sorbent with a suitable organic solvent, and after evaporation to dryness of the organic solvent the residue is dissolved in the eluent and an aliquot is analyzed using a suitable LC system.

As outlined in Chap. 1, pre- and postchromatographic derivatization have both pros and cons. An interesting development is the postchromatographic reaction of amines with ninhydrin or Peri, already present in the LC solvent. This system, developed by LePage and Rocha [87], combines the advantages of the pre- and postchromatographic labeling for a widespread and very important function. It seems worthwhile to optimize this method for amino labeling and to study the possibility of generalizing pre-/postchromatographic derivatization procedure.

The simultaneous analysis of analytes with different functional groups may be a necessity in some cases, e.g., profiling of amino acids in peptides and proteins. Nika described a postchromatographic system suitable for the detection of primary amino groups and sulfhydryl groups in peptides. After LC separation well-defined portions of the effluent are diverted, by means of an automatic splitting device, into one of the two reactors based on the ninhydrin and Ellman assay [dithioerythritol/sodium arsenite/5,5'-dithiobis(2-nitrobenzoic acid)], respectively [134].

## REFERENCES


1. Naish, P. J., *Analyst, 109*:809 (1984).
2. Parkin, J. E., and H. T. Lau, *J. Chromatogr., 314*:488 (1984).
3. Hulshoff, A., and A. D. Foerch, *J. Chromatogr., 220*:275 (1981).
4. Ahuja, S., *J. Chromatogr. Sci., 17*:168 (1979).
5. Frei, R. W., *Res. Dev., 28*:42 (1977).
6. Frei, R. W., *Methodol. Surv. Biochem., 7*:243 (1978).
7. Morozowich, W., and M. J. Cho, in *GLC and HPLC Determination of Therapeutic Agents*, Part I (T. Kiyoshi and W. Morozowich, eds.), Chromatogr. Sci. Series, Vol. 9, Marcel Dekker, New York, 1978.
8. Ross, M. S. F., *J. Chromatogr., 141*:107 (1977).
9. Durst, H. D., M. Milano, E. J. Kikta, Jr., S. A. Conelly, and E. Grushka, *Anal. Chem., 47*:1797 (1975).
10. Grushka, E., H. D. Durst, and E. J. Kikta, Jr., *J. Chromatogr., 112*:673 (1975).
11. Roorda, I., C. Gonnet, and J. L. Rocca, *Analyst, 18*:409 (1982).
12. Patience, R. L., and J. D. Thomas, *J. Chromatogr., 234*:225 (1982).
13. Desbene, P.-L., S. Coustal, and F. Frappier, *Anal. Biochem., 128*:359 (1983).
14. Lam, S., and E. Grushka, *J. Liq. Chromatogr., 1*:33 (1978).
15. Gupta, R. N., F. Eng, and M. L. Gupta, *Clin. Chem., 25*:1303 (1979).
16. Gupta, R. N., P. M. Keane, and M. L. Gupta, *Clin. Chem., 25*:1984 (1979).
17. Alric, R., M. Cociglio, J. P. Blayac, and R. Puech, *J. Chromatogr., 224*:289 (1981).
18. Kline, W. F., D. P. Enagonio, D. J. Reeder, and W. E. May, *J. Liq. Chromatogr., 5*:1697 (1982).
19. Moody, J. P., and S. M. Allan, *Clin. Chim. Acta, 127*:263 (1983).
20. Distler, W., *J. Chromatogr., 192*:240 (1980).
21. Borch, R. F., *Anal. Chem., 47*:2437 (1975).
22. Jordi, H. C., *J. Liq. Chromatogr., 1*:215 (1978).
23. Bussell, N. E., and R. A. Miller, *J. Liq. Chromatogr., 2*:697 (1979).
24. Miller, R. A., N. E. Bussell, and C. Ricketts, *J. Chromatogr., 1*:291 (1978).
25. Zoutendam, P. H., P. B. Bowman, T. M. Ryan, and J. L. Rumph, *J. Chromatogr., 283*:273 (1984).
26. Zoutendam, P. H., P. B. Bowman, J. L. Rumph, and T. M. Ryan, *J. Chromatogr., 283*:281 (1974).

27. Watkins, W. D., and M. B. Peterson, *Anal. Biochem.*, *125*:30 (1982).
28. Stellaard, F., D. L. Hachey, and P. D. Klein, *Anal. Biochem.*, *87*:359 (1978).
29. Nagels, L., C. Debeuf, and E. Esmans, *J. Chromatogr.*, *190*:411 (1980).
30. Clark, J. H., and J. M. Miller, *Tetrahedron Lett.*, 599 (1977).
31. Hulshoff, A., H. Roseboom, and J. Renema, *J. Chromatogr.*, *186*:535 (1979).
32. Shaikh, B., N. J. Pontzer, J. E. Molina, and M. I. Kelsey, *Anal. Biochem.*, *85*:47 (1978).
33. Lindner, W., and W. Santi, *J. Chromatogr.*, *176*:55 (1979).
34. Hoffman, N. E., and J. C. Liao, *Anal. Chem.*, *48*:1104 (1976).
35. Bissinger, J. M., K. T. Rullo, J. T. Stoklosa, C. M. Shearer, and N. J. De Angelis, *J. Chromatogr.*, *268*:102 (1983).
36. Ikeda, M., K. Shimada, and T. Sakaguchi, *Chem. Pharm. Bull.*, *30*:2258 (1982).
37. Ikeda, M., K. Shimada, and T. Sakaguchi, *J. Chromatogr.*, *272*:251 (1983).
38. Munson, J. W., *J. Pharm. Sci.*, *63*:252 (1974).
39. Munson, J. W., and R. Bilous, *J. Pharm. Sci.*, *66*:1403 (1977).
40. Momose, T., *Chem. Pharm. Bull.*, *26*:1627 (1980).
41. Miwa, H., M. Vamamoto, and T. Momose, *Chem. Pharm. Bull.*, *28*:599 (1980).
42. Westerlund, D., J. Carlqvist, and A. Theodorsen, *Acta Pharm. Suec.*, *16*:187 (1979).
43. Carlqvist, J., and D. Westerlund, *J. Chromatogr.*, *164*:373 (1979).
44. Pollitzer, I. R., G. W. Griffin, B. J. Dowty, and L. J. Laseter, *Anal. Lett.*, *6*:539 (1973).
45. Horikawa, R., and T. Tanimura, *Anal. Lett.*, *15*:1629 (1982).
46. Görög, S., A. Lauko, M. Renyei, and B. Hagedus, unpublished work.
47. Heinrikson, R. L., and S. C. Meredith, *Anal. Biochem.*, *136*: 65 (1984).
48. Biggs, H. G., and L. J. Gentilcore, *Clin. Chem.*, *30*:851 (1984).
49. Pucci, P., G. Sannia, and G. J. Marino, *J. Chromatogr.*, *270*: 371 (1983).
50. Brown, A. S., J. E. Mole, A. Weissenger, and J. C. Bennet, *J. Chromatogr.*, *148*:532 (1978).
51. Abrahamsson, M., K. Groningsson, and S. Castensson, *J. Chromatogr.*, *154*:313 (1978).
52. Moser, P. W., and E. E. Rickli, *J. Chromatogr.*, *176*:451 (1979).
53. Johnsson, N. D., M. W. Hunkapillar, and L. E. Hood, *Anal. Biochem.*, *100*:335 (1979).

54. Tarr, G. E., *Anal. Biochem.*, *111*:27 (1981).
55. Hawke, D., Y. Pau-Miau, and J. E. Shively, *Anal. Biochem.*, *120*:302(1982).
56. Zimmerman, C. L., E. Appella, and J. J. Pisano, *Anal. Biochem.*, *77*:569 (1977).
57. Lammens, J., and M. Verzele, *Chromatographia*, *11*:376 (1978).
58. Lin, J.-K., and C.-H. Wang, *Clin. Chem.*, *26*:579 (1980).
59. Chang, J.-Y., P. Martin, R. Bernasconi, and D. G. Braun, *FEBS Lett.*, *132*:117 (1981).
60. Chang, J.-Y., R. Knecht, and D. G. Braun, *Methods Enzymol.*, *91*:41 (1983).
61. Hoppel, C. L., D. E. Weir, A. P. Gibbons, S. T. Ingalls, A. T. Brittain, and F. M. Brown, *J. Chromatogr.*, *272*:43 (1983).
62. Caudill, W. L., and R. M. Wightman, *Anal. Chim. Acta*, *141*:269 (1978).
63. Wilkinson, J. M., *J. Chromatogr. Sci.*, *16*:547 (1978).
64. Furukawa, H., Y. Mori, Y. Takeuchi, and K. Ito, *J. Chromatogr.*, *136*:428 (1977).
65. Goto, J., M. Hasegawa, S. Nakamura, K. Shimada, and T. Nambara, *J. Chromatogr.*, *152*:413 (1978).
66. Nambara, T., S. Ikegawa, M. Hasegawa, and J. Goto, *Anal. Chim. Acta*, *101*:111 (1978).
67. Nachtman, F., *Int. J. Pharm.*, *4*:337 (1980).
68. Radjai, M. K., and R. T. Hatch, *J. Chromatogr.*, *196*:319 (1980).
69. Lawrence, J. F., and R. W. Frei, *Chemical Derivatization in Liquid Chromatography*, Elsevier, Amsterdam, 1976, p. 157.
70. Lou, M. F., and M. Siena, *Biochem. Med.*, *25*:309 (1981).
71. Hamilton, P. B., in *Handbook of Biochemistry* (H. A. Sober, ed.), CRC Press, Cleveland, 1968, B43.
72. LePage, J. N., and E. M. Rocha, *Anal. Chem.*, *55*:1360 (1980).
73. Lawrence, J. F., and R. W. Frei, *Chemical Derivatization in Liquid Chromatography*, Elsevier, Amsterdam, 1976, p. 113.
74. Gray, W. R., in *Methods in Enzymology*, Vol. 25 (C. H. W. Hirs and S. N. Timasheff, eds.), Academic Press, New York, 1972, p. 121.
75. Tsuji, K., J. F. Goetz, W. VanMeter, and K. A. Gusciora, *J. Chromatogr.*, *175*:141 (1979).
76. Barends, D. M., J. S. Blauw, C. W. Mijnsbergen, C. J. L. R. Govers, and A. Hulshoff, *J. Chromatogr.*, *322*:321 (1985).
77. Helboe, P., and S. Kryger, *J. Chromatogr.*, *235*:215 (1982).
78. Elrod, L. Jr., L. B. White, and C. W. Wong, *J. Chromatogr.*, *208*:357 (1981).
79. Wong, L. T., A. R. Beaubien, and A. P. Pakuts, *J. Chromatogr.*, *231*:145 (1982).

80. Barends, D. M., J. S. Blauw, H. M. Smits, and A. Hulshoff, *J. Chromatogr.*, *276*:385 (1983).
81. Barends, D. M., C. L. Zwaan, and A. Hulshoff, *J. Chromatogr.*, *182*:201 (1980).
82. Barends, D. M., C. L. Zwaan, and A. Hulshoff, *J. Chromatogr.*, *222*:316 (1981).
83. Barends, D. M., C. L. Zwaan, and A. Hulshoff, *J. Chromatogr.*, *225*:417 (1981).
84. Spragg, B. P., and A. D. Hutchings, *J. Chromatogr.*, *258*:289 (1983).
85. Kabra, P. M., P. K. Bhatnagar, M. A. Nelson, J. H. Wall, and L. J. Marton, *Clin. Chem.*, *29*:672 (1983).
86. Kabra, P. M., P. K. Bhatnagar, and M. A. Nelson, *J. Chromatogr.*, *307*:224 (1984).
87. Ross, M. S. F., *J. Chromatogr.*, *133*:408 (1977).
88. Wellons, S. L., and M. A. Carey, *J. Chromatogr.*, *154*:219 (1978).
89. Clark, C. R., and M. M. Wells, *J. Chromatogr. Sci.*, *16*:332 (1978).
90. Larsen, N. E., K. Marinelli, and A. M. Heilesen, *J. Chromatogr.*, *221*:182 (1980).
91. Edwards, D. J., in *Handbook of Derivatives for Chromatography* (K. Blau and G. S. King, eds.), Heyden, London, 1978, p. 394.
92. Iskierko, J., E. Soczewinsky, and B. Kanadys-Sobieraj, *Chem. Anal. (Warsaw)*, *25*:955 (1980).
93. Chen, Y.-Z., *Mikrochim. Acta*, *1*:343 (1980).
94. Bjorkqvist, B., *J. Chromatogr.*, *204*:109 (1981).
95. Mitra, A. K., C. L. Baustian, and T. J. Mikkelson, *J. Pharm. Sci.*, *69*:257 (1980).
96. Lingeman, H., H. A. van Munster, J. H. Beijnen, W. J. M. Underberg, and A. Hulshoff, *J. Chromatogr.*, *352*:261 (1986).
97. Kudoh, M., I. Matoh, and S. Fudano, *J. Chromatogr.*, *261*:293 (1983).
98. Seiler, N., and L. Demisch, in *Handbook of Derivatives for Chromatography* (K. Blau and G. S. King, eds.), Heyden, London, 1978, p. 370.
99. Lawrence, J. F., in *Chemical Derivatization in Analytical Chemistry*, Vol. 2, *Separation and Continuous Flow Techniques*, (R. W. Frei and J. F. Lawrence, eds.), Plenum Press, London, 1982, p. 220.
100. Tocksteinova, D., J. Churacek, J. Slosar, and L. Skalik, *Mikrochim. Acta*, *1*:507 (1978).
101. Bauer, M., L. Mailhe, and L. Nguyen, *J. Chromatogr.*, *292*:468 (1984).
102. Ohkuma, S., *J. Pharm. Soc. (Japan)*, *75*:1124 (1955).

103. Lehrfeld, J., *J. Chromatogr.*, *120*:141 (1976).
104. Nachtmann, F., and K. W. Budna, *J. Chromatogr.*, *136*:278 (1977).
105. White, C. A., J. F. Kennedy, and B. T. Golding, *Carbohydr. Res.*, *76*:1 (1979).
106. Nachtmann, F., H. Spitzy, and R. W. Frei, *Anal. Chem.*, *48*:1576 (1976).
107. Fujii, Y., R. Oguri, A. Mitsuhashi, and M. Yamazaki, *J. Chromatogr. Sci.*, *21*:495 (1983).
108. Besso, H., Y. Saruwatari, K. Futamura, T. Kunihiro, T. Fuwa, and O. Tanaka, *Planta Med.*, *37*:226 (1979).
109. White, C. A., S. W. Vass, J. F. Kennedy, and D. G. Large, *J. Chromatogr.*, *264*:99 (1983).
110. Bjorkqvist, B., and H. Toivonen, *J. Chromatogr.*, *153*:265 (1978).
111. Fitzpatrick, F. A., M. A. Wynalda, and D. G. Kaiser, *Anal. Chem.*, *49*:1032 (1977).
112. Reid, A. D., and P. R. Baker, *J. Chromatogr.*, *260*:115 (1983).
113. Ludden, T. M., L. K. Ludden, K. Wade, and S. R. B. Allerheiligen, *J. Pharm. Sci.*, *72*:693 (1983).
114. Rosenblum, M. G., G. N. Hortobagyi, W. Wingender, and E. M. Hersh, *J. Liq. Chromatogr.*, *7*:159 (1984).
115. Munger, D., L. A. Sternson, A. J. Repta, and T. Higuchi, *J. Chromatogr.*, *143*:375 (1977).
116. Gfeller, J. C., A. K. Beck, and D. Seebach, *Helv. Chim. Acta*, *63*:728 (1978).
117. Studebaker, J. F., S. A. Slocum, and E. L. Lewis, *Anal. Chem.*, *50*:1500 (1978).
118. Nishiyama, J., and T. Kuninori, *Anal. Biochem.*, *138*:95 (1984).
119. Borys, H. K., and R. Karler, *J. Chromatogr.*, *205*:303 (1981).
120. Kawamoto, T., E. Okada, and T. Fujita, *J. Chromatogr.*, *267*:414 (1983).
121. Clarke, E. G. C., *Isolation and Identification of Drugs*, The Pharmaceutical Press, London, 1978, p. 367.
122. Roder, E., and J. Merzhauser, *Fres. Z. Anal. Chem.*, *272*:34 (1974).
123. Edwards, D. J., and K. Blau, *Anal. Biochem.*, *45*:387 (1972).
124. Selim, S., *J. Chromatogr.*, *136*:271 (1977).
125. Helmcher, G., H. Voelter, and W. Schuhle, *Tetrahed. Lett.*, 1417 (1977).
126. Mathees, D. P., and W. C. Purdy, *Anal. Chim. Acta*, *109*:161 (1979).
127. Kuwata, K., M. Uebori, and Y. Yamazaki, *Anal. Chem.*, *52*:857 (1980).
128. Sango, S., *J. Liq. Chromatogr.*, *2*:763 (1979).

129. Fitzpatrick, F. A., M. A. Wynalda, and D. G. Kaiser, *Anal. Chem.*, *49*:1032 (1977).
130. Roder, E., I. Pigulla, and J. Troschuetz, *Fres. Z. Anal. Chem.*, *288*:56 (1977).
131. Abou, H. M., T. Medwick, and L. C. Bailey, *Anal. Chim. Acta*, *93*:221 (1977).
132. Sugiura, T., T. Hayashi, S. Kawai, and T. Ohno, *J. Chromatogr.*, *110*:385 (1975).
133. Rosenfeld, J. M., M. Mureika-Russel, and A. Phatak, *J. Chromatogr.*, *283*:127 (1984).
134. Nika, H., *J. Chromatogr.*, *297*:261 (1984).

# 8
# Electrochemical Derivatization

J. A. LISMAN and WILLY J. M. UNDERBERG *State University of Utrecht, Utrecht, The Netherlands*

HENK LINGEMAN* *Leiden University, Leiden, The Netherlands*

## INTRODUCTION

In this chapter an introduction will be presented on the principles of electrochemical detection. In particular, amperometric detection (AD) is discussed. The emphasis will be on the advantages of AD over some of the other widely applied detection techniques (e.g., absorbance, fluorescence), some of its inherent disadvantages, and the prerequisites for combining liquid chromatography (LC) with AD [1,2].

Since 1974, when the first commercially available LC electrochemical detector was introduced, an overwhelming number of articles on design, performance, theory, and application has appeared in the scientific literature and today LC-AD is widely accepted as a sensitive and selective technique.

Because nowadays a considerable number of amperometric detection devices are commercially available, all possessing specific advantages and disadvantages, this chapter presents randomly chosen detection devices as examples, along with some general remarks. A certain knowledge of the principles of electrochemistry is therefore of vital importance for the chromatographer who wants to achieve an optimum LC-AD performance. Therefore, some of the basic theoretical and practical information necessary for proper LC-AD system operation and understanding is discussed in the following subsections.

The main part of this chapter, however, will deal with the possibilities of converting nonelectroactive analytes into electroactive

---

**Present affiliation*: Free University, Amsterdam, The Netherlands

compounds by means of prechromatographic or postchromatographic derivatization procedures [3,4].

Furthermore, an introduction and some applications will be given for dual-electrode systems allowing the electrochemical conversion of the analyte to a product with favorable properties for AD.

In the last part of this chapter some indirect chemical derivatizations will be discussed. In these cases the analyte is not detected but rather some other component involved in the reaction. The decrease in the concentration of that particular solute is a measure for the concentration of the analyte. Furthermore, some biochemical procedures will be discussed.

## Electrochemical Detection in Liquid Chromatography

In addition to the absorbance, fluorescence, and chemiluminescence detection in LC, electrochemical detection—in particular AD—is gaining popularity. The combination of LC and AD allows the selective and sensitive detection of a considerable number of compounds in complex matrices. AD is selective because only compounds which possess an electrophore and which are active under the chosen conditions can be detected. In general, an advanced chromatographic system should be used in combination with AD to elevate the signal over the noise and the background [5].

The linear dynamic range and the accuracy of amperometric detectors is at least comparable to that of the other LC detectors. Detection limits in the low nanogram range can be obtained. An advantage is the possibility of manipulating selectivity by changing the potential of the working electrode (WE). The potential can be chosen in such a way that only the compound under investigation is detected [6]. Sensitive analytical procedures using LC-AD in complex matrices are described for catecholamines and the corresponding metabolites [7], the anticancer agents teniposide and etoposide [8], platinum-containing cytostatic agents [9], a number of phenolic compounds [10] such as Δ-9-tetrahydrocannabinol [11], amines [12], benzodiazepines [13], nitrophenols [14], and sulfur-containing compounds such as sulfonamides [15].

A disadvantage of AD, and of the other sensitive and selective detection principles, is the limited applicability of the technique. However, this disadvantage can be overcome by using a pre- or postchromatographic derivatization procedure [16].

A more pronounced disadvantage of AD is the fact that the amperometric signal is relatively sensitive to interferences, e.g., changes in the surface of the electrodes [16], which will influence the relation between the current and the working potential, or the influence of oxygen [17] when working in the reductive mode. Since AD is based on an electrochemical reaction of the analyte, it is more sensitive to changes in the LC conditions than the more stable processes of light absorption, emission, or refraction. The result is that for the successful application of AD some knowledge about the basic principles of amperometry is essential.

## PRINCIPLES OF ELECTROCHEMICAL DETECTION

### Detection Device

Normally amperometry is the applied technique in electroanalytical flow-through detection devices. In these devices a potentiostatic method, meaning that a potential is applied and the resulting current measured, is used. A precondition is that the applied potential be kept constant. For those electroanalytical techniques using a dropping mercury electrode the term polarography is in use, which may cause confusion.

Electrochemical detection of an analyte in the effluent of a LC column is based on electrochemical conversion (electrolysis) of the analyte at a measuring WE placed in the effluent stream (Fig. 8.1).

Since electrons cannot move freely in solution, only those analyte molecules staying close to the electrode surface can react, and the reaction yields are normally 3 to 5% [5]. However, the reaction yield strongly depends on the

- Surface of the electrodes
- Efficiency of the mass transport (geometry of the cell, flow rate, diffusion coefficients) [18]
- Number of involved electrons
- Value of the working potential in relation to the half-wave potential

An increase of the reaction yield can be established by increasing the electrode surface or by increasing the working potential. However, in general this will not result in an improved detection limit because the noise will also increase [5]. Furthermore, the volume of the detection cell cannot be increased unlimitedly because of band-broadening effects, which will destroy the chromatographic resolution [5]. In modern LC the analytes will be eluted in volumes less than 100 μl, meaning that flow cell volumes of 0.1 to 1.0 μl are required.

Applying wall-jet detection devices, where the flow is not parallel to the electrode surface, the cell volume is not determined by the cell geometry but by the combination of the thickness of the diffusion layer and the electrode surface [8].

Reaction yields of 100% can be achieved and in this case the term amperometry is replaced by coulometry. The advantage of coulometric detection over amperometric detection is that, because

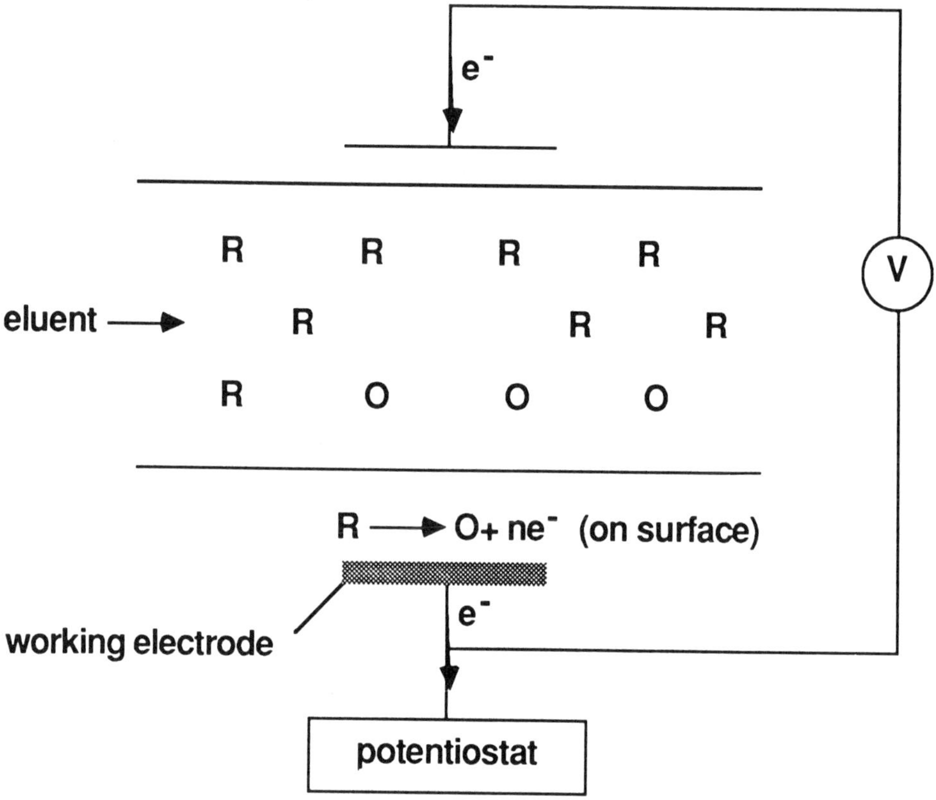


Figure 8.1 Amperometric detection principle.

it is an absolute measurement, no internal standard or calibration is necessary. However, as stated before, in general the application of a coulometric detection device will not be advantageous, except for the use of this detection device in a dual-electrode system. In these systems the detection selectivity is even further increased by using the combination of an oxidation and a reduction reaction [19].

The detection selectivity can also be increased by the application of fast-scanning potentiostatic techniques [20], e.g., for the separation of nitrobenzene compounds, possessing different electrochemical properties, which can be established by detecting them at various potentials.

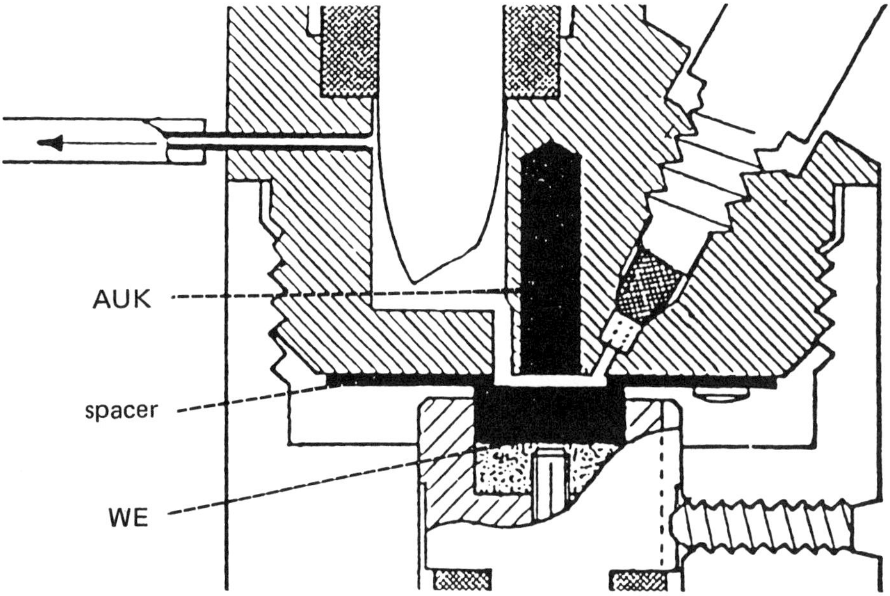


Figure 8.2 Three electrode detection system. REF, reference electrode; AUX, auxiliary electrode; WE, working electrode. A, AMOR; B, VT-03. [Reprinted with permission from Spark Holland BV, Emmen, The Netherlands (A) and ANTEC Instruments, Leiden, The Netherlands (B).]

*(Three)-Electrode System*

The WE current (i.e., detector response) is strongly influenced by voltage changes between the electrode and the mobile phase. A constant and known potential difference between WE and the mobile phase is therefore a vital requirement to obtain a stable, reproducible, and predictable detector response.

For this reason, electrochemical measuring cells are usually based on a three-electrode arrangement (Fig. 8.2). The WE is kept at "zero" potential by connecting it at the virtual ground of the electronics. A counterelectrode (auxiliary electrode, AUX) is used to apply the potential difference between the mobile phase and the WE which is necessary for electrolysis of the analyte.

A stable and predictable potential difference between mobile phase and WE is, however, impossible with a two-electrode system because of polarization effects in the electroactive/eluent boundary layer [21]. Therefore, a third electrode is needed to monitor the

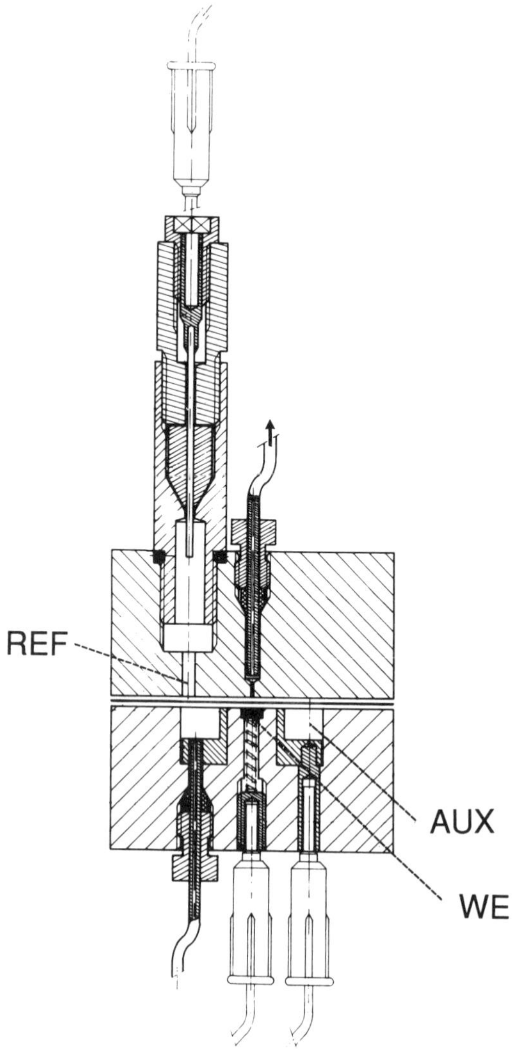


Figure 8.2 (continued)

potential of the mobile phase. This third electrode is called the reference electrode (REF). Its potential versus mobile phase is constant and known, and is invariant with respect to mobile phase composition and current between AUX and WE.

The potential selected for WE versus REF (i.e., versus mobile phase) is applied and controlled by AUX; if any difference is measured between the potential setting and the actual potential between WE and REF, AUX adapts its potential to eliminate the difference.

*Electrode Materials*

The normally applied electrode material for reactions in the reductive mode is mercury or mercury amalgam. The advantages of mercury are the large potential range over which the electrode can be used (-2.7 V versus saturated calomel to about 0.3-0.5 V), the possibility of renewing the electrode surface during the analysis (dropping mercury electrode), and the fact the application of mercury electrodes is well described in the literature [2,10]. The disadvantages of the dropping mercury electrode are that combination with a small-volume flow cell is relatively difficult and that for oxidation processes mercury is only applicable for easily oxidized compounds [5].

In this case electrodes of gold, platinum, or carbon (e.g., glassy carbon, carbon paste) are used. These materials can be used in a potential window of -1.0 V versus silver/silver chloride electrode to 1.3 V [22]. The main disadvantages of these materials is pollution of the surface resulting in a decreased sensitivity. Therefore, normally glassy carbon is applied because this material is less sensitive to these pollution problems [16]. An advantage of glassy carbon over carbon paste is the fact that glassy carbon is resistant to relatively high concentrations of organic solvents [8].

In summary, it can be stated that in case of oxidative detection, carbon is the best choice for the WE in nearly all circumstances. Therefore glassy carbon (an amorphous glasslike carbon) is the standard WE material in most cells for LC-AD nowadays.

Many different sorts of glassy carbon are commercially available with great differences in performance. Moreover, the performance of glassy carbon as electrode material also depends strongly on its "electrochemical history" and may be changed in some cases by electrochemical pretreatment [23]. Experiments with different kinds of glassy carbon may result in a considerable improvement of detector performance for a certain analyte. Gold/mercury amalgam needs a relatively high "potential" for reduction of oxygen and is therefore frequently used in reductive LC-AD.

The AUX electrode must be electrochemically and chemically inert and therefore carbon or stainless steel are the electrode materials chosen in the majority of the detection devices.

The ideal REF electrode has a stable and reproducible potential versus the eluent, invariant with mobile phase changes. The silver/silver chloride (Ag/AgCl) electrode is used in the majority of the published electrochemical flow cell configurations.

The geometry of the detection cell is also important for the degree of pollution during the analysis. For instance, by choosing a wall-jet-type cell (see below), the surface will be cleaned continuously [12]. Other methods to clean the electrode surface are described, such as thermal [24] and electrochemical [25] treatment of the surface, but polishing of the cell is still a reliable and simple way of cleaning the surface.

### *Cell Design Considerations*

Important aspects in amperometric cell design are the volume and the geometry of the applied flow cell. The geometry should allow a high sensitivity. In absorbance and fluorescence detection devices this is achieved by using flow cells with a low cell volume and a large optical pathlength. However, applying AD it is not so easy to predict the proper geometry of the cell, in order to obtain the most suitable signal-to-noise ratio, because this strongly depends on the applied cell type.

One of the most important parameters in amperometric cell design is the efficiency of mass transport. This efficiency strongly depends on the thickness of the diffusion layer, which is influenced by the cell geometry [26].

In addition, cell geometry also influences the linearity of the AD signal because of the occurring potential drop by the resistance of the eluent between the WE and the AUX electrode. In order to minimize this potential drop, the eluent should be a well-conducting solvent and the electrode configuration should be optimized (three-electrode configuration) [8].

A large number of alternative cell designs have been described in the literature [20]. In essence, however, three major design principles may be distilled out of all these different (often home-made) cells:

1. Thin-layer design
2. Wall-jet design
3. Porous flow-through design

The relevant characteristics of these designs will be described below. The construction of the flow cell strongly depends on the chosen electrode material. The shape of the mercury electrodes is determined by the surface of the mercury drop. The flow direction of the mercury, using a dropping mercury electrode, can be horizontal or vertical. The flow of liquid containing the analyte is usually perpendicular to the flow of mercury and in this way the transport efficiency of the analyte is as high as possible. However, the falling mercury drops will cause current oscillations, resulting in a relatively high background signal. The application of static mercury electrodes can solve this problem, but they have the disadvantage that the electrode surface is infrequently renewed [27].

The application of static electrode surfaces allows the application of all kinds of materials (e.g., mercury, carbon, gold, platinum) and different geometries (e.g., disc, bar, grate) are possible. In LC a fast response and a small detection volume is of major importance. Therefore, in particular thin-layer detection devices are popular. Two different types have been described. In the first type the liquid flow is perpendicular to the electrode surface (wall-jet cell) and in the second type the liquid flow is parallel to the electrode which is incorporated into the detection block (thin-layer cell). In addition to these two types, a few combination devices were recently described [21] as well as a porous flow-through cell which can be applied for coulometric detection. A last type of detection cell is the rotating electrode in which the flow of liquid is directed perpendicular to the electrode surface. The efficiency of mass transport (thickness of the diffusion layer) depends on the rotation speed and can be optimized in order to obtain the most suitable signal-to-noise ratio [28].

Diffusion and convection are the major processes responsible for analyte transport toward the WE.

The thin-layer cell design is based on reduction of the diffusion pathlength; the mobile phase is directed along the WE surface as a thin film of liquid. The wall-jet design is based on flow rate acceleration perpendicular to the WE to create strong convective motion of the analyte toward the WE. The porous electrodes may be regarded as a combination of thin-layer and wall-jet design: short diffusion pathlength and strong convective motion due to high liquid velocities in narrow pores.

Important features for choosing the proper cell design are:

WE surface area
Contribution to peak broadening
Linear dynamic range

With respect to the WE surface area it can be stated that the signal-to-noise ratio decreases at an increasing rate as a function of the WE surface and that the optimum electrolytic efficiency is 5-25%. This means that thin-layer and wall-jet cells usually have electrolytic efficiencies in the optimum range while porous cells have efficiencies in the order of 100% which normally result in an unfavorable signal-to-noise ratio.

Since the background (noise) decreases proportionally to the WE surface, miniaturization of the electrochemical cell does not affect the signal-to-noise ratio. Therefore, the standard cell volume can be chosen in such a way that microbore LC can be applied.

The linear dynamic range often extends the dynamic range of the operational amplifiers which are used to amplify the signal. However, in order to maintain a large linear dynamic range in poorly conducting mobile phases, the presence of the three electrodes is of great importance.

### *Selection of the Detection Potential*

In order to determine the optimum WE potential for a certain analysis, the relation between the detector response (electrolysis current) and the WE potential (voltage) should be known. Such a relationship (CV curve) provides practically all information necessary to determine the optimum detection potential for a certain analyte under given chromatographic conditions.

It seems evident that detection potential is chosen beyond $E_{opt}$ (Fig. 8.3) because of maximum response stability (independent of minor potential fluctuations due to change of temperature, reference potential, or WE surface). However, the influence of WE potential on "background current" (electrolysis current of mobile phase constituents), i.e., on baseline offset, stability, and noise, should also be taken into account.

Electrolysis of the mobile phase constituents will cause a continuous detector response (background current, comparable to "dark current" of an absorbance detector) resulting in a chromatographic baseline level that differs from the electrical detector zero response level. The difference, "baseline offset," is an important analysis parameter because baseline fluctuations (noise, drift) due to fluctuations in electrolysis conditions (potential, mobile phase flow rate, temperature) are proportional to baseline offset. Therefore, the background current should be kept as small as possible.

Besides using pure and nonelectroactive solvents and chemicals to reduce background current, the relationship between WE potential and background current should also be established to select the optimum WE potential for analysis of the analyte.

However, the background current increases exponentially with higher voltages due to oxidation of a major mobile phase constituent. Since a major mobile phase constituent is always present at the WE surface in very high "concentrations," a considerable electrolysis current is already obtained at potentials much lower than the half-wave potential for electrolysis of the specific constituent and the limiting current will be excessively high. The result is that, in order to obtain the best signal-to-noise ratio, an optimum in the applied WE potential should be found.

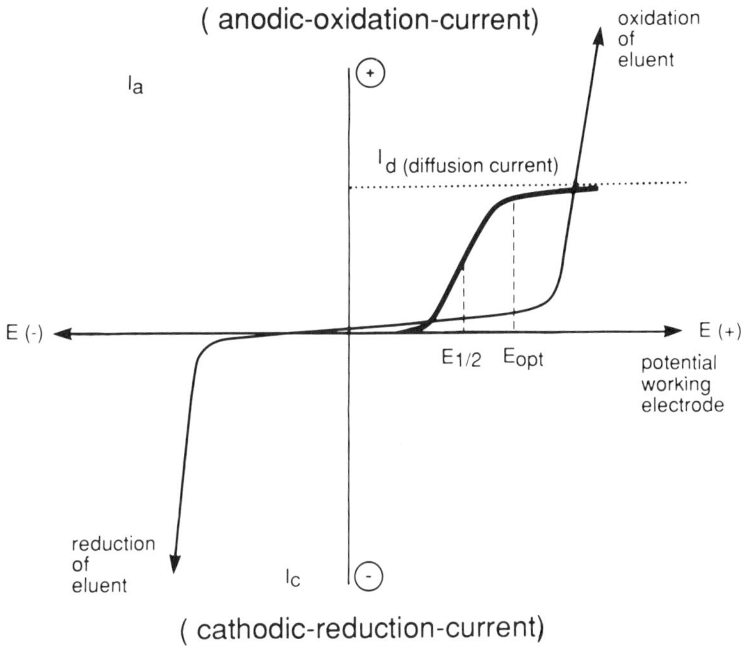


**Figure 8.3** Current-voltage relationship. $E_{opt}$, optimum potential; $E_{1/2}$, half-wave potential.

In addition to sensitivity, selectivity is another important feature of AD. Selective detection is based on the fact that different electroactive substances have different $E_{1/2}$ values and, consequently, have maximum detector response at different potentials. This means that any potential window is a compromise between sensitivity and selectivity. Since the number of electroactive substances that can be electrolyzed increases with increasing WE potential (either oxidative or reductive), the optimal selectivity is obtained at minimum potential (Fig. 8.3).

The final choice for the detection potential usually is a compromise between sensitivity, stability of response, and selectivity based on the fact that an increase of the potential will result in

- An increasing response of the analyte
- An increasing background current and, consequently, decreasing baseline stability
- An increasing response stability of the analyte (response plateau in CV curve)
- A decreasing selectivity

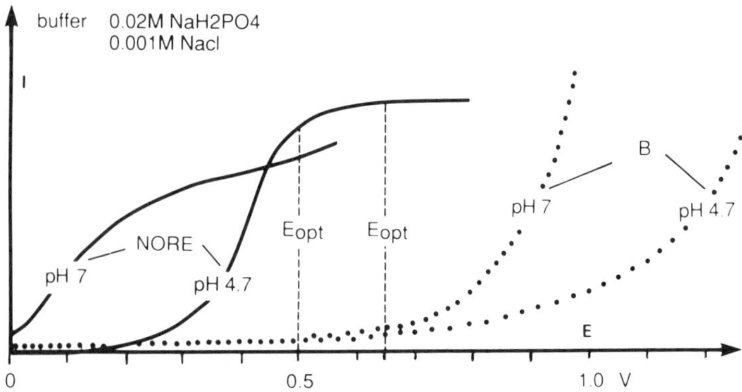


Figure 8.4 Influence of pH on current-voltage relationship for norepinephrine (NORE) in a phosphate buffer. (Reprinted with permission from Electrochemical Detection in HPLC, Spark Holland BV, Emmen, The Netherlands, 1987 [101].)

## Chromatographic Conditions

The mobile phase composition may strongly influence the value of the optimum detection potential as shown in Fig. 8.4. This means that optimum mobile phase composition, with respect to separation, may be incompatible with AD. For instance, pH variations may strongly influence the background current as well as the analyte electrolysis current. Therefore, the mobile phase should preferably be buffered.

For reversed phase (RP) chromatography, electrolysis of water usually limits the WE potential region at the oxidation side. The practical WE potential region in a 0.02 M phosphate buffer using a glassy carbon electrode related to the pH are listed in Table 8.1.

Obviously, the optimum potential for detection of analyte X in this mobile phase is a compromise: higher potentials will increase the peak height, but will also increase the baseline offset and thus baseline noise and drift. Lower potentials will decrease the peak height, but also decrease noise and drift. The maximum peak/noise ratio will probably be near $E_{opt}$ in Fig. 8.4.

### *Mobile Phase Components*

The most important prerequisite for LC eluents in combination with AD is conductivity, meaning that an electrolyte should always be present in eluent. This explains why RP chromatography using aqueous mobile phases is more popular than normal phase chromatography using organic solvents [29], although postchromatographic

Table 8.1 Practical Limits for Detection Potential (vs. Ag/AgCl) in Aqueous Phosphate Buffers at Various pH Values Using a Glassy Carbon Electrode

| pH | $E_{ox}$ (limit) | $E_{red}$ (limit) |
|---|---|---|
| 2 | +1.3 | |
| 4 | +1.2 | |
| 7 | +1.1 | -1.1 |
| 9 | +0.9 | -1.2 |
| 11 | | -1.3 |

addition of an electrolyte may be a solution. The lower and upper limits of the usually applied solvents in RP chromatography are listed in Table 8.2. The potential ranges of a number of electrolytes are given in Table 8.3.

It will be clear that the data in Table 8.2 and 8.3 are given only as guidelines because the exact values depend on the purity of the compounds, on the electrode material, and on the required sensitivity. Therefore, all the chemicals used must be electrochemically inert and free of electroactive contaminants. No chemicals dedicated to LC are available. Therefore, selection of mobile phase components with the highest available purity is advisable. A high background current at moderate potentials usually indicates the presence of an electroactive component in the mobile phase.

The prerequisites for some of the applied eluent components are:

Table 8.2 Potential Range of Solvents

| Solvent | Potential limits vs. saturated calomel electrode | | Ref. |
|---|---|---|---|
| Water | -2.7 | +1.5 | 80 |
| Methanol | -2.2 | +1.8 | 81 |
| Acetonitrile | -3.5 | +2.4 | 82 |
| Dimethylformamide | -3.5 | +1.5 | 83 |
| Tetrahydrofuran | -3.6 | +1.8 | 84 |
| Dichloromethane | -1.7 | +1.8 | 85 |
| Acetic acid | -1.7 | +2.0 | 86,87 |

Table 8.3 Potential Range in Dimethylsulfoxide in Presence of Electrolytes

| Electrolyte (0.1 M) | Potential range vs. saturated calomel electrode | |
|---|---|---|
| Lithium perchlorate | -2.68 | +2.10 |
| Potassium perchlorate | -2.33 | +2.10 |
| Sodium perchlorate | -2.08 | +2.10 |
| Potassium nitrate | -2.33 | +2.10 |
| Potassium tetrafluoroborate | -2.33 | +2.10 |
| Potassium persulfate | -2.33 | +2.10 |
| Lithium chloride | -2.68 | +1.52 |
| Tetraethylammonium perchlorate | -2.30 | +2.10 |
| Tetra-*n*-butylammonium bromide | -2.40 | +1.45 |

*Source:* Reprinted with permission from J. Courtot-Coupez and M. Le Demezet, *Bull. Soc. Chim. Fr.*, 4744 (1967) [88].

Water. The applied water should be of LC grade, distilled, or deionized in that order of preference.

Buffering salts. Phosphate, acetate, and citrate are frequently applied, but the use of amines should be avoided.

Ion-pairing agents. Sulfonic acids can be applied as anionic agents and quaternary ammonium salts as cationic agents.

Organic Modifiers. Acetonitrile and simple aliphatic alcohols are the most frequently used organic moderators. However, methanol can only be applied at potentials less than 1.0 V. Tetrahydrofuran may be contaminated with peroxides after prolonged exposure to air and is therefore not recommended. Amine compounds should be avoided as modifier.

Conductance. The mobile phase must be electrically conductive to support charge transfer between mobile phase and electrodes. Buffering the mobile phase usually is sufficient to provide enough electrolyte for conductance. For nonaqueous mobile phases ammonium perchlorate (0.05M) or sodium acetate (0.05 M) may be added as supporting electrolyte.

Degassing. For oxidative detection removal of dissolved air from the mobile phase is necessary to prevent air bubble formation at the column outlet which disturbs the electrolysis process. Vacuum filtration usually is sufficient to remove enough air for "bubble-free" operation. However, air readily redissolves in the mobile phase. Continuous helium bubbling is therefore the only effective degassing method for longer periods. If a small overpressure of helium is maintained in a closed mobile phase container with a small exhaust hole, very little helium flow is required to keep the air away after thorough bubbling for a few minutes. Helium degassing is a very effective and yet economic way of removing dissolved air. Oxygen can also be removed by the addition of sulfite to the mobile phase [1]. However, the use of sulfite is only successful at pH values over 7. A last possibility is the use of a "scrubber," a cell with electrodes of porous silver over which a negative potential is applied [30].

For reductive detection, thorough removal of oxygen is essential, since oxygen is easily reduced and moderate negative potential (beyond -0.4 to -0.5 V versus Ag/AgCl), causing an excessive background current. More than 99% dissolved oxygen should preferably be removed for very sensitive measurements.

## Detectability of the Analytes

Using absorbance or fluorescence detection, the optimum detection conditions are established by recording the corresponding spectra. Using AD voltammograms can be applied for this purpose.

### *Cyclic Voltammogram*

Cyclic voltammography gives rapidly useful information about the optimum detection conditions (Fig. 8.5A) [31]. In the "off-line" mode a scan is made over a certain potential range, using conditions comparable to the detection conditions (e.g., electrode material, electrolyte, pH) in the flow cell. First of all, a scan is made from a high potential to a low potential and subsequently vice versa. The measured limiting current and half-wave potential give information about the oxidative and reductive behavior of the analyte [30].

### *Hydrodynamic Voltammogram*

The next step is the recording of a hydrodynamic voltammogram (Fig. 8.5B). The difference between a cyclic voltammogram and a hydrodynamic voltammogram is that the latter is recorded "on-line" and provides information about the signal of the AD after LC separation.

In order to optimize the LC-AD conditions for an analyte, several hydrodynamic voltammograms should be recorded, e.g., to establish the influence of the pH on the amperometric behavior of the analyte [8,32,33].

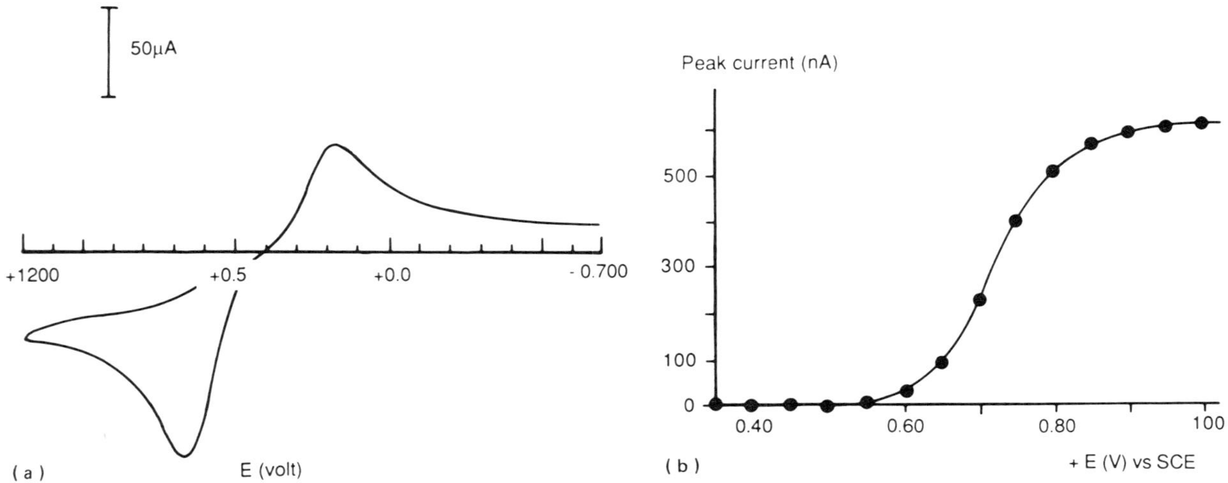


**Figure 8.5** (a) Cyclic voltammogram of epinephrine; (b) hydrodynamic voltammogram of epinephrine.

In case of fast electrode reactions, in which the number of involved protons is the same as the number of involved electrons, the shift in the half-wave potential is theoretically 59 mV for each pH unit (Nernst equation). Oxidation is in principle more difficult at lower pH values. Therefore, the pH of the eluent should be increased when phenols are oxidized [8]. The result is that a lower WE potential can be used resulting in a better signal-to-noise ratio. Another influence of the pH can be the occurrence of side reactions, meaning that more electrons are involved in the reactions. Examples are the oxidation of physostigmine [32] and epinephrine [33].

## DERIVATIZATION FOR ELECTROCHEMICAL DETECTION

Compounds with poor electrochemical (redox) properties can be detected using electrochemical detection after labeling with a suitable compound having better electrochemical properties. As in other detection-improving derivatization procedures, labeling may take place pre- or postchromatographically [34-39]. The advantages and disadvantages of both types of derivatization, such as changes in sensitivity and selectivity, (lack of) freedom in reaction conditions, improvement of chromatographic performance [35], possibility of separation of excess reagent and interfering side products in subsequent chromatography, etc., also apply in this case.

Prechromatographic derivatization can also be used to increase the stability of the analytes [34]. The half-lives for degradation of arylhydroxylamines are in the order of 8 min in phosphate buffer of pH 7.5 [37], meaning that these compounds, metabolites of arylamines and nitro compounds, will decompose partially during the LC analysis. Conversion of these analytes into the more stable hydroxyurea analogs can prevent this degradation. Even if the analytes are already electroactive, prechromatographic derivatization may be performed for this reason, such as in the case of thiols [38].

The advantages and disadvantages of pre- and postchromatographic procedures are emphasized in Chap. 1 of this book. In combination with electrochemical detection mainly the postchromatographic electrochemical reactors (vide infra) are of importance.

Apart from the labeling procedures, some nonlabeling reactions resulting in electroactive products are described. Examples of this kind will also appear in this chapter.

### Labeling Procedures

The majority of chemical reactions resulting in an increased electroactivity of the reaction product are labeling reactions, where as a result of a covalent binding to an active function in the substrate, an electrophore is introduced [2]. In most cases the purpose of these derivatizations is the improvement of detection sensitivity. However, as in all labeling procedures, a certain loss of selectivity also results since the label is at best function-specific rather than compound-specific. The use of modern LC techniques, with a high resolution power, usually compensates for this disadvantage. The advanced LC permits the choice of a small selection of labels with which a large number of functions can be derivatized. In laboratories with a variety of separation and detection problems occurring at a low frequency, the use of a limited number of these "broad-spectrum" labeling procedures is a great advantage.

The additional prerequisites (see Chap. 1 of this book) for an ideal electrochemical derivatization are [2,3] as follows:

The derivative should be detectable in the oxidative mode at relatively low WE potentials. However, in case the reductive mode is applied, potentials between 0 and -500 mV should be used.

The derivatization procedure should be applied in aqueous media.

Many of the labeling reactions used to improve electrochemical detection can also be found in other areas, such as in LC with absorbance (UV-Vis) and with fluorescence detection. Examples to illustrate this are the aromatic nitro compounds, used for UV-Vis as well as AD, and *o*-phthaldialdehyde (OPA) derivatives, which are detectable both electrochemically and fluorimetrically. More information about the chemical reactivity of these labels can be found elsewhere in this book (Chaps. 7 and 9).

*Electrochemical Probes*

The probes that can be applied for electrochemical derivatization can be divided into oxidizable and reducible probes. In case both probes are available an oxidizable probe will be preferred since the application of WE potentials of 0 V and higher means that oxygen will not interfere with detection. In case reducible probes are applied, time-consuming and complicated procedures to remove the oxygen from the eluent and the samples are unavoidable [17].

Electrochemical probes contain two different parts. One of them is the electrophore and the second is the reactive side chain which interacts with the analyte. It will be obvious that most of the normally applied electrochemical probes are borrowed from absorbance or fluorescence detection, e.g., OPA, which is one of the frequently applied fluorescence probes for the labeling of primary amines

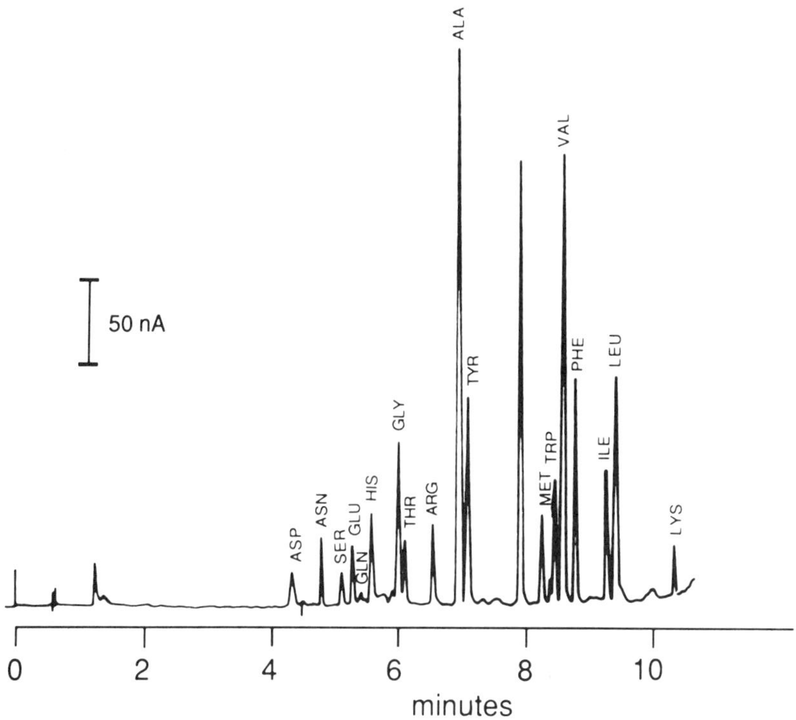


**Figure 8.6** Reversed phase analysis (gradient elution) of *o*-phthaldialdehyde derivatives of amino acids in beer applying amperometric detection at 700 mV versus Ag/AgCl. (Reprinted with permission from L. A. Allison, G. S. Mayer, and R. E. Shoup, *Anal. Chem.*, *56*:1089 (1984) [40].)

(Fig. 8.6). However, the resulting derivatives, the isoindoles, also possess favorable electrochemical properties; they can be oxidized at a potential of about 750 mV [40-42].

Reacting amines with OPA and various thiols circumvents the conventional problems of derivative stability and fluorescence quantum efficiencies. A representative example is the separation of 22 derivatized (prechromatographic mode) amino acids in about 10 min using gradient elution on a 3-μm octadecyl silica RP column. The detection limits are in the order of 500 fmol. The applied deriva-

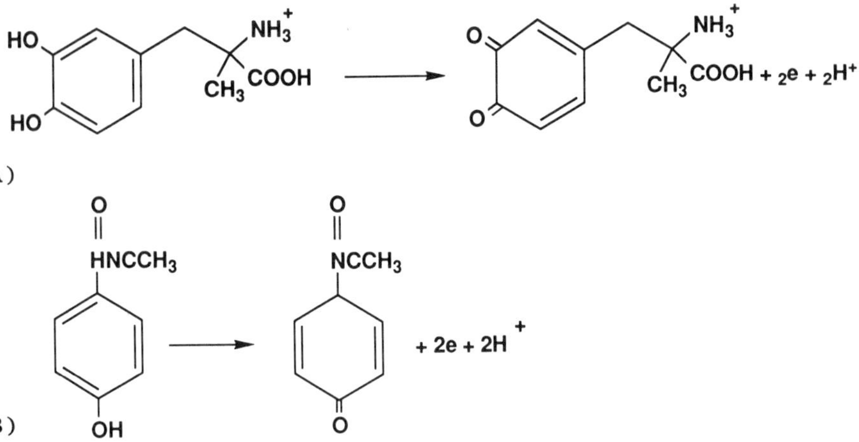


(A) (B)

Figure 8.7 Electrode reactions of α-methyldopa (A) and paracetamol (B).

tization procedure is as follows: a 20-μl sample is placed in a conical reaction vial and mixed with 100 μl of a buffered (pH 9.5) OPA/thiol solution. After exactly 2 min 1/2 μl is injected onto the chromatographic system. Detection is performed using a glassy carbon WE and a potential of 0.70 V versus Ag/AcCl [40]. This procedure is also applied to the determination of amino acids in beer and brain samples. The beer is sonicated before derivatization, while the brain samples are homogenized in cold methanol/water using a Polytron and ice bath [40].

Other oxidizable functions in electrochemical probes are the aromatic amino, phenol, or guaiacol functions. The electrochemical behavior of these functional groups is well described in the literature [5], e.g., the oxidation of α-methyldopa (Fig. 8.7A) and the oxidation of paracetamol (Fig. 8.7B).

In addition to the described probes, the ferrocenyl compounds possess favorable electrochemical properties. These tags have been developed in order to avoid the problems (interferences by the presence of oxygen) which are inherent to the use of reducible tags (Fig. 8.8). The detection limits that can be achieved with these probes are low in comparison with the aromatic nitro components. Until now only a few ferrocenyl derivatives have been applied, such as *N*-succinimidyl-3-ferrocenyl propionate for the labeling of amines

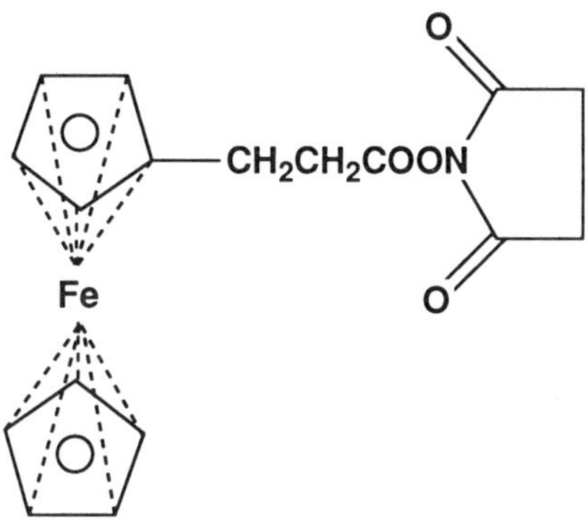


Figure 8.8 Chemical structure of *N*-succinimidyl-3-ferrocenyl propionate.

[43] and ferrocenoyl azide for the derivatization of alcohols [44], but recently ferrocenyl reagents have also been developed for the electrochemical derivatization of thiols and carboxylic acids [44].

The reason the ferrocenyl derivatives are becoming more and more popular is that by using the same electrophore a large number of analytes can be derivatized selectively by applying different ferrocenyl probes, favorable chromatographic properties, and the low half-wave potentials of the derivatives. The half-wave potential for the *N*-succinimidyl-3-ferrocenyl propionate derivatives is about 280 mV versus the Ag/AgCl REF electrode; which means, for instance, that these probes can be used in bioanalysis because the half-wave potential of the easily oxidizable catecholamines is about 400 mV using comparable conditions.

In addition to the ferrocenyl probes, a group of reagents with the same electrophore but different reactive functions to allow selective reaction with the analytes, the isothiocyanates, can be applied. Isothiocyanates are reagents used in absorbance and fluorescence derivatization studies in order to improve the detectability of both primary and secondary amine functions and the *p*-*N*,*N*-dimethylaminophenyl derivative is used as an oxidative probe (+500 mV) for the labeling of arylhydroxylamines in, for instance, liver homogenates [45]. The maleimides are in principle fluorescence probes developed for the conversion of thiols into fluorescent products, but *N*-(4-anilinophenyl)maleimide can be applied for the electrochemical derivatization of cysteine, glutathione, and penicillamide at a WE potential of around 1100 mV [38].

The derivatization of thiols with *N*-(4-anilinophenyl)maleimide is normally performed in the prechromatographic mode. A chilled solution of the malemide (1 ml containing 50 μg) in 0.033 M phosphate buffer (pH 6.85) is mixed with 0.1-4.0 μg of the thiol under investigation. The reaction mixture is kept for 10-90 min at 273 K and subsequently the excess of reagent is removed by a double extraction with 2-ml portions of diethyl ether. The aqueous phase is heated at 323 K for 20 min and an aliquot is subjected to isocratic RP ($C_{18}$) chromatography. Detection is performed with a glassy carbon WE and an Ag/AgCl reference electrode. The detection limits of glutathione and penicillamine derivatives are about 20 pg at 4-nA full scale [38].

The derivatization conditions are not treated in this section because they are discussed extensively in the chapters on absorbance and fluorescence detection of this book.

The most applied examples of reducible probes are the aromatic nitro compounds. These reagents are normally applied to improve the absorbance detection of compounds containing an amine function but the derivatives can also be reduced at WE potentials of -700 to -1100 mV. Although the electrochemistry of this function is yet to be understood in detail, and may even vary depending on the pH, the general reaction sequence for the reducible nitro function is likely to occur according to Eq. 1 [33]:

$$\begin{aligned} &\text{R-}\emptyset\text{-NO}_2 + 4e^- + 4H^+ \longrightarrow \text{R-}\emptyset\text{-NHOH} + H_2O \\ &\text{R-}\emptyset\text{-NHOH} + 2e^- + 2H^+ \longrightarrow \text{R-}\emptyset\text{-NH}_2 + H_2O \end{aligned} \tag{1}$$

($\emptyset$, aromatic nucleus)

As mentioned during the discussion of the oxidizable probes the majority of tags used in electrochemical derivatization are derived from the field of fluorescence derivatization, e.g., the application of hydrazines for the labeling of the carbonyl function. 2,4-Dinitrophenylhydrazine can be applied for the prechromatographic derivatization of aldehydes and ketones and the detection of the derivatives is performed at a WE potential at -750 to -1100 mV [46,47].

The prechromatographic determination of 17-ketosteroid sulfates (i.e., dehydroepiandrosterone sulfate) in human serum (0.1 ml) is performed by extraction of the analyte with acetonitrile (2 ml). The mixture is allowed to stand at 298 K for 5 min and after centrifugation for 5 min (1000 g) and addition of the internal standard the mixture is evaporated to dryness. To the residue 10 μl *p*-nitrophe-

Table 8.4 Electrochemical Derivatization Reagents and Reactive Functional Groups

| Reagent | Abbreviation | Functional group | Mode | Ref. |
|---|---|---|---|---|
| *o*-Phthaldialdehyde | OPA | Amine (prim.) | Ox | 17, 40-42 |
| *p*-Hydroxysuccinimide esters (vanillic acid/homovanillic acid) | | Amine (prim. + sec.) | Ox | 89 |
| *p*-*N*,*N*-Dimethylaminophenylisothiocyanate | DMAPI | Amine (prim.) | Ox | 45, 90 |
| *N*-Succinimidyl-3-ferrocenyl propionate | | Amine (prim.) | Ox | 5 |
| Ferrocenoyl azide | | Hydroxyl | Ox | 44 |
| Salicylamide | | Hydrazine | Ox | 91 |
| *p*-Aminophenol | | Carboxyl | Ox | 3 |
| *N*-(4-Anilinophenyl)maleimide | APM | Thiol | Ox | 38 |
| 2-Cyanoacetamide | | Reducing carbohydrates | Ox | 92 |
| 2,4,6-Trinitrobenzenesulfonic acid | TNBS | Amine | Red | 18, 93, 94 |
| 1-Fluoro-2,4-dinitrobenzene | FDNB | Amine | Red | 18 |
| 2-Chloro-3,5-dinitropyridine | DNCP | Amine | Red | 18 |
| 3,6-Dinitrophthalic acid anhydride | DPNT | Amine | Red | 95 |
| *p*-Nitrophenylhydrazine | | Carbonyl | Red | 14, 96 |
| Semicarbazide hydrochloride | | Carbonyl | Red | 34 |
| 2,4-Dinitrophenylhydrazine | DNPH | Carbonyl | Red | 14, 46, 47 |
| *p*-*N*itrobenzyl bromide | pNBB | Carboxyl | Red | 97 |
| *p*-*N*itrophenacyl bromide | pNPB | Carboxyl | Red | 97 |
| 3,5-Dinitrobenzoyl bromide | DNBB | Carboxyl | Red | 97 |

Abbreviations: Red, reductive mode; Ox, oxidative mode.

Table 8.5 Applications of Electrochemical Derivatizations

| Analyte | Matrix | Derivatization reagent | Pre/post | Pot. | Ref. |
|---|---|---|---|---|---|
| Amines | | Vanillic acid ester of *p*-hydroxysuccinimide | Pre | +800 | 89 |
| | | TNBS, FDNB | Pre | -850 | 18 |
| Amino acids | | TNBS | Pre | -600 | 94 |
| | | DMAPI | Pre | +850 | 90 |
| | | *N*-Succinimidyl-3-ferrocenyl propionate | Pre | +400 | 5 |
| Peptide | | DNPT | Pre | -240 | 95 |
| Amino acid | Plasma, Brain | OPA | Pre | +400, +600 | 42 |
| | Beer | OPA | Pre | +150 | 40 |
| γ-Aminobutyric acid | Brain | TNBS | Pre | -800 | 93 |
| Sympaticomimetics | | OPA | Pre | +900 | 41 |
| Arylhydroxylamines | Liver | DMAPI | Pre | +500 | 45 |
| Hydrazines | Wastewater | Salicylaldehyde | Pre | +1000 | 91 |

| | | | | | |
|---|---|---|---|---|---|
| Carboxylic acids | | *p*NBB, DNBB, *p*NPB | Pre | -1100 | 97 |
| Fatty acids | Plasma | *p*-Aminophenol | Pre | +750 | 3 |
| Bile acids | Bile | *p*-Aminophenol | Pre | +750 | 3 |
| Prostaglandins | | *p*-Aminophenol | Pre | +750 | 3 |
| Cysteine, *N*-acetylcysteine, glutathione, D-penicillamide | | APM | Pre | +1100 | 38 |
| Aldehydes, ketones | | DNPH | Pre | -750 | 46 |
| Ketones | | Semicarbazide hydrochloride | Post | -1100 | 34 |
| Aldehydes | | DNPH | Pre | +1100 | 47 |
| Reducing carbohydrates | | 2-Cyanoacetamide | Post | +400 | 92 |
| 17-Ketosteroids | Brain | *p*-Nitrophenylhydrazine | Pre | +800 | 14 |
| | Serum | *p*-Nitrophenylhydrazine | Pre | +800 | 96 |
| Hydroxyl-containing steroids | | Ferrocenoylazide | Pre | +400 | 44 |

Abbreviations: Pre, prechromatographic; Post, postchromatographic; Pot., potential in mV vs. Ag/AgCl electrode. The abbreviations in the column "Derivatization reagent" are explained in Table 8.4.

nylhydrazine (50 µg/µl) and 100 µl of a trichloroacetic acid-benzene (30 mg/10 ml) solution is added. After subsequent heating for 20 min at 333 K the mixture is evaporated to dryness. The derivatives are separated on a $C_{18}$ RP column using 80% methanol and 20% 0.5% phosphate buffer (pH 3.0) as the eluent. Detection is performed using 0.8 V versus an Ag/AgCl reference electrode. The detection limit is approximately 80 ng/ml with a coefficient of variation of 4%.

In Tables 8.4 and 8.5 an overview is given of the electrochemical reagents that can be applied, of the selectivity of these reagents for certain functional groups, of the application of these probes in the oxidative or reductive mode, of the necessary potentials, and of the possibilities of pre- or postchromatographic labeling.

### Nonlabeling Procedures

Improvement of electrochemical detectability can also be obtained by a number of nonlabeling procedures. For instance, in the case of epinephrine analysis the number of electrons involved in the oxidation reaction can be increased from 2 to 4 when the pH is changed from about 7 to higher values. However, in this particular case the stationary phase, based on silica, will dissolve in the eluent using this drastic pH increase, causing a decisive decrease in chromatographic performance. Therefore, the pH of the eluent is changed by means of a postchromatographic addition of a suitable base.

In addition to the postchromatographic addition of hydroxyl ions it is sometimes necessary to add electrolyte ions after the LC separation has taken place. An example is the LC-AD determination of misonidazole and corresponding metabolites [34]. Postchromatographic addition of the electrolyte solution is also applied in the case of normal phase chromatography.

Another example is the conversion of a number of benzylalcohols (e.g., 3-methoxy-4-hydroxymandelic acid, metanephrines, 3-methoxy-4-hydroxyphenylglycol) by periodate to vanillin. Vanillin can be oxidized (+800 mV) and detected by means of AD [33].

## DUAL-ELECTRODE REACTION DETECTION SYSTEMS

In addition to chemical derivatization in order to introduce or improve the amperometric detectability of the analyte, some other principles can be applied. One of these principles is the electrochemical derivatization applicable when using dual-electrode systems.

A relatively new approach in AD is the application of two or more WE, with the possibility of independently establishing their

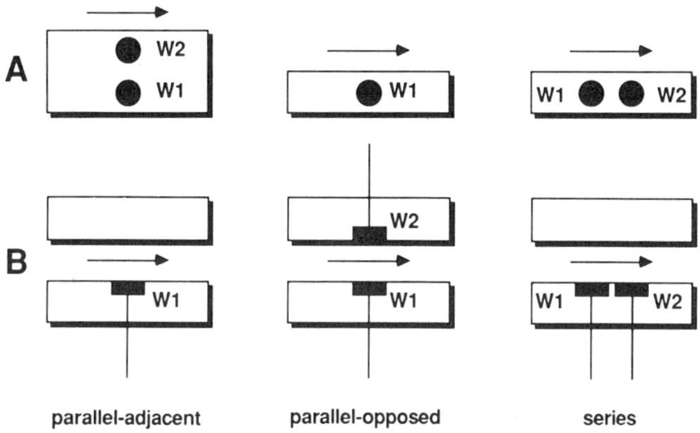


Figure 8.9 Dual-electrode configuration. (A) front sight, and (B) side sight; $W_1$, $W_2$, working electrode; →, Flow direction.

potentials. The dual-electrode systems can be divided into three types (Fig. 8.9).

In the "parallel-adjacent" (P-A) configuration the electrodes are placed next to each other on the same side of a rectangular channel. In the "series" (S) configuration the eluent passes first $WE_1$ and thereafter $WE_2$. In the "parallel-opposed" (P-O) configuration the electrodes are placed next to each other. All three configurations allow the simultaneous registration of two signals of a chromatogram, e.g., one WE in the oxidative mode and the other in the reductive mode. Two electrodes of the same material or two electrodes of different material can be applied. Normally glassy carbon is used.

## Dual-Electrode Detection Devices

### *Parallel-Adjacent Configuration*

This configuration can be compared with a dual-wavelength absorbance detector. The selective and simultaneous detection of an easily reducible (or oxidable) compound and a not easily reducible (or oxidable) solute present in the same matrix can be established applying this configuration, e.g., for the simultaneous determination of glutathione and metabolites of benzene [48]. Another possibility is the simultaneous determination of a reducible and an oxidable analyte such as several phtherine derivatives [49].

In addition to the chromatography peak identity can also be established by recording hydrodynamic voltammograms [50]. However, using this dual-electrode configuration instead of the complete voltammograms, a constant ratio between the measured currents at different potentials is also a confirmation of the identity of the analyte. Examples are the analysis of the phtherines [49] and platinum-containing anticancer drugs [33].

Finally, the signal-to-noise ratio can be improved by using the difference signals of the two potentials which differ only slightly [51]. The registered chromatogram is composed of the signal of $WE_1$ corrected for the signal of $WE_2$. The interferences, which are responsible for the noise on the baseline, will be equally present at both electrodes. By subtracting the two signals the noise will be diminished and the signal-to-noise ratio improved. However, when the difference between the two potentials is too small, the signal will also decrease meaning that an optimum should be found. A second effect of measuring the differences in electrode response will be an increased selectivity because only analytes which will generate different currents at these potentials will be detected. This means that in this case the potential of the analyte should be in the steep part of the hydrodynamic voltammogram. A nice example is the analysis of phenolic acids in beer [51].

### *Parallel-Opposed Configuration*

This detection configuration can be compared with a photomultiplier tube. The reaction product formed at one of the electrodes diffuses to the other electrode where it reacts back to its original form, whereafter it can be detected again at the first electrode. The signal which is produced by the detector is proportional with the number of electrons involved in the electrode reaction multiplied by the number of cycles the molecule is going through. In this way a reaction yield of over 100% can be obtained, meaning that the sensitivity for reversible and semireversible couples in these systems will be higher than with classical AD [52]. The sensitivity is limited by the background current, which is relatively high in this type of system due to the low impedance. The background current can be reduced by switching the voltage on and off [52]. In principle, relatively low flow rates will be applied allowing a relatively large residence time of the analytes in the flow cell without the need for large-volume flow cells and the inherent problem of band broadening. For 2,4-toluenediamine [52] and catecholamine derivatives [53] detection limits in the picogram range are reported.

This cell type can only be applied for reversible or semireversible redox couples, meaning that the oxidized and the reduced forms are chemically stable and react at the applied electrode potentials [54]. Another feature of this configuration is that the background signal can be diminished by applying the same potential onto both electrodes and by measuring the signal of only one of the electrodes. The most important advantage of this detection device is its selectivity.

### *Series Configuration*

The S-type dual-electrode device may be compared with a fluorescence detection device. In this configuration the reaction product of the first electrode is determined at the second electrode. In this case the dual-electrode device can be applied as a reactor, a detector, and for electrochemical derivatization. Furthermore, this configuration results in an improved selectivity, based on the possibility of detecting the analytes with respect to their chemical and electrochemical reversibilities [55,56]. Relatively low detection limits can be obtained, since as a result of the decreased noise and background signals of $WE_2$, a less extreme potential can be applied.

The series configuration is also applied for the separation of analytes that cannot be separated chromatographically but possess different amperometric properties [57]. In this case, first of all the most easily oxidable (or reducible) solutes are determined onto $WE_1$ and the analytes with higher oxidation (or reduction) potentials are measured onto $WE_2$.

Another feature of this configuration is that disturbance of electroactive interferences can be avoided by coulometric conversion of these solutes [58]. In this case $WE_1$ shields $WE_2$, meaning that $WE_2$ can be applied at a more extreme potential. This means that applying the reductive mode this configuration can also be applied to eliminate interferences by oxygen [17].

Whenever gradient elution techniques are applied, the S-type configuration of the dual-electrode system can be applied to correct for the changes in the baseline signal. Detection is performed at the first electrode (coulometrically), at a suitable potential, and the signal of the second electrode, set at the same potential, is subtracted electronically [60].

The series configuration can be characterized by two parameters: shielding efficiency (SE) and collection efficiency (CE) [61]. SE is a measure of the amount of conversion at $WE_1$ and CE for the amount of conversion at $WE_2$. This means that CE is the fraction of the compound formed at $WE_1$ and converted again at $WE_2$. This value strongly depends on the geometry of the flow cell. SE is the value of the current at $WE_2$, when $WE_1$ is set at the same potential, divided by the current of $WE_2$ when $WE_1$ is switched off. Therefore, SE depends on CE.

Table 8.6 Electrochemical Derivatization by Means of Dual-Electrode Systems[a]

| Analyte(s) | Matrix | $WE_1$ (mV) | $WE_2$ (mV) | Ref. |
|---|---|---|---|---|
| Biogenic amines | Urine | +800 | -50 | 60 |
| Phenolic acids | Soda | +1000 | 0 | 98 |
| Paracetamol metabolites | Urine | +900 | 0 | 98 |
| K vitamins | Liver | -1200 | +50 | 99 |
| Dichlone | | -1000 | +200 | 55 |
| Organo-Sn compounds | | -1200 | +200 | 55 |
| Catecholamines | Brain | +850 | -150 | 100 |
| Phtherine derivatives | | -600 | +1000 | 101 |
| Cystine, cysteine | | -1100 | +50 | 102 |
| Penicillamine | | -1100 | +50 | 102 |

[a]The potentials are measured against a silver-silver chloride reference electrode.

### Electrochemical Reactors

The most important application of the dual-electrode systems is electrochemical derivatization. In this particular case it is called postchromatographic electrochemical derivatization. An overview of the analytes determined in this way is presented in Table 8.6. The thiols mentioned in the table can be reduced at a mercury electrode. The mercury itself is oxidized during this reaction (Eq. 2):

$$2RSH + Hg \longrightarrow Hg(RS)_2 + 2e + 2H^+ \qquad (2)$$

#### *Electrochemical Derivatization for Other Detection Devices*

Electrochemical conversions to facilitate the absorbance or fluorescence detection of the analyte are described in the literature, and both the on-line prechromatographic as well as the postchromatographic mode have been applied [62-64].

The prechromatographic mode can be applied when, in addition to an improvement of the detectability, the chromatographic performance should be improved. A nice application of this technique is the elucidation of (side) products of electrochemical reactions by means of LC [62]. One of the negative features of prechromatographic electrochemical derivatization in combination with LC is the problem of how to perform this reaction between the injection device and the column because of the relatively high pressures in this part of the system. This means that specially constructed reactors are needed to obtain a coulometric conversion of the analyte.

An example is the oxidation of catecholamines into their strongly fluorescent oxidation products [63].

Electrochemical reduction is an attractive method for the conversion of K vitamins to fluorescent products [64]. In principle, this derivatization can be performed in the prechromatographic as well as the postchromatographic mode. Both AD and coulometric detectors can be applied for this purpose, but the latter are to be preferred in view of their higher conversion yield.

Prechromatographic reduction cannot be applied in the assay of vitamin K epoxide, a product of the metabolic cycle, in the presence of vitamin K itself since both solutes are converted to the same hydroquinone.

Chromatography of the formed hydroquinones is performed with a RP system. In this particular case some sodium perchlorate is added to the eluent in order to have sufficient conductivity. Applying reduction at a potential of -700 mV, the minimum detectable amounts are about 5 pg. This procedure can be applied for the determination of K vitamins in blood serum samples.

### *Postchromatographic Electrochemical Derivatization*

Coulometric titrations are a nice way to detect nonelectroactive compounds [65-68]. Coulometric titration is a well-known analytical method in pharmaceutical analysis [65]. An oxidation or addition reaction is executed with a halogen ion generated in situ. The main advantage of this titration is that no unstable titration liquids are necessary and, moreover, coulometry is an absolute determination method. During the titrations the halogen is formed from a halogen salt by means of electrolysis and the formed halogen reacts in the following step with the analyte.

Coulometric titrations are also combined with LC [66]. A solution of a halogen salt is added to the eluent and, depending on the reactivity of the applied halogen ion, it is mixed with the eluent stock or added postchromatographically. The halogen is generated electrochemically and some of it reacts with the analyte (e.g., fatty acids, prostaglandins, vitamin A, D, or E, phenols, thiols). The amount of halogen used during this reaction can be determined by means of AD.

The on-line electrochemical production of bromine is described by Kok et al. [68]. The produced bromine reacts with the analytes in the effluent of the analytical column and the excess bromine can be measured by means of AD. The system is applied for phenolic ethers (Fig. 8.10) [68] and for alkyl amines [67].

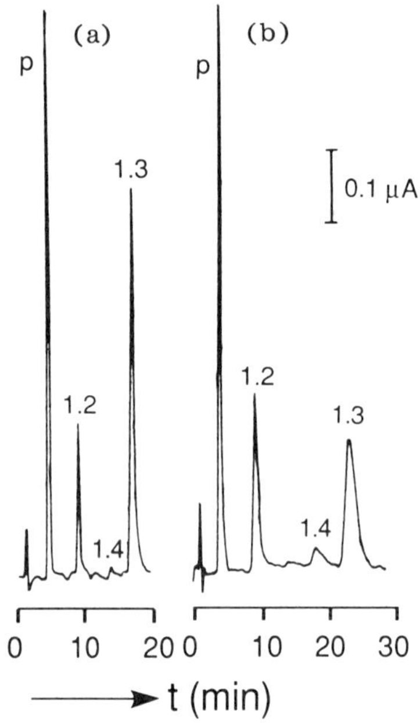


Figure 8.10 Reversed phase analysis of phenol (P) and dimethoxybenzenes (1.2, 1.3, 1.4) applying amperometric detection based on the on-line electrochemical generation of bromine. Stationary phase: RP-8 (a) and RP-2 (b). (Reprinted with permission from W. Th. Kok, U. A. Th. Brinkman, and R. W. Frei, *Anal. Chim. Acta, 162*:19 (1984) [68].)

## INDIRECT DETECTION OF ELECTROCHEMICALLY INACTIVE COMPOUNDS

In addition to the covalent labeling procedures for AD and coulometric titrations (see above), a number of other methods exist to detect electrochemically inactive compounds by means of AD. In these cases a reagent is added which interacts with the analyte. The prerequisite is that this reagent is electrochemically active, meaning that the excess of the reagent can be measured. A second possibility is the use of reagents forming an electroactive product during the interaction of the analyte with the reagent.

A considerable number of compounds are electrochemically nondetectable, although they can be oxidized or reduced. The reason is that the needed overvoltage to achieve an acceptable electron transfer rate is too high. Under these circumstances components of the eluent will be oxidized or reduced resulting in an unacceptably high background signal. Homogeneous redox reactions are normally faster than electrode reactions and can be applied to reduce the required voltage of the detector electrode [33].

Examples are the detection of reducing sugars, amino acids, and metal ions by the addition of reagents leading to the formation of electroactive products [69]. The sugars are oxidized with basic hexacyanoferrate(III) at 453 K and the reaction product, hexacyanoferrate(II), is detected coulometrically. The sugars were separated by means of ion exchange chromatography applying a strong cation exchange resin [33]. The amino acids can be detected due to the following electrode reaction (Eq. 3):

$$[Cu(SCN)]_2^- + 2(\text{amino acid})^- \longrightarrow [\text{Cu-amino Acid}_2] + 2SCN^- + e^- \qquad (3)$$

In order to detect the metal ions, the chromatographic effluent is mixed with a complex of mercury and diethylenetriaminepentaacetate ($[Hg\text{-}DTPA]^{3-}$) and the following reaction scheme can be given (Eq. 4):

$$[Hg\text{-}DTPA]^{3-} + M^{n+} + 2e^- \longrightarrow [M\text{-}DTPA]^{(5-n)-} + Hg \qquad (4)$$

in which $M^{n+}$ is metal cation. In case a metal is present (alkaline earth group) with a lower oxidation potential than mercury, this metal will be bound to the mercury and reduced on the electrode. After combination of this detection technique with cation exchange chromatography, detection limits of $10^{-7}$ mol can be achieved [33].

A last example is the determination of muramic acid in blood serum samples [70]. After the addition of bis(1,10-phenanthroline)-copper(II), in a postchromatographic mode (Eq. 5), the resulting product can be oxidized at a relatively low potential:

$$Cu(phen)_2^{2+} \xrightarrow[\text{pH 11, 368 K}]{\text{muramic acid}} Cu(phen)_2^{+} \xrightarrow[\text{40 mV}]{\text{oxidation}} Cu(phen_2^{2+} \qquad (5)$$

The blood serum samples are deproteinated by means of a protein precipitation with acetonitrile and muramic acid is chromatographed on a strong cation exchanger. The determination limit in serum is about 4 pmol. The advantage of applying a relatively low oxidation potential (40 mV) is a low background signal together with high stability of electrode system.

## Biochemical Methods

### *Prechromatographic Enzyme Reactions*

A number of compounds of clinical interest, such as ethanol, lactate, and maleate, are electrochemically inactive. However, these solutes can be oxidized in the presence of enzymes. These enzymes, the oxidoreductases, are able to oxidize a substrate and to transfer the liberated electrons to acceptor molecules such as nicotinamide adenine dinucleotide (NAD). The reduced form of this so-called cofactor, NADH, can be oxidized easily at a glassy carbon electrode [71].

After the incubation of a mixture of the substrate (e.g., ethanol, Eq. 6) and the enzyme under the proper conditions, a certain amount of NADH is obtained which is proportional to the amount of substrate originally present.

$$CH_3CH_2OH + NAD^+ \xrightarrow[\text{pH } 8.8]{\text{enzyme}} NADH + CH_3CHO + H^+ \tag{6}$$

Ethanol, for instance, can be measured in 100-μl blood samples by using alcohol dehydrogenase as the enzyme and by separating the resulting mixture on the reversed phase LC system [33]. The minimum detectable amount, applying this type of enzyme reaction and AD, is in the low nanogram range [72]. The produced NADH can be detected selectively by using an appropriate LC-AD system. Instead of NAD, other cofactors (e.g., NADPH, pyridoxal phosphate) can be applied which will enhance the applicability of the procedure [33].

Another possibility is the determination of enzyme activity by means of a procedure in which the substrate is converted by an enzyme into an electroactive product, where in this case the enzyme is the analyte.

An example is the analysis of the enzyme dopamine-β-hydroxylase [73]. This enzyme is able to convert dopamine to norepinephrine. Subsequently, these two compounds are separated by a RP-LC system and quantitated with AD.

### *Postchromatographic Enzyme Reactions*

Enzyme reactions can also be used in the postchromatographic mode in order to allow the AD of nonelectroactive compounds. In this case the enzyme is added to the LC eluent and after the reaction has taken place in a suitable reactor, the product is detected electrochemically.

This technique has been applied, for instance, to the analysis of acetylcholine in brain tissue. The enzymes acetylcholinesterase and choline oxidase are added to the mobile phase and acetylcholine is converted to choline, which is subsequently oxidized to betaine. During this last reaction step hydrogen peroxide is generated, which is detected with a platinum electrode [74].

Interesting developments in the near future will be the construction of reactors containing immobilized enzymes or other catalyzing agents [5]. An example is the adsorption of enzymes onto an ion exchanger for the determination of acetylcholine [75]. A second example is the analysis of phenolic glycosides [76]. β-Glucuronidase is immobilized on porous glass beads in a postchromatographic reactor. The phenolic products formed by the enzymatic reaction can be detected amperometrically.

A last example is the determination of bile acids [77,78]. The compounds are separated on a RP-LC system but because they are poorly detectable an enzyme reactor with immobilized 3-α-hydroxysteroid dehydrogenase has been applied postchromatographically. The enzyme catalyzes the oxidation of bile acids to the corresponding ketones in the presence of a coreactant, NAD.

The advantage of this method is that the necessary amount of enzyme is significantly lower. In these reactors the electroactive compound is formed and, again, detected by means of AD.

## FUTURE PERSPECTIVES

Because the intrinsic possibilities of AD are high selectivity and high sensitivity, this detection mode will maintain its position in the future and become even more important when the technique will become less sensitive for interferences and when automated procedures are developed for routine bioanalysis.

The applicability of this technique will improve when the development of, for example, ferrocene-based labels has progressed. The advantages of these labels are that they are oxidative reagents and that a number of different side chains can be attached to the ferrocene nucleus allowing the derivatization of a wide variety of analytes possessing different functional groups.

Another interesting development will be the use of dual-electrode systems because these systems will increase the applicability of reductive labels (e.g., aromatic nitro compounds). In this particular case the detection reaction is an oxidation at a relatively low potential at the second electrode. This means that interferences by the presence of oxygen can be avoided.

The dual-electrode systems will, in principle, improve the applicability of AD in bioanalysis, since the first electrode serves as screening for the second electrode. An example is the application of a voltammetric/amperometric detection device [79]. In this device the advantages of voltammetric and amperometric detection are combined because the three-dimensional data array of current response as a function of time and potential can be obtained. Nice features of this device are the possibility of recording hydrodynamic voltammograms using small amounts of the analytes and the possibility of obtaining peak identification of poorly resolved peaks.

Further development of the technique is also likely to occur in clinical chemistry, where LC-AD can be an interesting addition to the existing enzymatic methods by applying immobilized enzyme reactors and glass fiberoptics.

## REFERENCES


1. van Oort, W. J., J. den Hartigh, and R. J. Driebergen, in *Electrochemical Detectors* (T. H. Ryan, ed.), Plenum Press, New York, 1984.
2. Lingeman, H., W. J. M. Underberg, A. Takadate, and A. Hulshoff, *J. Liq. Chromatogr.*, *8*:792 (1985).
3. Ikenoya, S., O. Hiroshima, M. Ohmae, and K. Kawaba, *Chem. Pharm. Bull.*, *28*:2941 (1980).
4. Lawrence, J. F., in *Chemical Derivatization in Analytical Chemistry* (R. W. Frei and J. F. Lawrence, eds.), Plenum Press, New York, 1982.
5. Lavrich, C., and P. T. Kissinger, in *Therapeutic Drug Monitoring and Toxicology by Liquid Chromatography* (S. H. Y. Wong, ed.), Marcel Dekker, New York, 1985.
6. Richards, D. A., in *Electrochemical Detectors* (T. H. Ryan, ed.), Plenum Press, New York, 1984.
7. Mefford, F. I., *J. Neurosc. Methods*, *3*:207 (1981).
8. Holthuis, J. J. M., Bioanalysis, electrochemistry and pharmacokinetics of etoposide, Thesis, State University Utrecht, Utrecht, 1985.
9. Ding, X.-P., and I. S. Krull, *J. Liq. Chromatogr.*, *12*:2173 (1983).
10. Akerblom, M., and B. Lundgren, *J. Chromatogr.*, *258*:302 (1983).
11. Nakahara, Y., and H. Sekine, *J. Anal. Toxicol.*, *9*:121 (1985).
12. Suckow, R. F., and T. B. Cooper, *J. Pharm. Sci.*, *70*:257 (1981).
13. Cullen, L. F., M. P. Brindle, and G. J. Papariello, *J. Pharm. Sci.*, *62*:1708 (1973).
14. Shimada, K., M. Tanaka, and T. Nambara, *Anal. Lett.*, *13*:1129 (1980).
15. Alawi, M. A., and H. A. Russel, *Fres. Z. Anal. Chem.*, *307*:382 (1981).
16. van Bennekom, W. P., Contributions to pulse polarography and electrochemical flow-through detection, Thesis, State University, Leiden, 1981.
17. Bratin, K., and P. T. Kissinger, *J. Liq. Chromatogr.*, *4*:321 (1981).
18. Jacobs, W. A., and P. T. Kissinger, *J. Liq. Chromatogr.*, *5*:881 (1982).
19. Langeberg, J. P., and U. R. Tjaden, *J. Chromatogr.*, *5*:881 (1984).
20. Trojanek, A., in *Electrochemical Detectors, Fundamental Aspects and Analytical Applications* (T. H. Ryan, ed.), Plenum Press, New York, 1984.

21. *Electrochemical Detection in HPLC*, Spark Holland B. V., Emmen, The Netherlands, 1987.
22. Adams, R. F. (ed.), *Electrochemistry at Solid Electrodes*, Marcel Dekker, New York, 1969.
23. Reeuwijk, H. J. E. M., H. Lingeman, J. de Jong, U. R. Tjaden, and J. van der Greef, *Anal. Proc.*, *24*:301 (1987).
24. Tenygl, J., Proceedings 2, J. Heyrovsky Memorial Congress on Polarography, August 25-29, Prague, 1980.
25. Tenygl, J., and B. Fleet (eds.), *Automatic Electrochemical Analysis*, Marcel Dekker, New York (in press).
26. Elbicki, J. M., D. M. Morgan, and S. G. Weber, *Anal. Chem.*, *56*:978 (1984).
27. Alexander, P. W., and S. H. Qureshi, *J. Electroanal. Chem.*, *71*:235 (1976).
28. Oosterhuis, B., K. Brunt, B. H. C. Westerink, and D. A. Doornbos, *Anal. Chem.*, *52*:203 (1980).
29. Lemar, H., and M. Porthault, *J. Chromatogr.*, *130*:372 (1977).
30. Hanekamp, H. B., W. H. Voogt, P. Box, and R. W. Frei, *Anal. Chim. Acta*, *118*:81 (1980).
31. Kissinger, P. T., and W. R. Heineman, *J. Chem. Educat.*, *60*:702 (1983).
32. Whelpton, R., and T. Moore, *J. Chromatogr.*, *341*:361 (1985).
33. Kissinger, P. T., K. Bratin, G. C. Davis, and L. A. Pachla, *J. Chromatogr. Sci.*, *17*:137 (1979).
34. Little, C. J., J. A. Whatley, and A. D. Dale, *J. Chromatogr.*, *171*:63 (1979).
35. Lindroth, P., and K. Mopper, *Anal. Chem.*, *51*:1667 (1979).
36. Abdou, H. M., T. Medwick, and L. C. Bailey, *Anal. Chim. Acta*, *93*:221 (1977).
37. Sternson, L. A., W. J. DeWitte, and J. G. Stevens, *J. Chromatogr.*, *153*:481 (1978).
38. Shimada, K., M. Tanaka, and T. Nambara, *Anal. Chim. Acta*, *147*:375 (1983).
39. Meyer, S. D., and D. E. Tallman, *Anal. Chim. Acta*, *146*:227 (1983).
40. Allison, L. A., G. S. Mayer, and R. E. Shoup, *Anal. Chem.*, *56*:1089 (1984).
41. Leroy, P., A. Nicolas, and A. Moreau, *J. Chromatogr.*, *282*:561 (1983).
42. Joseph, H. M., and P. Davies, *J. Chromatogr.*, *277*:125 (1983).
43. Tanaka, M., K. Shimada, and T. Nambara, *J. Chromatogr.*, *309*:170 (1984).
44. Shimada, K., S. Orii, M. Tanaka, and T. Nambara, *9th Int. Symp. Column Liq. Chromatogr.*, Edinburgh, 1985, p. 309.
45. Musson, D. G., and L. A. Sternson, *J. Chromatogr.*, *188*:159 (1980).

46. Jacobs, W. A., and P. T. Kissinger, *J. Liq. Chromatogr.*, *5*:669 (1982).
47. Chiavara, G., and C. Bergamini, *J. Chromatogr.*, *318*:427 (1985).
48. Lunte, S. M., and P. T. Kissinger, *J. Chromatogr.*, *317*:579 (1984).
49. Lunte, C. E., and P. T. Kissinger, *Anal. Chem.*, *55*:1458 (1983).
50. Roston, D. A., and P. T. Kissinger, *Anal. Chem.*, *53*:1695 (1981).
51. Lunte, C. E., P. T. Kissinger, and R. E. Shoup, *Anal. Chem.*, *57*:1541 (1985).
52. Weber, S. G., and W. C. Purdy, *Anal. Chem.*, *54*:1757 (1982).
53. Fenn, R. J., S. Siggi, and D. J. Curran, *Anal. Chem.*, *50*: 1067 (1978).
54. Kurahashi, T., *The Yanaco News*, *11*:8 (1982).
55. McCrehan, W. A., and R. A. Durst, *Anal. Chem.*, *53*:1700 (1981).
56. Allison, L. A., and R. E. Shoup, *Anal. Chem.*, *55*:8 (1983).
57. Blank, C. L., *J. Chromatogr.*, *117*:35 (1976).
58. Schieffer, G. W., *Anal. Chem.*, *52*:1994 (1980).
59. Tjaden, U. R., J. de Jong, and C. F. M. van Valkenburg, *J. Liq. Chromatogr.*, *6*:2255 (1983).
60. Goto, M., E. Sakurai, and D. Ishii, *J. Liq. Chromatogr.*, *6*: 1907 (1983).
61. Hoogvliet, J. C., Electrochemical detection in liquid chromatography, Thesis, State University Leiden, Leiden, 1985.
62. Heineman, W. R., and P. T. Kissinger, *Anal. Chem.*, *52*:151R (1980).
63. Schieffer, G. W., *Anal. Chem.*, *53*:126 (1981).
64. Langenberg, J. P., Bioanalysis of ultra-trace levels of K vitamins using electrofluorimetric detection in HPLC, Thesis, State University Leiden, Leiden, 1985.
65. Kalinowska, Z.-E., *Die Pharmazie*, *22*:1 (1967).
66. King, W. P., and P. T. Kissinger, *Clin. Chem.*, *26*:1484 (1980).
67. Isaksson, K., J. Lindquist, and K. Lundström, *J. Chromatogr.*, *324*:333 (1985).
68. Kok, W. Th., U. A. Th. Brinkman, and R. W. Frei, *Anal. Chim. Acta*, *162*:19 (1984).
69. Takata, Y., and G. Muto, *Anal. Chem.*, *45*:1864 (1973).
70. Watanabe, N., *J. Chromatogr.*, *316*:495 (1984).
71. Blaedel, W. J., and R. A. Jenkins, *Anal. Chem.*, *47*:1337 (1975).
72. Davis, G. C., K. S. Holland, and P. T. Kissinger, *J. Liq. Chromatogr.*, *2*:663 (1979).
73. Davis, G. C., and P. T. Kissinger, *Anal. Chem.*, *51*:1960 (1979).

74. Potter, P. E., J. L. Meek, and N. H. Neff, *J. Neurochem.*, *41*:188 (1983).
75. Meek, J. L., and C. Eva, *J. Chromatogr.*, *317*:343 (1984).
76. Dalgaard, L., L. Nordholm, and L. Brimer, *J. Chromatogr.*, *265*:183 (1983).
77. Hayashi, M., Y. Imai, Y. Minami, S. Kawata, Y. Matsuzawa, S. Tarui, and K. Uchida, *J. Chromatogr.*, *338*:195 (1985).
78. Kamada, S., M. Maeda, A. Tsuji, and T. Kurahashi, *J. Chromatogr.*, *239* :773 (1982).
79. Lunte, C. E., J. F. Wheeler, and W. R. Heineman, *Anal. Chim. Acta*, *200*:101 (1987).
80. Baizer, M. M. (ed.), *Organic Electrochemistry*, Marcel Dekker, New York, 1973.
81. Roger, W., and S. M. Kipnes, *Anal. Chem.*, *27*:1916 (1955).
82. Billon, J. P., *J. Electroanal. Chem.*, *27*:486 (1955).
83. Breant, M., M. Bazoin, C. Buisson, M. Dupin, and J. M. Rebattu, *Bull. Soc. Chim. Fr.*, 5065 (1968).
84. Perichon, J., and R. Buvet, *Electrochim. Acta*, *9*:567 (1964).
85. Bard, A. J. (ed.), *Electroanalytical Chemistry*, Vol. 12, Marcel Dekker, New York, 1969.
86. Eberson, L., and K. Nybeg, *J. Am. Chem. Soc.*, *88*:1686 (1966).
87. Bachmann, G. B., and M. J. Astle, *J. Am. Chem. Soc.*, *64*:1303 (1942).
88. Courtot-Coupez, J., and M. Le Demezet, *Bull. Soc. Chim. Fr.*, 4744 (1967).
89. Shimada, K., M. Tanaka, and T. Nambara, *Chem. Pharm. Bull.*, *27*:2259 (1979).
90. Mahachi, T. J., R. M. Carlson, and D. P. Poe, *J. Chromatogr.*, *298*:279 (1984).
91. Kester, P. E. and N. D. Danielson, *Chromatographia*, *18*:125 (1984).
92. Honda, S., T. Konishi, and S. Suzuki, *J. Chromatogr.*, *299*: 245 (1984).
93. Caudill, W. L., G. P. Houck, and R. M. Wightman, *J. Chromatogr.*, *227*:331 (1982).
94. Caudill, W. L., and R. M. Wightman, *Anal. Chim. Acta*, *141*: 269 (1982).
95. Meek, J. L., *J. Chromatogr.*, *266*:401 (1983).
96. Shimada, K., M. Tanaka, and T. Nambara, *J. Chromatogr.*, *307*:23 (1984).
97. Roorda, I., C. Gonnet, and J. L. Rocca, *Analysis*, *10*:409 (1982).
98. Roston, D. A., and P. T. Kissinger, *Anal. Chem.*, *54*:429 (1982).

99. Harron, Y., C. A. W. Schubert, and P. V. Hauschka, *J. Chromatogr. Sci.*, *22*:89 (1984).
100. Mayer, G. S., and R. E. Shoup, *J. Chromatogr.*, *255*:533 (1983).
101. Lunte, C. E., and P. T. Kissinger, *Anal. Chem.*, *56*:658 (1984).
102. Eggli, R., and R. Asper, *Anal. Chim. Acta*, *101*:253 (1978).

# 9

# Fluorescence Derivatization

JUNICHI GOTO *Tohoku University, Aobayama, Sendai, Japan*

## INTRODUCTION

Liquid chromatography (LC) is well recognized as an ideal tool for the analysis of a variety of compounds, in particular compounds which are polar and unstable. The great success of this methodology is ascribable to the development of excellent software for the detection system as well as fine instrumentation and column technology hardware. Among numerous detection modes, fluorimetric detection (FD) has particularly seen growth in the last decade [1-8].

Fluorimetry is now established as a powerful analytical method with high sensitivity and selectivity. By this technique a minute amount ($10^{-12}$-$10^{-15}$ g) of the analyte can be determined with high reliability. Accordingly, on-line fluorescence detection combined with highly efficient columns serves for the characterization and determination of trace components in a complex matrix. In this chapter FD for use in LC will be discussed.

A marked increase in utilization of FD can be ascribed to the development of novel reagents and excellent reactions for derivatization. It is to be noted that the characteristics of fluorogenic reagents and their derivatives are influenced more or less by the conditions suitable for chromatographic separation.

Of a large number of known organic compounds, relatively few exhibit intense fluorescence. It is sufficiently substantiated that the rigid structure consisting of planar aromatic ring exhibits fluorescence in either the visible or the ultraviolet region. For instance, fluorescein shows intense fluorescence whereas phenolphthalein, with a similar structure, is not fluorescent. A marked difference between these two is explained in terms of rigidity due to the presence of an ether linkage in the molecule. Certain functional groups on an aromatic ring exert definite influences, predictable on both the energy and intensity of emission. Substitution of electron-donating groups such as $-OR$ or$-NR_2$ on an aromatic nucleus enhances the fluorescence intensity unless steric factors are involved, while electron-withdrawing groups such as $-NO_2$, $-COOH$, $-CN$, or halogen cause a substantial depression of the fluorescence intensity. In actuality, eosin being tetrabromofluorescein exhibits less intense fluorescence than fluorescein itself.

The fluorescence intensity is significantly affected by interactions between the solvent and solute molecule. The solvent is therefore a very important variable in FD. One very specific solvent-solute interaction is hydrogen bonding. The molecules having lowest excited singlet states of the $n$-$\pi^*$ type, such as quinoline or acridine, rarely fluoresce in nonpolar solvents. However, in polar solvents such as ethanol these molecules have the lowest excited singlet states of the $\pi$-$\pi^*$ type due to the formation of hydrogen bonding and become fluorescent.

The use of heavy-atom solvents (e.g., ethyl iodide), or solvents having heavy atom ions, causes a decrease in the fluorescence intensity. In aqueous media, ionic compounds exist in different species depending on pH due to dissociation or protonation of functional groups. It is generally desirable to avoid solvent-solute pairs which exhibit pronounced hydrogen-bonding tendencies.

The fluorescence efficiency in solution is a function of temperature. As the temperature is decreased, fluorescence efficiency of an organic compound usually (but not always) increases. A 1°C increase in temperature results in a 1-2% decrease in fluorescence intensity. In certain cases, such as tryptophan or indole-3-acetic acid, however, a 1°C rise in temperature exerts a 5-10% depression of the fluorescence intensity.

## FLUORESCENCE DERIVATIZATION OF NONFLUORESCENT COMPOUNDS

The LC method combined with on-line FD is particularly favorable for the analysis of trace components in a complex matrix, such as a biological specimen. A major drawback of this technique, however, is the limited number of compounds exhibiting sufficiently intense fluorescence. This inherent disadvantage can be overcome by the use of chemical reactions converting non- or weakly fluorescent compounds into highly fluorescent derivatives before or after LC separation. Fluorescence labeling of the analyte through a covalent bond will be discussed in the next section.

Various types of derivatization methods have been developed. Basic drugs like alkaloids can be detected employing the formation of fluorescent ion pairs with 9,10-dimethoxyanthracene-2-sulfonate [9-13]. In similar fashion, acidic analytes can be detected with basic fluorogenic reagents.

Postchromatographic oxidation is also useful for FD of biologically important substances. The highly sensitive detection of catecholamines and thiamine is readily attained by oxidation with potassium hexacyanoferrate(III) [14,15]. The phenothiazine drugs are similarly detected after permangate oxidation [16]. In some cases reduction in the eluate is also effective for sensitive detection. For instance, conversion of phylloquinone and menaquinone into hydroquinones is practically useful for fluorescence monitoring. Upon irradiation with UV light, cannabinol is converted to a fluorescent phenanthrene derivative [17] and catecholamines to fluorescent derivatives in the presence of ethylenediamine.

Digitalis glycosides [18] and cortisol [19] are transformed into fluorogenic products when heated with concentrated hydrochloric acid or ethanol-sulfuric acid, respectively. Indomethacin is easily converted to a fluorescent product by deacylation prior to LC [20]. The analytes having pertinent ligand structures are capable of forming fluorescent metal chelate compounds. 8-Hydroxyquinoline [21] and isonicotinoylhydrazone derivatives [22] after elution can be easily detected by the addition of Mg(II) and Al(III), respectively. In addition, enhancement of the inherent fluorescence is also applicable to a sensitive detection. A fluorescent probe, 1-anilinonaphthalene-8-sulfonic acid, exhibits enhanced fluorescence in the presence of lipids in aqueous solution, permitting the sensitive detection of lipids due to the formation of a hydrophobic complex [23]. Some fluorescent substrates are enhanced by the formation of inclusion compounds with β-cyclodextrin [24]. The pH alteration of the effluent also serves to enhance the fluorescence intensity [25-27].

Detection methods using nonfluorescent reagents which are capable of converting the eluates into fluorescent products in the flow system are also available. Upon oxidation of compounds, such as carbohydrates, organic acids, or acetaminophen, cerium(IV) is reduced to highly fluorescent cerium(III) [28,29]. The palladium(II)-calcein complex is decomposed by a ligand exchange reaction with organosulfur compounds, resulting in the release of fluorescent calcein [30]. As is well known, coenzyme nicotinamide adenosine dinicleotide (NADH), formed from $NAD^+$ along with the enzymatic reaction, is fluorescent [31,32]. The detection system utilizing enzymatic reactions will be discussed in Chap. 6.

### Fluorescence Labeling

One of the most important approaches to extending the applicability of LC is the development of highly sensitive and selective labeling reactions. Many investigations have been carried out for this purpose, and various fluorescence labeling reagents and reactions are now available. Almost all of the labeling reactions involve the use of pertinent reagents which possess both specific reaction functions and sensitive fluorophores as signal groups. These reagents are utilized in two ways, for pre- or postchromatographic derivatization. The fluorescence labeling reagents used for the analysis of biologically important substances are listed in Table 9.1.

The use of these reagents does not always permit selective labeling of the analytes owing to the side reactions with coexisting substances having similar partial structures. An efficient clean-up prior to LC is therefore needed to obtain reliable results. The design of suitable labeling reagents has to fulfill the following requirements: the fluorophore should have a large molar absorptivity and a large quantum yield of fluorescence to achieve high sensitivity. The excitation wavelength of a fluorophore should be longer than 300 nm to exclude possible interferences. The reagent should have adequate purity and be stable for a long period. The reagent itself and its degradation products formed during the labeling procedure should be nonfluorescent or be readily separable from the desired derivative. The analyte should react rapidly with the reagent under mild conditions to yield the derivative quantitatively as a single product. The structure of a fluorescent derivative should be unambiguously established. The derivative must be stable and possess favorable chromatographic properties. Furthermore, the reagent must be nontoxic.

To attain efficient labeling with a specific reagent, optimal reaction conditions should be established. The reaction solvent, concentration of reagents, use of catalysts, reaction temperature, and time become dominant factors for providing satisfactory results with high reproducibility. In addition, the suitable properties of a labeled product for LC with FD are also to be clarified.

### Prechromatographic Derivatization

Prechromatographic derivatization is widely carried out in labeling, with a fluorophore as well as with a chromophore or an electrophore, because it is methodologically of greater advantage. Any other solvents than the mobile phase in LC can be used for the derivatization reaction. The unchanged reagent and side reaction products which may possibly interfere with the detection are to be eliminated on a separation column and/or by the clean-up procedure. In other words, the appropriate chromatographic conditions for resolving the

Table 9.1 Fluorescence Derivatization Reagents and Reacting Functional Group

| Reagent | Functional group | Ref. |
|---|---|---|
| 9-Acridinylmaleimide | Sulfhydryl | 194,195 |
| 9-Aminophenanthrene | Carboxyl | 167 |
| 1-Anilinonaphthylmaleimide | Sulfhydryl | 196 |
| 1- and 9-Anthroylnitrile | Hydroxyl | 205-208, 211,212 |
| 9-Anthryldiazomethane | Carboxyl | 159-161 |
| *p*-(2-Benzimidazolyl)phenylmaleimide | Sulfhydryl | 190 |
| Benzoine | Guanidino | 216,217 |
| *p*-(2-Benzoxazolyl)phenylmaleimide | Sulfhydryl | 198 |
| BOC-aminomethylphenyl and -aminophenylisothiocyanate | Amine | 131 |
| 1-Bromoacetylpyrene | Carboxyl | 156 |
| 4-Bromomethyl-7-acetoxycoumarin | See Br-Mmc | 149,150 |
| 4-Bromomethyl-6,7-dimethoxycoumarin | See Br-Mmc | 148 |
| 3-Bromomethyl-6,7-dimethoxy-1-methyl-2(1H)-quinoxalinone | Carboxyl | 151-153 |
| Bromobimane | Sulfhydryl | 222 |
| 4-Bromomethyl-7-methoxycoumarin (Br-Mmc) | Carboxyl, imide, sulfhydryl, phenol | 137-145 |
| *N*-Chlorodansylamide | Sulfhydryl | 200,201 |
| 3-Chloroformyl-7-methoxycoumarin | Hydroxyl | 224,225 |
| 9-(Chloromethyl)anthracene | Carboxyl | 157 |
| 4-Chloro-7-nitrobenzo-2-oxa-1,3-diazole | Amine (prim. and sec.), sulfhydryl, phenol | 115-119 |
| α-Cyanoacetamide | Uronic acid, reducing sugar | 227-230 |
| 1,2-Diamino-4,5-dimethoxybenzene | α-Keto acid | 220 |
| 9,10-Diaminophenenthrene | Carboxyl | 168-179 |
| 4-Diazomethyl-7-methoxycoumarin | Carboxyl | 158 |
| 5-Dibutylaminonaphthalene-1-sulfonyl chloride | See Dns-Cl | 51 |
| *N*,*N*′-Dicyclohexyl-O-(7-methoxycoumarin-4-yl)methylisourea | Carboxyl | 147,162, 163 |

Table 9.1 (continued)

| Reagent | Functional group | Ref. |
|---|---|---|
| *N,N'*-Diisopropyl-O-(7-methoxycoumarin-4-yl)methylisourea | Carboxyl | 147,162, 163 |
| 9,10-Dimethoxyanthracene-2-sulfonic acid | Amine | 13 |
| 7-Dimethylamino-4-methylcoumarin-maleimide | Sulfhydryl | 193,197 |
| 5-Dimethylaminonaphthalene-1-sulfonylaziridine | Sulfhydryl | 199 |
| 5-Dimethylaminonaphthalene-1-sulfonyl chloride (Dns-Cl) | Amine (prim. and sec.), sulhydryl, phenol, hydroxyl | 33-50 |
| 5-Dimethylaminonaphthalene-1-sulfonylhydrazine | Carbonyl | 173-183 |
| 4-Dimethylamino-1-naphthoylnitrile | Hydroxyl | 210 |
| 4-Dimethylamino-1-naphthylisothiocyanate | Amine | 137 |
| 5,5-Dimethyl-1,3-cyclohexanedione | Aldehyde | 221 |
| 1,2-Diphenylethylenediamine | Catecholamine | 214 |
| 9-Fluorenylmethylchloroformate | Amine | 126 |
| Fluoresceinisothiocyanate | Amine | 128 |
| 4-Fluorobenzo-2-oxa-1,3-diazole-7-sulfonate | Sulfhydryl | 202,203 |
| 4-Fluoro-7-nitrobenzo-2-oxa-1,3-diazole | Amine (prim. and sec.), sulfhydryl, phenol | 120-125 |
| 4-Fluoro-7-sulfamoylbenzo-2-oxa-1,3-diazole | Sulfhydryl | 202,203 |
| Glycylglycine | Catecholamine | 101 |
| 4-Hydrazino-2-oxa-1,3-diazole | Carbonyl | 186 |
| 4'-Hydrazino-2-stilbazole | Carbonyl | 184,185 |
| 4-Hydroxy-7-methoxycoumarin | See Br-Mmc | 141,146, 147 |
| 9-(Hydroxymethyl)anthracene | Carboxyl | 166 |
| 9-Isothiocyanatoacridine | Amino | 136 |
| 7-Methoxycoumarin-3- and -4-carbonylazide | Hydroxyl | 213 |

Table 9.1 (continued)

| | | |
|---|---|---|
| 2-Methoxy-2,4-diphenyl-3-(2H)-furanone | Amine | 74,75 |
| *N'*-Methylnicotinamide chloride | α-Oxomethylene | 187-189 |
| 5-Methylphenylaminonaphthalene-1-sulfonyl chloride | See Dns-Cl | 52 |
| Naphthacyl bromide | Carboxyl | 154 |
| 2-Naphthylchloroformate | Amine | 127 |
| Naphthylisocyanate | Amine, hydroxyl | |
| Ninhydrin | Amine | 218,219 |
| Panacyl bromide | Carboxyl | 155 |
| 9,10-Phenanthrenequinone | Guanidino | 135,136 |
| *o*-Phenylenediamine | α-Keto acid | 169-172 |
| 4-Phenylspiro[furan-2(3H)-1'-phthalan]-3,3-dione (fluorescamine) | Amine (prim. and sec.) | 53-74 |
| *o*-Phthaldialdehyde | Amine (prim. and sec.) | 17,46,57, 72,76-114 |
| Pyrene-1-carbonylnitrile | Hydroxyl | 209 |
| 1-Pyrenylmaleimide | Sulfhydryl | 191-193 |
| *N*-Succinimidyl-2-naphthoxyacetate | Amine | 132,133 |

derivatives can be chosen extensively. In certain cases, however, the difficulty is the removal of interfering peaks from the chromatogram. It is recommended that the purity of reagents, solvents, and materials used not only for LC but also for pretreatment be checked prior to use. Although in a few cases the reaction mixture can be directly injected into the chromatograph without any pretreatment, removal of excess reagents and solvents in the reaction mixture is usually required. The stability of the derivatives against hydrolytic cleavage and thermal decomposition is also very important.

### Postchromatographic Derivatization

In recent years, considerable efforts have been directed to the development of postchromatographic derivatization methods. This technique has several advantages over prechromatographic derivatization. The postchromatographic derivatization needs a rapid and reproducible reaction but not necessarily the quantitative formation of a sole product. The eluate can be monitored simultaneously by methods other than fluorescence detection prior to derivatization. The detection systems with postchromatographic derivatization can be automated with more ease than that with prechromatographic derivatization.

On the other hand, the postchromatographic derivatization method has inevitable disadvantages. Diffusion of the eluate or its derivative makes the bands broad and, hence, decreases the resolution. The choice of suitable conditions for the separation and derivatization is somewhat restricted. Occasionally, unchanged reagent, decomposed products, and impurities derived from the reagent interfere with the detection of the derivatized analyte.

## FLUORESCENCE DERIVATIZATION REAGENTS, METHODS, AND APPLICATIONS

LC with fluorescence labeling has been widely applied to the determination of trace components in biological materials. As previously mentioned, the derivatization reaction is influenced by the reagent concentration, solvent, and reaction temperature. Therefore, modifications of the conditions are often needed.

Although various reactions have been developed for the preparation of derivatives, the replacement reaction of active hydrogens such as those in OH, SH, or $NH_2$ groups accounts for a large portion. Derivatization reactions are discussed below in general, according to the representative functional groups.

### Amino Group

5-Dimethylaminonaphthalene-1-sulfonyl (dansyl; Dns) chloride has been proposed as a pre- or postchromatographic derivatization reagent for introducing a fluorophore into primary and secondary amino compounds [33-48]. This reagent is also capable of reacting with phenolic and alcoholic hydroxyl groups and thiols [49,50]. The derivatization reaction is performed in either aqueous or nonaqueous media in the presence of base such as sodium carbonate or triethylamine. The latter reaction condition, however, is preferred because of no formation of unfavorable hydrolyzate and the shorter reaction time. Therefore, this method is usually employed for prechromatographic derivatization. The dansylated products are relatively stable and show high sensitivity, the detection limit being in the subnanogram to picogram level. Similar types of reagents, 5-dibutylaminonaphthalene-1-sulfonyl (Bns) chloride [51] and 5-methylphenylaminonaphthalene-1-sulfonyl (Mns) chloride [52], are also available. The derivatization procedures with these reagents are almost the same as with Dns-Cl. The derivatives obtainable with the former two are somewhat more lipophilic.

4-Phenylspiro[furan-2(3H)-1'-phthalan]-3'-dione, also called fluorescamine, is nonfluorescent itself. Among alcohols, primary

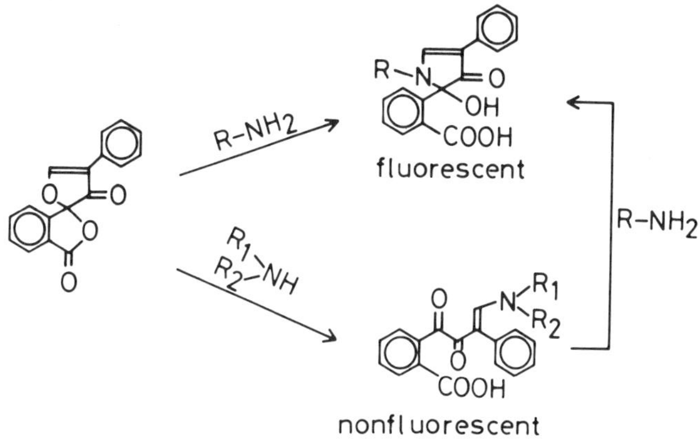


Figure 9.1 Derivatization of primary and secondary amines with fluorescamine.

and secondary amines, only primary amines provide highly fluorescent pyrrolinones (Fig. 9.1) [53-71]. Accordingly, fluorescamine is usually employed for the detection of primary amines as a postchromatographic derivatization reagent, although it is also applicable to prechromatographic derivatization. The derivatization reaction is attained by addition of a fluorescamine solution in ethanol, acetonitrile, or acetone into a test solution of amines in borate or phosphate buffer (pH 7.5-9.0). The use of acetone as a solvent is more suitable because of the resulting Schiff base linkage. Owing to the formation of an equilibrated mixture of two fluorescent products, this derivatization technique is unfavorable for prechromatographic labeling of α-amino acids (Fig. 9.2).

Usually, secondary amines are detected after conversion to primary amines by oxidation with *N*-chlorosuccinimide followed by derivatization with fluorescamine [72,73]. It is also reported that the nonfluorescent aminoenones formed from fluorescamine and secondary

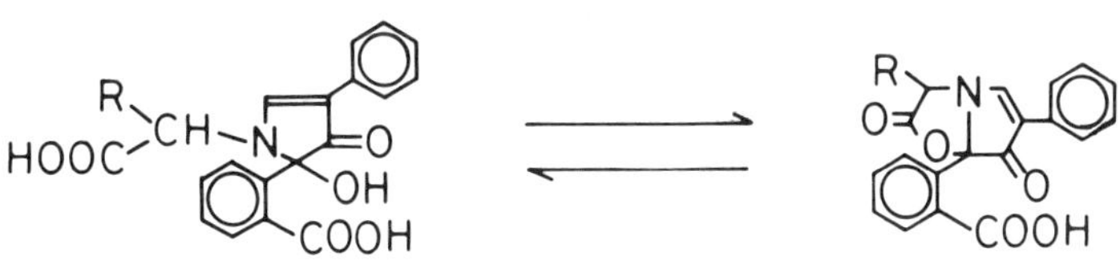

Figure 9.2 Formation of two products during derivatization of amines with fluorescamine.

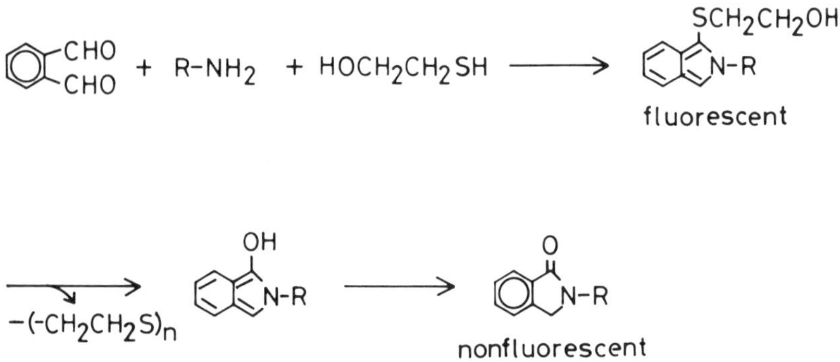


**Figure 9.3** Derivatization of primary amines with *o*-phthaldialdehyde in the presence of 2-mercaptoethanol.

amines can be transformed into fluorescent products by treatment with primary amines (Fig. 9.1) [74]. 2-Methoxy-2,4-diphenyl-(2H)-furanone (MDPF), whose structure is closely related to that of fluorescamine, also acts with primary amines to yield fluorescent pyrrolinones [74].

Secondary amines are converted to aminodienones which in turn are converted to fluorescent products with ethanolamine [75]. *o*-Phthaldialdehyde (OPA) reacts with primary amines in the presence of a thiol compound such as ethanethiol, 3-mercapto-1-propanol, 2-mercaptoethanol, or *N*-acetyl-L-cysteine, providing fluorescent isoindole derivatives [17,46,76-104]. This reaction is utilized for pre- or postchromatographic derivatization. The sensitivity obtainable by the OPA method is somewhat higher than that with fluorescamine. Secondary amines also become detectable upon prior oxidation with chloramine T [105], *N*-chlorosuccinimide [57,72], or sodium hypochlorite [106,107]. The fluorescent isoindoles formed with primary amines and thiols undergo spontaneous rearrangement, yielding the nonfluorescent derivatives (Fig. 9.3). The unfavorable instability of the isoindole derivatives can be avoided by the use of enthanethiol [53,108-111] or 3-mercapto-1-propanol [112] instead of 2-mercaptoethanol. It is reported that derivatization of histidine with OPA provides three peaks on the chromatogram [113]. The use of *N*-acetyl-L-cysteine with OPA for derivatization of amino acid enantiomers affords diastereomers which are readily resolved by LC on a conventional column [114].

A nonfluorescent reagent, 4-chloro-7-nitrobenzo-2-oxa-1,3-diazole (NBD-Cl), furnishes fluorescent derivatives with primary and secondary amines [115-119]. This sensitive reaction is used for prechromatographic as well as postchromatographic derivatization.

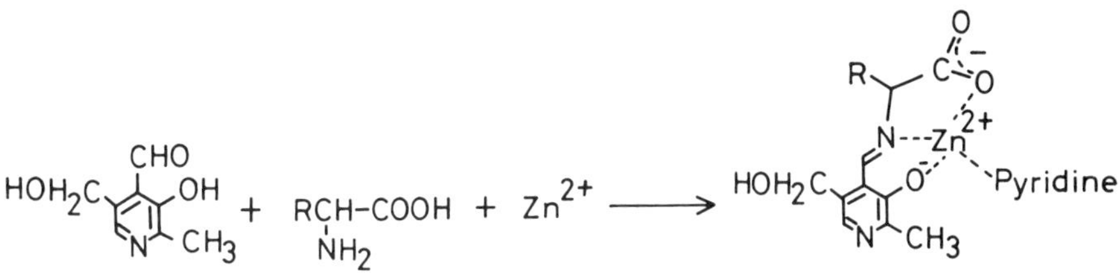


Figure 9.4 Chelation of amino acids with Zn(II) and pyridoxal.

Although this reagent is also capable of reacting with phenolic hydroxyl and sulfhydryl groups, its higher reactivity toward amino groups is effective for selective derivatization. 4-Fluoro-7-nitrobenzo-2-oxa-1,3-diazole (NBD-F), with a more electronegative fluoro substituent, is even more reactive than NBD-Cl [120-125].

The chloroformate-type reagents such as 9-fluorenylmethylchloroformate (FMOCCl) [126] of 2-naphthylchloroformate (NCF) [127] are also available for prechromatographic derivatization of primary and secondary amines. 1,2-Naphthoylenebenzimidazole-6-sulfonyl chloride is applicable to the prechromatographic derivatization of aliphatic primary and secondary amines.

So-called Edman-type isothiocyanate reagents, fluoresceinisothiocyanate [128], 9-isothiocyanatoacridine [129], 4-dimethylamino-1-naphthylisothiocyanate [130], or BOC-aminophenyl- and BOC-aminoethylphenylisothiocyanates [131] react with primary amines to give fluorescent thioureido derivatives. Condensation of *N*-succinimidyl-2-naphthoxyacetate with amino acids provides fluorescent derivatives [132,133]. The amino groups adjacent to carboxyl, carbonyl, or hydroxyl functions react with pyridoxal and Zn(II) to form fluorescent metal chelate compounds (Fig. 9.4) [134].

9,10-Phenanthrenequinone, which is used as a postchromatographic derivatization reagent, condenses with guanidino compounds, providing fluorescent 2-aminophenanthrimidazoles [135,136].

A number of representative examples of derivatizing amino compounds are given below.

*Derivatization of Alkaloids with Dansyl Chloride [33]*

An aqueous solution of alkaloids (10 μl, <30 nmol) is mixed with 0.1% Dns-Cl in acetone (50 μl) and 0.1 M sodium carbonate (50 μl), and the mixture is kept at 318 K for 20 min. The reaction mixture is extracted with benzene (0.5 ml) and an aliquot of the organic layer is injected onto a normal phase (NP) column. The derivatized alkaloids are monitored by FD, the limit for cephalin, emetine, and ephedrine being 1 ng.

### *Derivatization of 7-Aminoflunitrazepam and 7-Aminodesmethylflunitrazepam with Fluorescamine [65]*

To a plasma sample (0.5-4 ml) containing amines (5-50 ng) is added buffer solution (pH 10, 2 ml) and 1% isoamyl alcohol in ether (10 ml). After centrifugation, the amines in the organic layer are reextracted into 0.5 M sulfuric acid (2 ml). The acid phase is adjusted to pH 9-10 with 1 M sodium hydroxide solution and then extracted with ether (10 ml). After evaporation to dryness under a stream of nitrogen, the residue is dissolved in buffer solution (pH 8)/acetonitrile (3:1, 100 μl), followed by the addition of 0.5% fluorescamine in acetone (20 μl). The mixture is shaken well for 60 sec and an aliquot is injected onto a reversed phase (RP) column. The detection is carried out with a fluorometer (excitation wavelength 390 nm; emission wavelength 470 nm), the limit of detection being 0.5-1 ng/ml plasma.

### *Derivatization of Primary and Secondary Amines with 2-Methoxy-2,4-diphenyl-3(2H)-furanone [75]*

To an aqueous sample solution (100 μl) is successively added 50 mM sodium borate buffer (pH 9.6, 100 μl) and 10 mM MDPF in acetonitrile (100 μl), and the whole is equilibrated at 293 K for 30 min. An aliquot of the solution is injected onto a RP column. The effluent is led to the first fluorometer to detect fluorescent pyrrolinones formed from primary amines. The eluate from the flow cell is mixed with 5 M ethanolamine solution (pH 10, adjusted with concentrated hydrochloric acid) and the combined flows are passed through a reaction coil at 333 K. During this procedure, nonfluorescent aminodienones formed from secondary amines are converted to fluorescent pyrrolinones. The pyrrolinone derivatives are monitored by FD (excitation wavelength 390 nm; emission wavelength 480 nm), the limit of detection being 3 pmol for primary amines and 50 pmol for secondary amines.

### *Derivatization of Catecholamines with o-Phthaldialdehyde [102]*

A plasma sample is deproteinized with perchloric acid, adjusted to pH 7.0 with 0.5 M potassium hydroxide and immediately derivatized with OPA (350 μg) in buffer solution (pH 10.4, 400 μl). After addition of sodium chloride (2 g), the reaction mixture is extracted with ethyl acetate and the organic layer concentrated to about 100 μl under a stream of nitrogen. An aliquot of the solution is injected onto a RP column. The eluate is monitored by FD (excitation wavelength 340 nm; emission wavelength 480 nm), the limit of detection being 100 pg.

*Postchromatographic Derivatization of Cystamine with o-Phthaldialdehyde [98]*

Cystamine in biological materials is separated on a cation exchange resin column. The eluate is mixed with 0.08% sodium hypochlorite followed by the addition of 0.08% OPA solution containing 0.2% 2-mercaptoethanol, and the reaction product is monitored by FD (excitation wavelength 348 nm; emission wavelength 450 nm), the limit of detection being 2 pmol.

*Derivatization of Amines, Amino Acids, and Imino Acids with 4-Fluoro-7-nitrobenzo-2-oxa-1,3-diazole [124]*

To a solution of amines, amino acids, or imino acids (about 30 nmol) in phosphate buffer (pH 7.0, 0.2 ml) is added a freshly prepared solution of NBD-F (about 300 nmol) in ethanol (0.2 ml). The mixture is heated at 323 K for 90 min and after addition of 0.5 M hydrochloric acid (25 μl) to stop the reaction, an aliquot of the solution is injected onto a column packed with polyethylene glycol dimethacrylate gel. The derivatives are monitored by FD (excitation wavelength 467-472 nm; emission wavelength 524-541 nm), the limits of detection for putrescine and spermidine being 76 and 430 fmol, respectively. The reaction rates for secondary amine, imino acid, primary amine, and amino acid decrease in this order. Phenolic compounds also react with this reagent to form weakly fluorescent derivatives.

*Derivatization of Pirlimycin with Fluorenylmethylchloroformate [126]*

Pirlimycin in serum (1 ml) is extracted with chloroform. To the extract is added 0.25 mM sodium hydroxide (10 μl) and FMOCC1 (6.4 μl) in acetonitrile (10 μl). The whole mixture is allowed to stand at room temperature for 2.5 hr and an aliquot of the solution is injected onto a RP column. The carbamate derivative is monitored by FD (excitation wavelength 270 nm; emission wavelength 300 nm), the limit of detection being 0.1 ng/ml serum.

*Derivatization of Tertiary Amines with 2-Naphthylchloroformate [127]*

The tertiary amines (10 pmol-5 nmol) are heated with a 10-fold molar excess of NCF potassium carbonate (10 mg) in dry benzene (100 μl) at 373 K for 1 hr. The excess reagent is removed by extracting with saturated methanolic potassium hydroxide solution (300 μl) for 1 min. After addition of water (1 ml), an aliquot of the benzene layer is injected onto a RP column. The resulting carbamates are monitored by FD (excitation wavelength 275 nm; emission wavelength 335 nm), the limit of detection for diphenhydramine being 200-300 pg.

The absolute yield of the carbamate produced from diphenhydramine is 81%. This method is applicable to the tertiary amines having the *N*-methyl, -ethyl, or -benzyl group. Primary and secondary amines are easily converted to the corresponding carbamates under the mild conditions.

*Derivatization of Catecholamines with 1,2-Diphenylethylenediamine [214]*

To an aqueous solution (150 μl) of catecholamine adjusted to pH 6.0-6.5 is added 20 mM potassium hexacyanoferrate(III) (10 μl), acetonitrile (150 μl), and 0.1 M 1,2-diphenylethylenediamine in 0.1 M hydrochloric acid (50 μl), and the mixture is allowed to stand at 310 K for 40 min. An aliquot is injected onto a RP column. The derivatives are monitored by FD (excitation wavelength 345 nm; emission wavelength 486 nm), the limit of detection being 2 fmol.

*Postchromatographic Derivatization of Catecholamines with Glycylglycine [101]*

The catecholamines are separated on an ion exchange resin column. The eluate is mixed with 1.5% glycylglycine in 0.5 M boric acid solution containing 2 mM zinc sulfate and 0.01% hexacyanoferrate(III) in 0.5 M potassium borate buffer (pH 9.4). The combined solution is passed through a reaction coil at 368 K and, after cooling, the fluorescent derivatives are monitored by FD (excitation wavelength 350 nm; emission wavelength 500 nm), the limit of detection being 1 ng.

*Fluorescence Detection of Methadone, Phencyclidine, and Their Metabolites by Postchromatographic Ion Pair Extraction Using 9,10-Dimethoxyanthracene-2-sulfonic Acid [13]*

The amines are separated by ion pair chromatography on a RP column using 9,10-dimethoxyanthracene-2-sulfonic acid (30 mg/liter) as a counterion and the eluate is mixed with chloroform. The combined flows are passed through an extraction coil and then a phase separator. The organic layer is monitored by FD (excitation wavelength 380 nm; emission wavelength 445 nm). The detection limit is 1.0-6.0 ng/ml plasma.

*Derivatization of Guanidino Compounds with Benzoin [216]*

To a 200-μl aliquot of a sample solution is added 4 mM benzoin in methylcellosolve (100 μl), 0.1 M 2-mercaptoethanol and 0.2 mM sodium sulfite (100 μl), and 2 M potassium hydroxide (200 μl) under ice cooling, and the solution is heated at 373 K for 5 min. Hydrochloric acid-Tris-hydrochloric acid buffer (pH 9.2, 200 μl) is added and an aliquot of the mixture is injected onto a RP column. The 2-substituted amino-4,5-diphenylimidazoles formed are monitored by FD (excitation wavelength 325 nm; emission wavelength 425 nm), the limit of detection being 20-100 fmol.

### Carboxyl Group

A number of alkylating agents with active halogens as leaving groups have been developed for the fluorescence labeling of the carboxyl group. The use of crown ether which is capable of forming a soluble cation complex with basic catalysts such as potassium carbonate or potassium hydroxide accelerates the derivatization reactions. This type of reagent reacts not only with the carboxyl group but also with other acidic functional groups.

4-Bromomethyl-7-methoxycoumarin (Br-Mmc) was first developed as a prechromatographic derivatization reagent for the carboxyl group and has been widely used [137-145]. The detection limit for fatty acids is lower than 10 pmol. 4-Hydroxy-7-methoxycoumarin (Hy-Mmc) has been newly developed to improve the instability of Br-Mmc [141,146,147]. 4-Bromomethyl-6,7-dimethoxycoumarin (Br-Mdmc) provides esters having higher quantum yields than the corresponding Mmc esters [148].

The esters formed from 4-bromomethyl-7-acetoxycoumarin (Br-Mac) undergo facile hydrolysis with alkali to yield highly fluorescent 7-hydroxycoumarin. Accordingly, the method involving chromatographic separation of Mmc esters followed by alkaline hydrolysis of the eluate is recommended [149,150].

Recently, 3-bromomethyl-6,7-dimethyoxy-1-methyl-1(1H)-quinoxalinone was developed as a prechromatographic labeling reagent [151-153]. The detection limit obtainable for fatty acids is approximately 1-3 fmol. Similar alkylating agents, naphthacyl bromide [154], panacyl bromide [155], 1-bromoacetylpyrene [156], and 9-(chloromethyl)-anthracene [157], are also available.

The diazoalkane derivatives such as 4-diazomethyl-7-methoxycoumarin (D-Mmc) [158] or 9-anthryldiazomethane [159-161] are highly reactive toward the carboxyl group without catalyst. The major disadvantage is that most of these reagents lack stability.

The *O*-alkyl isourea derivatives, which are readily prepared from equimolar amounts of alcohol and carbodiimide, are well known as labeling reagents for the carboxyl group. *N,N'*-Dicyclohexyl- and *N,N'*-diisopropyl-*O*-(7-methoxycoumarin-4-yl)-methylisourea (DCCI, DICI) produce fluorescent esters [147,162, 163].

Fluorescent amines can also be utilized for the labeling of carboxyl groups. In general, it is necessary that the carboxylic acid function be activated prior to coupling with amines. For this purpose carbodiimide derivatives such as *N,N'*-dicyclohexylcarbodiimide or *N*-ethyl-*N'*-(3-dimethylaminopropyl)carbodiimide hydrochloride are

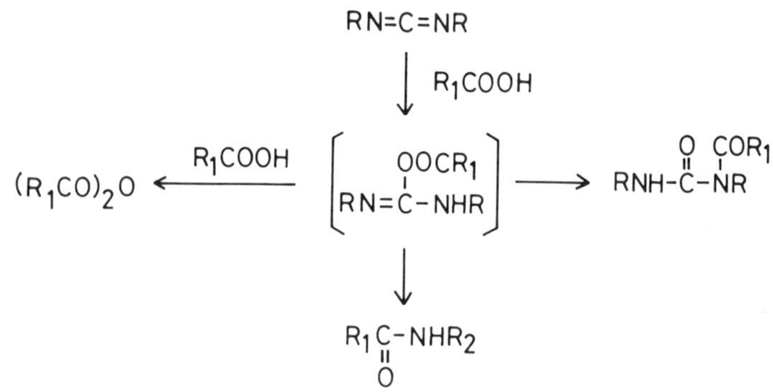


Figure 9.5 Reaction of carboxylic acids with carbodiimide derivatives.

widely used [164-166]. The addition of 1-hydroxybenzotriazole is effective for suppressing the formation of *N*-acylisourea derivatives (Fig. 9.5) [165,166]. Activation of carboxylic acids is also attained by the use of 2-bromo-1-methylpyridinium iodide and *N,N'*-carbonyldiimidazole [166], although oxalyl chloride or thionyl chloride [167] can be used to prepare acid chlorides. 9,10-Diaminophenanthrene (DAP) condenses with carboxylic acids, resulting in fluorescent phenanthroimidazole derivatives (Fig. 9.6) [179]. In this condensation reaction, methyl polyphosphate is used for dissolving DAP. Derivatization of α-keto acids with *o*-phenylenediamine produces 2-quinoxalinol [169-172].

A number of representative examples of derivatizing carboxylic acids are given below.

*Derivatization of Fatty Acids with 4-Bromomethyl-6,7-dimethoxycoumarin [148]*

The fatty acids are dissolved in anhydrous acetone (100 μl) and neutralized with 10% methanolic potassium hydroxide. A threefold molar excess of Br-Mdmc in acetone, 0.2 M 18-crown-6 in acetonitrile (100 μl), and anhydrous potassium carbonate (2 mg) is added to the

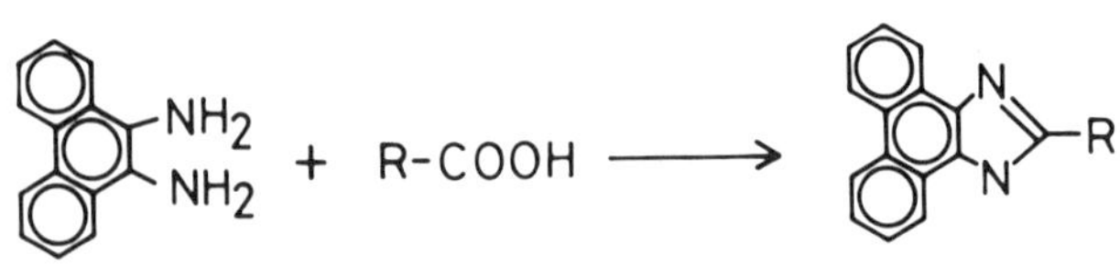


Figure 9.6 Derivatization of carboxylic acids with 9,10-diaminophenanthrene.

solution and heated at 343 K for 30 min. An aliquot of the reaction mixture is injected onto a RP column. The fatty acid esters are monitored by FD (excitation wavelength 340 nm; emission wavelength 425 nm), the limit of detection being 0.5 pmol.

*Derivatization of Fatty Acids with 4-Bromomethyl-7-acetoxycoumarin [149]*

The fatty acids (0.1-20 nmol), a 2.5-fold molar excess of Br-Mac, and an equimolar amount of dibenzo-18-crown-6 are dissolved in acetone (50 µl) and placed in an ampule containing a finely powdered mixture of potassium hydrogen carbonate and sodium sulfate (1:1, 3 mg). The ampule is sealed, heated at 323 K for 30 min in the dark, and an aliquot of the solution is injected onto a RP column. After separation, the esters are hydrolyzed into the highly fluorescent compound in the continuous flow as follows: The eluate from the separation column is mixed with 0.1 M borate buffer (pH 11.0) and the combined flows are passed through a short coil at 323 K. The liberated 4-hydroxymethyl-7-hydroxycoumarin is monitored by FD (excitation wavelength 365 nm; emission wavelength 460 nm), the limit of detection being 10 fmol.

*Derivatization of Ibuprofen with 9-(Hydroxymethyl)anthracene [166]*

A plasma sample (150 µl) is acidified with 4 M hydrochloric acid (10 µl) and then extracted with dichloromethane (1 ml). The organic layer is evaporated under a stream of nitrogen and the residue redissolved in chloroform (100 µl) containing 1-hydroxybenzotriazole (5 g). After evaporation to dryness under a stream of nitrogen, *N*-ethyl-*N'*-(3-dimethylaminopropyl)carbodiimide hydrochloride (30 µg) in chloroform (30 µl) and triethylamine (5 µl) are added. The resulting solution is kept at 273 K for 30 min. Lower yield may be obtained when the reaction is run at higher temperatures. A solution of 9-(hydroxymethyl)anthracene (20 µg) in chloroform (100 µl) is then added and the mixture is allowed to stand at room temperature for 10 min. After evaporation, the residue obtained is redissolved in LC eluent (1 ml) and an aliquot of the solution is injected onto a RP column. 2-Bromo-1-methylpyridinium iodide and *N,N'*-carbonyldiimidazole instead of *N*-ethyl-*N'*-C3-dimethylaminopropyl)carbodiimide are also useful as activators for esterification of carboxylic acids.

*Derivatization of Bile Acids with 1-Bromoacetylpyrene [156]*

To a solution of bile acids in methanol (100 µl) is added 0.01% potassium hydroxide in methanol (50 µl) and the solution is evaporated under a stream of nitrogen. The residue is dissolved in acetonitrile (200 µl) containing 0.05% dicyclohexyl-18-crown-6 and 12.5 mM

1-bromoacetylpyrene and heated at 313 K for 30 min. After cooling, an aliquot of the solution is injected onto a RP column. Bile acid esters are monitored by FD (excitation wavelength 370 nm; emission wavelength 440 nm), the limit of detection being 5 pmol. Since taurine-conjugated bile acids are unable to be directly esterified, group separation into the free, glycine-, and taurine-conjugated fractions followed by deconjugation with alkali is necessary prior to derivatization.

*Derivatization of Nicotinic Acid with N,N'-Dicyclohexyl-O-(7-methoxycoumarin-4-yl)-methylisourea [163]*

To a solution of nicotinic acid in acetone (50 μl) is added 5 mM DCCI in acetone (50 μl), and the whole is heated at 373 K for 15 min. An aliquot of the reaction mixture is injected onto a RP column. The resulting ester is monitored by FD, the limit of detection being 1 pmol. In general, aliphatic and aromatic carboxylic acids are derivatized more easily in tetrahydrofuran of benzene. A similar derivatization reagent, *N*,*N'*-diisopropyl-*O*-(7-methoxycoumarin-4-yl)-methylisourea (DICI), is also available.

*Derivatization of Prostaglandins with 9-Anthryldiazomethane [160]*

A solution of prostaglandins in ethyl acetate is put into a 2-ml test tube and the ethyl acetate is removed by a stream of nitrogen. To the residue obtained is added 0.1% 9-anthryldiazomethane in ethyl acetate (50 μl), and the solution is kept at 313 K for 30 min. The reaction mixture is diluted with ethyl acetate (70 μl) and a 100-μl aliquot of the solution is applied to a gel permeation column to remove the interfering substances derived from the reagent. The fraction containing fluorescent prostaglandin esters is collected and evaporated in vacuo. The residue is dissolved in acetonitrile (180 μl) and an aliquot of the solution is injected onto a RP column. The 9-anthrylmethyl esters are monitored by FD (excitation wavelength 365 nm; emission wavelength 418 nm), the limit of detection being 30-120 pg. When a large quantity of prostaglandin (1 μg) is derivatized, the reaction mixture may be directly applicable to LC without further purification. Derivatization with this reagent can be widely used for carboxylic acids.

*Derivatization of Naproxen Enantiomers with 1-(4-Dimethylamino-1-naphthyl)ethylamine [165]*

A serum sample containing naproxen enantiomers (5-200 ng) is subjected to clean-up with an octadecylsilyl-bonded silica cartridge. To the residue obtained is added 1-hydroxybenzotriazole (5 μg), (-)- or (+)-1-(4-dimethylamino-1-naphthyl)ethylamine in pyridine (20 μl), and 1-ethyl-3-(3-dimethylaminopropyl)carbodiimide hydrochloride

(100 μg) in dichloromethane (200 μl), and the mixture is allowed to stand at 277 K for 45 min. No racemization of naproxen or the reagent takes place during the condensation reaction. A lower yield is obtained when an excess of 1-hydroxybenzotriazole is used. After removal of the solvent under a stream of nitrogen, the residue is purified by minicolumn chromatography on silica and then subjected to LC analysis on a NP column. Elution of the diastereomeric amides is monitored by FD (excitation wavelength 320 nm; emission wavelength 410 nm), the limit of quantitation being 100 pg.

### *Derivatization of Fatty Acids with 9,10-Diaminophenanthrene [168]*

To a solution of fatty acids in toluene or chloroform (10 μl) is added DAP (1 mg) in methyl polyphosphate (0.5 ml). The mixture is heated at 358 K for 3-6 min and then injected directly onto a RP column. Quantitative derivatization takes place for all fatty acids except formic acid, which undergoes degradation under these reaction conditions. If a large volume of the sample solution has to be analyzed, 5 M aqueous ammonium hydroxide (0.5 ml) is added to the reaction mixture with rapid cooling; the resulting solution is mixed thoroughly, and then an aliquot of the chloroform layer is subjected to LC. The 2-substituted phenanthrimidazole derivatives formed are monitored by FD (excitation wavelength 255 nm; emission wavelength 385 nm), the limit of detection being 10 pg.

### *Derivatization of α-Keto Acids with 1,2-Diamino-4,5-dimethoxybenzene [220]*

To a serum or urine sample (50 μl) is added 5 mM 1,2-diamino-4,5-dimethoxybenzene monohydrochloride in 0.5 hydrochloric acid (50 μl) containing 0.7 M 2-mercaptoethanol, heated at 373 K for 2.5 hr in the dark, and an aliquot of the reaction mixture is injected onto a RP column. The 3-substituted 6,7-dimethoxy-2(1H)-quinoxalinone derivatives formed are monitored by FD (excitation wavelength 368; emission wavelength 455 nm), the limit of detection being 10-300 fmol.

### *Derivatization of α-Keto Acids with o-Phenylenediamine [170]*

After deproteinization with methanol, a plasma sample containing α-keto acids is mixed with three drops of 0.1 M acetic acid and 0.1 M sodium chloride (1 ml), and then applied to a column of hydrazide gel (0.3 ml). After washing with 0.1 M sodium chloride (5 ml), the gel is transferred to a test tube and 0.1% *o*-phenylenediamine sulfate in 2 M hydrochloric acid (2 ml) containing 2-mercaptoethanol (5 μl) is added. The mixture is heated at 353 K for 2 hr and then diluted with a sodium sulfate-saturated solution (8 ml). The quinoxalinol derivatives thus formed are extracted with ethyl acetate and the extract

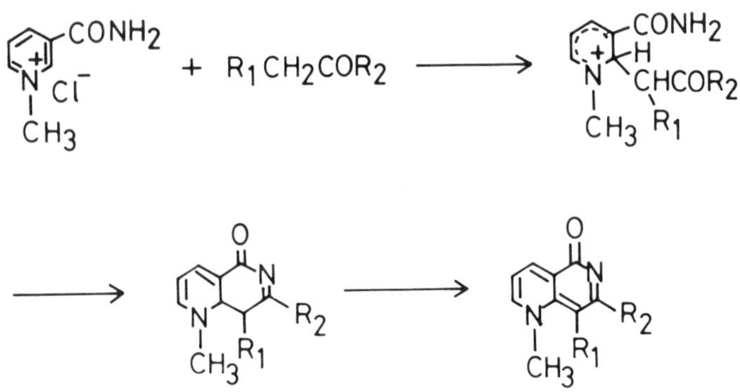

Figure 9.7 Reaction of *N*-methylnicotinamide with α-oxomethylene derivatives.

is evaporated to dryness. The residue is redissolved in two drops of dimethylformamide and five drops of water, and an aliquot of the solution is injected onto a RP column. The derivatives are monitored by FD (excitation wavelength 350 nm; emission wavelength 410 nm). The quinoxalinol derivative from *p*-hydroxyphenylpyruvic acid exhibits no fluorescence.

## Carbonyl Group

Dansylhydrazine (Dns-H) reacts with the carbonyl function in the presence of trichloroacetic acid to provide the highly fluorescent hydrazone [173-183]. The dansylhydrazones should be submitted to LC analysis as soon as possible due to their limited stability. 4'-Hydrazino-2-stilbazole [184-185] and 4-hydrazino-2-oxa-1, 3-diazole (NBD-H) [186] are also reactive toward the carbonyl compounds to afford fluorescent hydrazones.

*N'*-Methylnicotinamide chloride reacts selectively with the oxomethylene group ($-CH_2CO-$), providing fluorescent derivatives (Fig. 9.7) [187-198]. This reaction is applicable to pre- and postchromatographic derivatization where the detection limit of about 500 fmol is obtainable for acetophenone.

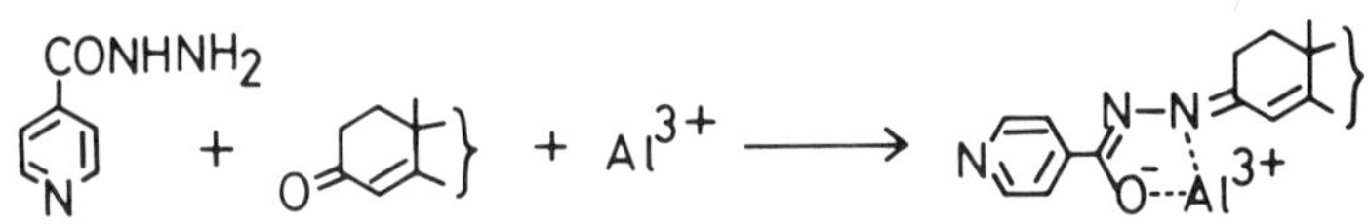

Figure 9.8 Chelation of $\Delta^4$-3-ketosteroids with Al(III) and isonicotinoyl hydrazide.

Isonicotinoyl hydrazide (INH) is capable of forming the fluorescent metal chelates with $\Delta^4$-3-ketosteroids in the presence of Al(III) (Fig. 9.8) [22], and this reaction is employed for postchromatographic derivatization.

A number of representative examples of derivatizing aldehydes and ketones are given below.

### *Derivatization of 3-Oxo Bile Acids with Dansyl Hydrazine [180]*

3-Oxo bile acids formed from common bile acids with 3α-hydroxysteroid dehydrogenase in the presence of $NAD^+$ are dissolved in benzene (200 μl) containing 0.1% trichloroacetic acid and 0.2% Dns-H, and allowed to stand at 303 K for 10 min. After evaporation to dryness under a stream of nitrogen, the residue is redissolved in methanol and an aliquot of the solution injected onto a RP column. The hydrazones are monitored by FD (excitation wavelength 365 nm; emission wavelength 520 nm), the limit of detection being 0.5 pmol.

### *Derivatization of Phenylpyruvic Acid with 4'-Hydrazino-2-stilbazole [185]*

Phenylpyruvic acid in urine or serum is extracted with ethyl acetate and the extract evaporated to dryness in vacuo at room temperature To the residue are added water (1 ml), 0.5 M ammonium chloride (pH 4.0, adjusted with 0.1 M hydrochloric acid) (0.5 ml), and 1.2 mM 4'-hydrazino-2-stilbazole dihydrochloride in methanol (0.5 ml). This mixture is heated at 323 K for 10 min in the dark and then cooled in ice. Within 2 hr, an aliquot of the reaction mixture is injected onto a RP column and the eluate is monitored by FD, the limit of detection being 25 pmol/0.2 ml urine and 32 pmol/0.2 ml serum.

### *Derivatization of Aldehydes with 5,5-Dimethyl-1,3-Cyclohexanedione(Dimedone) [221]*

To an aqueous sample (2 ml) containing aldehydes are added extraction reagent (1 ml) and 9 M sulfuric acid (0.25 ml), and the solution is heated at 373 K for 20 min. An aliquot of the mixture is injected directly onto a RP column. The derivatives are monitored by FD (excitation wavelength 385 nm; emission wavelength 460 nm), the limit of detection being 30 fmol. The extraction reagent consists of ammonium acetate (60 g) and dimedone (2.1 g) in isopropanol (17.5 ml) made up to 100 ml with water. To eliminate aldehyde contamination, the reagent is heated at 373 K for 30 min in a sealed vessel, cooled to room temperature, and extracted with two 20-ml portions of carbonyl-free dichloromethane.

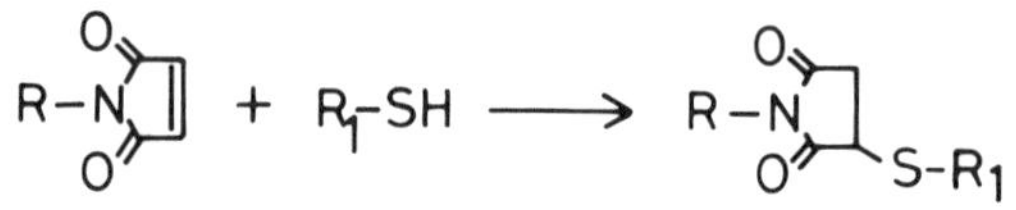

Figure 9.9 Reaction of thiols with maleimide derivatives.

## Sulfhydryl Group

Maleimide derivatives with potential fluorophores, such as *p*-(2-benzimidazolyl)phenyl (BIPM) [190], 1-pyrenyl (PM) [191-193], 9-acridinyl (NAM) [194,195], 1-anilinonaphthyl (ANM) [196], 7-dimethylamino-4-methylcoumarin (DACM) [193,197], or *p*-(2-benzoxazolyl)-phenyl (BOPM) groups [198], have been developed as prechromatographic derivatization reagents for the analysis of thiols. These reagents react selectively with the sulfhydryl groups under mild conditions to form Michael adducts (Fig. 9.9). In almost all cases the reagents themselves do not show any significant fluorescence while the resulting adducts are highly fluorescent. This type of derivatization reaction proceeds readily in biological fluids. A postchromatographic derivatization method using NAM is also available.

Dansyl derivatives have also been proposed for prechromatographic labeling of thiols. Dansylaziridine (Dns-A) reacts selectively with sulfhydryl groups (Fig. 9.10) [199] and a nonfluorescent reagent, *N*-chlorodansylamide (NCDA) [200-201], furnishes the highly fluorescent dansylamide with sulfhydryl groups (Fig. 9.11). However, this process is not a regular derivatization. 4-Fluorobenzo-2-oxa-1,3-diazole-7-sulfonate (SBD-F) and 4-fluoro-7-sulfamoylbenzo-2-oxa-1,3-diazole ($NH_2$-SBD-F) are used for prechromatographic derivatization of thiols, detection limits being in the pmol level [202,203], OPA reacts with thiols in the presence of a primary amine such as taurine, providing fluorescent derivatives.

A number of representative examples of derivatizing sulfhydryl groups are given below.

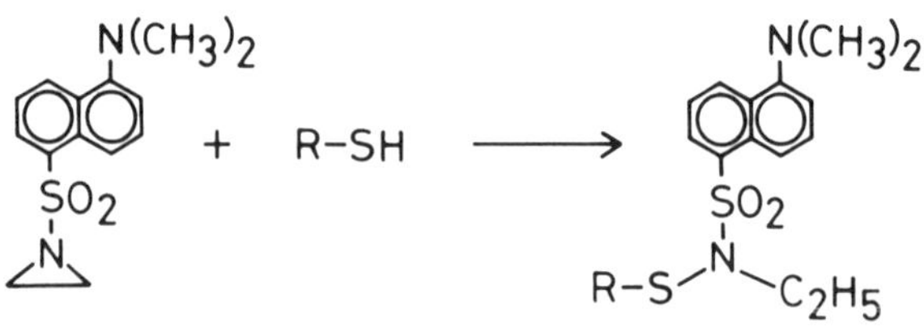

Figure 9.10 Derivatization of thiols with dansylaziridine.

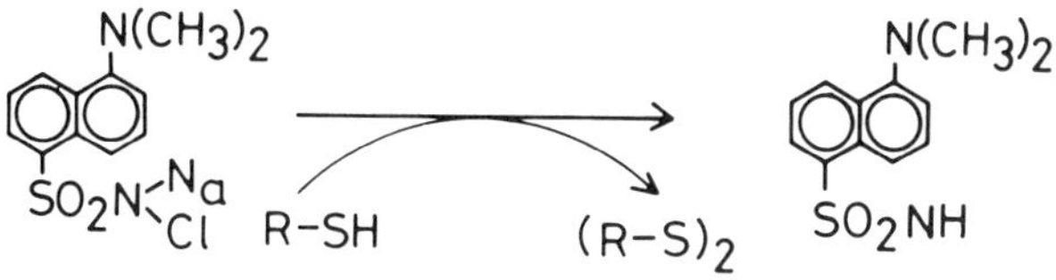


Figure 9.11 Derivatization of thiols with *N*-chlorodansylamide.

*Derivatization of Captopril with N-(1-Pyrene)maleimide [192]*

A plasma sample (1 ml) is deprotoneinized with perchloric acid and the supernatant adjusted to pH 6.5 with 1 M potassium phosphate. After addition of 2 mM *N*-(1-pyrene)maleimide in acetone (25 µl), the solution is allowed to stand at room temperature for 15 min and then extracted with ethyl acetate-benzene (1:1, 3 ml). The extract is dried, redissolved in methanol, and an aliquot of the solution is injected onto a RP column. The adduct is monitored by FD (excitation wavelength 340 nm; emission wavelength 390 nm), the limit of detection being 150 pmol/ml plasma.

*Derivatization of Thiols with Ammonium 4-Fluorobenzo-2-oxa-1,3-diazole-7-Sulfonate [202]*

To a fresh blood sample (1 ml) is added an equal volume of chilled 10% trichloroacetic acid containing 1 mM disodium EDTA and then centrifuged. To the supernatant (10 µl) is added 1 mM disodium EDTA (80 µl), 2.5 M borate buffer (pH 9.5, 0.2 ml) containing 4 mM disodium EDTA, and ammonium SBD-F (200 µg) in 2.5 M borate buffer (pH 9.5, 0.1 ml), and the mixture is allowed to stand at 333 K for 60 min. An aliquot is injected onto a RP column. The derivatives are monitored by FD (excitation wavelength 385 nm; fluorescence wavelength 515 nm), the limit of detection being 0.07-1.1 pmol. The derivatives are stable for at least a week when kept at pH 9.5 in a refrigerator.

*Derivatization of Thiols with Bromobimane [222]*

To a solution of 0.1 M ammonium bicarbonate (1.9 ml) containing 40 mM bromobimane in acetonitrile (50 µl) is added in aqueous solution of thiols (0.1 ml). The mixture is shaken at ambient temperature in the dark for 5 min. After extraction of the excess reagent with dichloromethane, the aqueous phase is adjusted to pH 7 and then injected onto a RP column. The derivatives are monitored by FD (excitation wavelength 356 nm; emission wavelength 450 ± 20 nm), the limit of detection for 2,3-dimercaptopropane-1-sulfonic acid being 10 pmol.

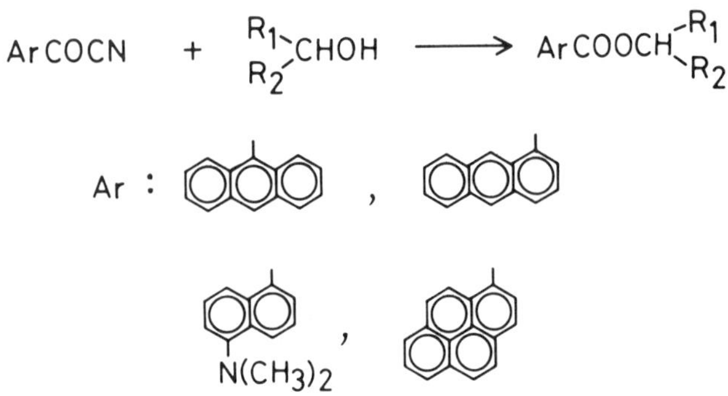


Figure 9.12 Reaction of primary and secondary alcohols with nitrile derivatives.

*Postchromatographic Derivatization of Thiols and Disulfides with o-Phthaldialdehyde [204]*

The thiols and disulfides are separated by ion exchange chromatography. The eluate is mixed with a sulfite-taurine reagent containing 25 mM taurine, 1 mM disodium EDTA, 15 mM sodium sulfite, and 1 mM L-ascorbic acid in 0.75 M phosphate buffer (pH 7.0), and the mixture is passed through a short coil at 323 K to reduce the disulfide to the corresponding thiol. The eluate and sulfite-taurine reagent mixture reacts with a fluorogenic reagent containing 25 mM OPA and 1 mM disodium EDTA in carbonate buffer (pH 10.5). The isoindole derivatives formed are monitored by FD, the limit of detecting being 250-750 pmol.

### Hydroxyl Group

Carboxylic acid chlorides are commonly used for derivatization of hydroxyl compounds. These derivatization reagents, however, are not necessarily satisfactory with respect to stability and feasibility. Recently, fluorescent reagents with carbonyl nitrile as a reacting group were developed to overcome these problems. 1-Anthroylnitrile (1-AN) [205-208] and pyrene-1-carbonylnitrile [209] as well as 4-dimethylamino-1-naphthoylnitrile [210] are capable of reacting with primary and secondary hydroxyl functions under mild conditions, while 9-anthroylnitrile (9-AN) [205,211,212] reacts with primary hydroxyl groups, providing the highly fluorescent esters (Fig. 9.12). For instance, the detection limit for bile acids with 1-AN is estimated to be 20 fmol.

7-Methoxycoumarin-3- and -4-carbonylazides (3-MCCA, 4-MCCA) are also proposed as prelabeling reagents for primary and secondary alcohols [213]. 3-MCCA provides better fluorescent derivatives than 4-MCCA.

A number of representative examples of derivatizing hydroxyl containing compounds are given below.

*Derivatization of Bile Acids with 1-Anthroylnitrile [206]*

To a solution of bile acids (1-100 pmol each) in acetonitrile (100 μl) containing 0.16% quinuclidine is added 1-AN (200 μg) in acetonitrile (100 μl), and the mixture is kept at 333 K for 20 min. After addition of methanol (50 μl) to decompose the excess reagent, the fluorescent bile acid derivatives are separated into the free, glycine and taurine conjugate fractions by ion exchange chromatography on the lipophilic gel piperidinohydroxypropyl Sephadex LH-20. Subsequent resolution of each fraction into individual bile acid derivatives is attained on a RP column. The 3-(1-anthroyl) bile acids are monitored by FD (excitation wavelength 370 nm; emission wavelength 470 nm), the limit of detection being 20 fmol. This derivatization method is also applicable to other hydroxyl compounds having sterically unhindered secondary alcohol function such as bile acid 7- and 12-sulfates [207], and prostaglandins [208]. In similar fashion, derivatization with pyrene-1-carbonylnitrile provides the corresponding esters which show a better detector response [209].

*Derivatization of Corticosteroids with 9-Anthroylnitrile [211,212]*

A serum sample (100 μl) is diluted with 0.5 M monobasic sodium phosphate (1 ml) and passed through an octadecylsilyl bonded silica cartridge. After successive washing with water (2 ml) and 10% ethanol (3 ml), cortisol and related compounds are eluted with ethanol (3 ml). To the dried eluate is added 9-AN (100 μg) in 1% triethylamineacetonitrile (100 μl), and the mixture is allowed to stand at room temperature for 30 min. After evaporation to dryness under a stream of nitrogen, the residue is redissolved in 5% acetone-hexane (2 ml) and subjected to a Clin-Elut cartridge to remove the excess reagent. The derivatized corticosteroids are separated on a RP or NP column and monitored by FD (excitation wavelength 360 nm; emission wavelength 460 nm), the limit of detection being 20 fmol. Derivatization of corticosteroids with 9-AN occurs exclusively at the primary hydroxyl group.

*Derivatization of Alcohols with Naphthylisocyanate [223]*

To a toluene solution (1 ml) containing the alcoholic compound (<400 ng) are added a 1.5-fold molar excess of naphthylisocyanate in toluene (0.5 ml) and 1 M triethylamine in toluene (0.5 ml), and the mixture is heated (primary and secondary alcohols, 60 min at 368 K; tertiary alcohols, 120 min at 413 K). Then a threefold molar excess of diethylamine in toluene (1 ml) is added to the hot solution. The reaction mixture is centrifuged and an aliquot of the supernatant is injected onto a RP or NP column. The naphthylurethane derivatives formed are monitored by FD (excitation wavelength 308-318 nm; emission wavelength 350-356 nm), the limit of detection being 10 pmol.

*Derivatization of Primary and Secondary Alcohols with 3-Chloroformyl-7-methoxycoumarin [224,225]*

A solution of alcohols in benzene or acetone (50 μl) is treated with 10 mM 3-chloroformyl-7-methoxycoumarin in benzene (50 μl) and heated at 373 K for 20 min. The reaction mixture is diluted with methanol (900 μl) and an aliquot of the solution is injected onto a RP column. The resulting esters are monitored by FD (excitation wavelength 355 nm; emission wavelength 400 nm), the limit of detection for cholesterol being 15 pmol.

## REFERENCES

1. Blau, K., and G. S. King, *Handbook of Derivatization for Chromatography*, Heyden and Son, London, 1977.
2. Schulman, S. G., *Fluorescence and Phosphorescence Spectroscopy: Physicochemical Principles and Practice*, Pergamon Press, Oxford, 1977.
3. Pryde, A., and M. T. Gilbert, *Applications of High Performance Liquid Chromatography*, Chapman and Hall, London, 1979.
4. Knapp, D. R., *Handbook of Analytical Derivatization Reactions*, John Wiley and Sons, New York, 1979.
5. Kabra, P. M., and L. J. Marton, *Liquid Chromatography in Clinical Analysis*, Human Press, Clifton, NJ, 1981.
6. Frei, R. W., and J. F. Lawrence, *Chemical Derivatization in Analytical Chemistry*, Vol. 2, *Separation and Continuous Flow Techniques*, Plenum Press, New York, 1982.
7. Vickrey, T. M., *Liquid Chromatography Detectors*, Marcel Dekker, New York, 1983.
8. Lingeman, H., W. J. M. Underberg, A. Takadate, and A. Hulshoff, *J. Liq. Chromatogr.*, *8*:789 (1985).
9. Gfeller, J. C., G. Frey, J. M. Huen, and J. P. Thevenin, *J. Chromatogr.*, *172*:141 (1979).
10. Lawrence, J. F., U. A. Th. Brinkman, and R. W. Frei, *J. Chromatogr.*, *185*:473 (1979).
11. Buuren, C. van, J. F. Lawrence, U. A. Th. Brinkman, I. L. Honigberg, and R. W. Frei, *Anal. Chem.*, *52*:700 (1980).
12. Reddingius, R. J., G. J. de Jong, U. A. Th. Brinkman, and R. W. Frei, *J. Chromatogr.*, *205*:77 (1981).

13. Derendorf, H., and E. R. Garret, *J. Pharm. Sci.*, *72*:630 (1983).
14. Causon, R. C., and M. E. Carruthers, *J. Chromatogr.*, *229*: 301 (1982).
15. Kishimoto, Y., S. Ohgitani, A. Yamatodani, M. Kuro, and F. Okumura, *J. Chromatogr.*, *231*:121 (1982).
16. Muusze, R. G., and J. F. K. Huber, *J. Chromatogr. Sci.*, *12*:779 (1974).
17. Cociglio, M., D. Sauvaire, and A. Alric, *J. Chromatogr.*, *307*:351 (1984).
18. Gfeller, J. C., G. Frey, and R. W. Frei, *J. Chromatogr.*, *142*:271 (1977).
19. Gotelli, G. R., J. H. Wall, P. M. Kabra, and L. J. Marton, *Clin. Chem.*, *27*:441 (1981).
20. Bernstein, M. S., and M. A. Evans, *J. Chromatogr.*, *229*:179 (1982).
21. Miura, K., H. Nakamura, H. Tanaka, and Z. Tamura, *J. Chromatogr.*, *210*:536 (1981).
22. Horikawa, R., T. Tanimura, and Z. Tamura, Method for fluorescence detection in the high-performance liquid chromatography of $\Delta^4$-3-ketosteroids, *J. Chromatogr.*, *168*:526 (1979).
23. Asmus, P. A., J. W. Jorgenson, and M. Novotny, Fluorescence enhancement: New selective detection principle for liquic chromatography, *J. Chromatogr.*, *126*:317 (1976).
24. Nakamura, H., and Z. Tamura, Fluorometric determination of thiols by liquid chromatography with postcolumn derivatization, *Anal. Chem.*, *53*:2190 (1981).
25. Lawrence, J. F., and R. W. Frei, Fluorigenic labelling of carbamates using dansyl chloride. II. Fluorescence phenomena of the derivatives, *J. Chromatogr.*, *66*:93 (1972).
26. Lee, S. H., L. R. Field, W. H. Howald, and W. F. Trager, High-performance liquid chromatographic separation and fluorescence detection of warfarin and its metabolites by postcolumn acid/base manipulation, *Anal. Chem.*, *53*:467 (1981).
27. Assenza, S. P., and P. R. Brown, Ultraviolet and fluorescence characterization of purines and pyrimidines by post-column pH manipulation, *J. Chromatogr.*, *289*:355 (1984).
28. Katz, S., W. W. Pitt, Jr., J. E. Mrochek, and S. Dinsmore, Sensitive fluorescence monitoring of carbohydrates eluted by a borate mobile phase from an anion-exchange column, *J. Chromatogr.*, *101*:193 (1974).
29. Wolkoff, A. W., and R. H. Larose, A highly sensitive technique for the liquid chromatographic analysis of phenols and other environmental pollutants, *J. Chromatogr.*, *99*:731 (1974).
30. Werkhoven-Goewie, C. E., W. M. A. Niessen, U. A. Th. Brinkman, and R. W. Frei, Liquid chromatographic detector

for organosulfur compounds based on a ligand-exchange reaction, *J. Chromatogr.*, *203*:165 (1981).
31. Okuyama, S., N. Kokubun, S. Higashidate, D. Uemura, and Y. Hirata, A new analytical method of individual bile acids using high performance liquid chromatography and immobilized 3α-hydroxysteroid dehydrogenase in column form, *Chem. Lett.*, *1979*:1443.
32. Kamada, S., M. Maeda, and A. Tsuji, Separation and determination of bile acids by high-performance liquid chromatography using immobilized 3α-hydroxysteroid dehydrogenase and an electrochemical detector, *J. Chromatogr.*, *239*:773 (1982).
33. Frei, R. W., W. Santi, and M. Thomas, Liquid chromatography of dansyl derivatives of some alkaloids and the application to the analysis of pharmaceuticals, *J. Chromatogr.*, *116*:365 (1976).
34. Bayer, E., E. Grom, B. Kaltenegger, and R. Uhlmann, Separation of amino acids by high performance liquid chromatography, *Anal. Chem.*, *48*:1106 (1976).
35. Meffin, P. J., S. R. Harapat, and D. C. Harrison, High-pressure liquid chromatographic analysis of drugs in biological fluids. II. Determination of an Antiarrhythmic drug, tocainide, as its dansyl derivative using a fluorescence detector, *J. Pharm. Sci.*, *66*:583 (1977).
36. Saeki, Y., N. Uehara, and S. Shirakawa, Sensitive fluorimetric method for the determination of putrescine, spermidine and spermine by high-performance liquid chromatography and its application to human blood, *J. Chromatogr.*, *145*:221 (1978).
37. Brown, N. D., R. B. Sweett, J. A. Kintzios, H. D. Cox, and B. P. Doctor, Determination of 5-dimethylaminonaphthalene-1-sulfonyl derivatives of urinary polyamines by ion-pair high-performance liquid chromatography, *J. Chromatogr.*, *164*:35 (1979).
38. Werkhoven-Goewie, C. E., U. A. Th. Brinkman, and R. W. Frei, The use of solvent segmentation in continuous-flow system, and fluorescence labelling by derivatization, *Anal. Chim. Acta*, *114*:147 (1980).
39. Tapuhi, Y., N. Miller, and B. L. Karger, Practical considerations in the chiral separation of Dns-amino acids by reversed-phase liquid chromatography using metal chelate additives, *J. Chromatogr.*, *205*:325 (1981).
40. Courte, S., and N. Bromet, Trace determination of trimetazidine in plasma by high-performance liquid chromatography using fluorescence detection, *J. Chromatogr.*, *224*:162 (1981).
41. Griesmann, G. E., W.-Y. Chan, and O. M. Rennert, Determination of γ-Aminobutyric acid by reversed-phase high-performance liquid chromatography and pre-column labeling for fluorescence detection, *J. Chromatogr.*, *230*:121 (1982).

42. Sommadossi, J. P., M. Lemar, J. Necciari, Y. Sumirtapura, J. P. Cano, and J. Gaillot, High-performance liquid chromatographic method for the determination of plasma and urine metapramine after dansylation, *J. Chromatogr.*, *228*:205 (1982).
43. Farid, N. A., and S. M. White, Determination of mexiletine and its metabolites in serum by liquid chromatography with fluorescence detection, *J. Chromatogr.*, *275*:458 (1983).
44. Kneifel, H., and A.-S. Jandel, Rapid amino acid analysis in biotechnology: Determination of ananine and aspartic acid, *J. Liq. Chromatogr.*, *6*:1395 (1983).
45. Kågedal, B., M. Kaellberg, and B. Soerbo, Determination of S-sulfocysteine in urine by high-performance liquid chromatography, *J. Chromatogr.*, *276*:418 (1983).
46. Shih, F. F., and A. D. Kalmar, Determination of glutamine and asparagine by isocratic elution reverse phase liquid chromatography with fluorescent detection, *J. Liq. Chromatogr.*, *7*:1169 (1984).
47. Parris, N., and D. Gallelli, Dansylation of amino acids and by-product formation, *J. Liq. Chromatogr.*, *7*:917 (1984).
48. Bontemps, J., J. Laschet, and G. Dandifrosse, *J. Chromatogr.*, *311*:59 (1984).
49. Schmidt, G. J., F. L. Vandemark, and W. Slavin, *Anal. Biochem.*, *91*:636 (1978).
50. Yamada, K., E. Kayama, Y. Aizawa, and K. Oka, *J. Chromatogr.*, *223*:176 (1981).
51. Kamimura, H., H. Sasaki, and S. Kawamura, *J. Chromatogr.*, *225*:115 (1981).
52. Cory, R. P., R. R. Becker, R. Rosenbluth, and I. Isenberg, *J. Am. Chem. Soc.*, *90*:1643 (1968).
53. Udenfriend, S., S. Stein, P. Boehlen, W. Dairman, W. Leimgruber, and M. Weigele, *Science*, *178*:871 (1972).
54. Walker, S. E., and P. E. J. Coates, *J. Chromatogr.*, *223*:131 (1981).
55. Sedman, A. J., and J. Gal., *J. Chromatogr.*, *232*:315 (1982).
56. Stein, S., P. Boehlen, J. Stone, W. Dairman, and S. Udenfriend, *Arch. Biochem. Biophys.*, *155*:203 (1973).
57. Felix, A. M., and G. Terkelson, *Arch. Biochem. Biophys.*, *157*:177 (1973).
58. Frei, R. W., L. Michel, and W. Santi, *J. Chromatogr.*, *126*:665 (1976).
59. McHugh, W., R. A. Sandman, W. G. Haney, and S. P. Sood, *J. Chromatogr.*, *124*:376 (1976).
60. Lewis, R. V., and D. DeWald, *J. Liq. Chromatogr.*, *5*:1367 (1982).
61. Yuan, P. M., H. Pande, B. R. Clark, and J. E. Shively, *Anal. Chem.*, *120*:289 (1982).

62. Szókán, G., *J. Liq. Chromatogr.*, *5*:1493 (1982).
63. Farid, N. A., *J. Pharm. Sci.*, *68*:249 (1979).
64. Licht, A., R. L. Bowman, and S. Stein, *J. Liq. Chromatogr.*, *4*:825 (1981).
65. Sumirtapura, Y. C., C. Aubert, Ph. Coassolo, and J. P. Cano, *J. Chomatogr.*, *232*:111 (1982).
66. Narayanan, T. K., and L. M. Greenbaum, *J. Chromatogr.*, *306*:109 (1984).
67. Imai, K., *J. Chromatogr.*, *105*:135 (1975).
68. Schwedt, G., *J. Chromatogr.*, *118*:429 (1976).
69. Samejima, K., *J. Chromatogr.*, *96*:250 (1974).
70. Samejima, K., M. Kawase, S. Sakamoto, and Y. Endo, *Anal. Biochem.*, *76*:392 (1976).
71. Kai, M., T. Ogata, K. Haraguchi, and Y. Ohkura, *J. Chromatogr.*, *163*:151 (1979).
72. Weigele, M., S. DeBernardo, and W. Leimgruber, *Biochem. Biophys. Res. Commun.*, *50*:352 (1973).
73. Felix, A. M., and G. Terkelson, *Anal. Biochem.*, *56*:610 (1973).
74. Nakamura, H., and Z. Tamura, *Anal. Chem.*, *52*:2087 (1980).
75. Nakamura, H., K. Takagi, Z. Tamura, R. Yoda, and Y Yamamoto, *Anal. Chem.*, *56*:919 (1984).
76. Hodgin, J. L., *J. Liq. Chromatogr.*, *2*:1047 (1979).
77. Lindroth, P., and K. Mopper, *Anal. Chem.*, *51*:1667 (1979).
78. Joys, T. M., and H. Kim, *Anal. Biochem.*, *94*:371 (1979).
79. Boehlen, P., and M. Mellet, *Anal. Biochem.*, *94*:313 (1979).
80. Cooper, J. D. H., and D. C. Turnell, *J. Chromatogr.*, *227*:158 (1982).
81. Boehlen, P., and R. Schroeder, *Anal. Biochem.*, *126*:144 (1982).
82. Kucera, P., and H. Umagat, *J. Chromatogr.*, *255*:563 (1983).
83. Joseph, M. H., and P. Davies, *J. Chromatogr.*, *227*:125 (1983).
84. Cronin, J. R., S. Pizzarello, and W. E. Gandy, *Anal. Chem.*, *93*:174 (1979).
85. Ishida, Y., T. Fujita, and K. Asai, *J. Chromatogr.*, *204*:143 (1981).
86. Huges, G. J., K. J. Winterhalten, E. Boller, and K. T. Wilson, *J. Chromatogr.*, *235*:417 (1982).
87. Sato, M., and K. Yagi, *J. Chromatogr.*, *242*:185 (1982).
88. Schlabach, T. D., and T. C.Wehr, *Anal. Biochem.*, *127*:222 (1982).
89. Cooper, J. D. H., M. T. Lewis, and D. C. Turnell, *J. Chromatogr.*, *285*:490 (1984).
90. Hogan, D. L., K. L. Kraemer, and J. I. Isenberg, *Anal. Biochem.*, *127*:17 (1982).
91. Wassner, S. J., and J. B. Li, *J. Chromatogr.*, *227*:497 (1982).
92. Radjai, M. K., and R. T. Hatch, *J. Chromatogr.*, *196*:319 (1980).

93. Kuwada, M., and K. Katayama, *Anal. Biochem.*, *131*:173 (1983).
94. Eslami, M., and J. D. Stuart, *J. Liq. Chromatogr.*, *7*:1117 (1974).
95. Davies, A. T., S. T. Ingalls, and C. L. Hoppel, *J. Chromatogr.*, *306*:79 (1984).
96. Hare, T. A., and N. V. Bala Manyam, *Anal. Biochem.*, *101*: 349 (1980).
97. Grove, J., R. G. Alken, and P. G. Schechter, *J. Chromatogr.*, *306*:383 (1984).
98. Ida, S., Y. Tanaka, S. Ohkuma, and K. Kuriyama, *Anal. Biochem.*, *136*:352 (1984).
99. Brodie, R. R., L. F. Chasseaud, L. Rooney, S. Darragh, and R. F. Lambe, *J. Chromatogr.*, *274*:179 (1983).
100. Dye, D., T. East, and W. F. Bayne, *J. Chromatogr.*, *284*: 457 (1984).
101. Seki, T., and Y. Yamaguchi, *J. Chromatogr.*, *284*:457 (1984).
102. Davies, T. P., C. W. Gehrke Jr., C. H. Williams, C. W. Gehrke, and K. O. Gerhardt, *J. Chromatogr.*, *228*:113 (1982).
103. Essers, L., *J. Chromatogr.*, *305*:345 (1984).
104. Kawamoto, T., I. Mashimo, S. Yamaguchi, and M. Watanabe, *J. Chromatogr.*, *305*:373 (1984).
105. Drescher, D. G., and K. S. Lee, *Anal. Biochem.*, *84*:559 (1979).
106. Roth, M., and A. Hampai, *J. Chromatogr.*, *83*:353 (1973).
107. Himuro, A., H. Nakamura, and Z. Tamura, *Anal. Chim. Acta*, *147*:317 (1983).
108. Roth, M., *Anal. Chem.*, *43*:880 (1971).
109. Simons, S. S. Jr., and D. F. Johnson, *Anal. Biochem.*, *82*: 250 (1977).
110. Wheler, G. H. T., and J. T. Russel, *J. Liq. Chromatogr.*, *4*:1281 (1981).
111. Burbach, J. P. H., A. Prins, J. L. M. Lebouille, J. Verhoef, and A. Witter, *J. Chromatogr.*, *237*:339 (1982).
112. Stobaugh, J. F., A. J. Repta, L. A. Sternson, and K. W. Garren, *Anal. Biochem.*, *135*:495 (1983).
113. Gardner, W. S., and W. H. Miller, *Anal. Biochem.*, *101*:61 (1980).
114. Nimura, N., and T. Kinoshita, *J. Chromatogr.*, *352*:169 (1986).
115. Ghost, P. B., and M. W. Whitehouse, *Biochem. J.*, *108*:155 (1968).
116. Wolfram, J. H., J. I. Feinberg, R. C. Doerr, and W. Fiddler, *J. Chromatogr.*, *132*:37 (1977).
117. Krol, G. J., J. M. Banovsky, C. A. Mannan, R. E. Pickering, and B. T. Kho, *J. Chromatogr.*, *163*:383 (1979).

118. Ahnoff, M., I. Grundevik, A. Arfwidsson, J. Fonselius, and B.-A. Persson, *Anal. Chem.*, *53*:485 (1981).
119. Johnson, L., S. Lagerkvist, P. Lindroth, M. Ahnhoff, and K. Martinsson, *Anal. Chem.*, *54*:939 (1982).
120. Imai, K., and Y. Watanabe, *Anal. Chim. Acta*, *130*:377 (1981).
121. Watanabe, Y., and K. Imai, *Anal. Biochem.*, *116*:471 (1981).
122. Watanabe, Y., and K. Imai, *J. Chromatogr.*, *239*:723 (1982).
123. Watanabe, Y., and K. Imai, *Anal. Chem.*, *55*:1786 (1983).
124 Toyo'oka, T., Y. Watanabe, and K. Imai, *Anal. Chim. Acta*, *149*:305 (1983).
125. Watanabe, Y., and K. Imai, *J. Chromatogr.*, *309*:279 (1984).
126. Shah, J. A., and D. J. Weber, *J. Chromatogr.*, *309*:95 (1984).
127. Gübitz, G., R. Wintersteiger, and A. Hartinger, *J. Chromatogr.*, *218*:51 (1981).
128. Maeda, H., N. Ishida, H. Kawauchi, and K. Tuzimura, *J. Biochem.*, *65*:777 (1969).
129. Sinsheimer, J. E., D. D. Hong, J. T. Stewart, M. L. Fink, and J. H. Burckhalter, *J. Pharm. Sci.*, *60*:141 (1971).
130. Ichikawa, H., T. Tanimura, T. Nakajima, and Z. Tamura, *Chem. Pharm. Bull.*, *18*:1493 (1970).
131. L'Italien, J. J., and S. B. H. Kent, *J. Chromatogr.*, *283*:149 (1984).
132. Falter, H., K. Jayasimhula, and R. A. Day, *Anal. Biochem.*, *67*:359 (1975).
133. Chen, S. S.-H., A. Y. Kou, and H.-H. Y. Chen, *J. Chromatogr.*, *276*:37 (1983).
134. Maeda, M., A. Tsuji, S. Ganno, and Y. Ohnishi, *J. Chromatogr.*, *77*:434 (1973).
135. Yamamoto, Y., T. Manji, A. Saito, K. Maeda, and K. Ohta, *J. Chromatogr.*, *162*:327 (1979).
136. Baker, M. D., H. Y. Mohammed, and H. Veening, *Anal. Chem.*, *53*:1658 (1981).
137. Grushka, E., S. Lam, and J. Chassin, *Anal. Chem.*, *50*:1398 (1978).
138. Hayashi, K., J. Kawase, K. Yoshimura, K. Ara, and K. Tsuji, *Anal. Biochem.*, *136*:314 (1984).
139. Lam, S., and E. Grushka, *J. Chromatogr.*, *158*:207 (1978).
140. Crozier, A., J. B. Zaerr, and R. O. Morris, *J. Chromatogr.*, *238*:157 (1982).
141. Dünges, W., and N. Seiler, *J. Chromatogr.*, *145*:483 (1978).
142. Gandelman, M. S., and J. W. Birks, *Anal. Chim. Acta*, *155*:159 (1983).
143. Kimura, K., and T. Shono, *J. Liq. Chromatogr.*, *5*:223 (1982).
144. Desbene, P.-L., S. Coustal, and F. Frappier, *Anal. Biochem.*, *128*:359 (1983).

145. Yoshida, S., S. Hirose, and M. Iwamoto, *J. Chromatogr.*, *383*:61 (1986).
146. Lloyd, J. B. F., *J. Chromatogr.*, *178*:249 (1979).
147. Goya, S., A. Takadate, H. Fujino, and M. Irikura, *Yakugaku Zasshi*, *100*:744 (1980).
148. Farinotti, R., Ph. Siard, J. Bourson, S. Kirkiacharian, B. Valeur, and G. Mahuzier, *J. Chromatogr.*, *269*:81 (1983).
149. Tsuchiya, H., T. Hayashi, H. Naruse, and N. Takagi, *J. Chromatogr.*, *234*:121 (1982).
150. Tsuchiya, H., T. Hayashi, M. Sato, M. Tatsumi, and N. Takagi, *J. Chromatogr.*, *309*:43 (1984).
151. Yamaguchi, M., S. Hara, R. Matsunaga, M. Nakamura, and Y. Ohkura, *Anal. Sci.*, *1*:295 (1985).
152. Yamaguchi, M., R. Matsunaga, S. Hara, M. Nakamura, and Y. Ohkura, *J. Chromatogr.*, *375*:27 (1986).
153. Yamaguchi, M., K. Fukuda, S. Hara, M. Nakamura, and Y. Ohkura, *J. Chromatogr.*, *380*:257 (1986).
154. Distler, W., *J. Chromatogr.*, *192*:240 (1980).
155. Watkins, W. D., and B. M. Peterson, *Anal. Biochem.*, *125*:30 (1982).
156. Kamada, S., M. Maeda, and A. Tsuji, *J. Chromatogr.*, *272*:29 (1983).
157. Korte, W. D., *J. Chromatogr.*, *243*:153 (1982).
158. Takadate, A., T. Tahara, H. Fujino, and S. Goya, *Chem. Pharm. Bull.*, *30*:4120 (1982).
159. Barker, S. A., G. A. Monti, S. T. Christian, F. Benington, and R. D. Morin, *Anal. Biochem.*, *107*:116 (1980).
160. Hatsumi, M., S.-I. Kimata, and K. Hirosawa, *J. Chromatogr.*, *253*:271 (1982).
161. Imaoka, S., Y. Funae, T. Sigumoto, N. Hayahara, and M. M. Maekawa, *Anal. Biochem.*, *128*:459 (1983).
162. Goya, S., A. Takadate, and H. Fujino, *Yakugaku Zasshi*, *102*:63 (1982).
163. Tsuruta, Y., and K. Kohashi, *J. Chromatogr.*, *309*:309 (1984).
164. Goto, J., N. Goto, A. Hikichi, T. Nishimaki, and T. Nambara, *Anal. Chim. Acta*, *120*:187 (1980).
165. Goto, J., N. Goto, and T. Nambara, *J. Chromatogr.*, *239*:559 (1982).
166. Lingeman, H., A. Hulshoff, W. J. M. Underberg, and F. B. J. M. Offermann, *J. Chromatogr.*, *290*:215 (1984).
167. Ikeda, M., K. Shimada, T. Sakaguchi, and U. Matsumoto, *J. Chromatogr.*, *305*:261 (1984).
168. Lloyd, J. B. F., *J. Chromatogr.*, *189*:359 (1980).
169. Hayashi, T., H. Tsuchiya, H. Todoriki, and H. Naruse, *Anal. Biochem.*, *122*:173 (1982).
170. Hayashi, T., H. Tsuchiya, and H. Naruse, *J. Chromatogr.*, *273*:245 (1983).

171. Kieber, D. J., and K. Mopper, *J. Chromatogr.*, *281*:135 (1983).
172. Speek, A. J., J. Schrijver, and W. H. P. Scheurs, *J. Chromatogr.*, *305*:53 (1984).
173. Chayen, R., R. Dvir, S. Gould, and A. Harell, *Anal. Biochem.*, *42*:283 (1971).
174. Kawasaki, T., M. Maeda, and A. Tsuji, *J. Chromatogr.*, *163*:143 (1979).
175. Kawasaki, T., M. Maeda, and A. Tsuji, *J. Chromatogr.*, *226*:1 (1981).
176. Alpenfels, W. F., *Anal. Biochem.*, *114*:153 (1981).
177. Toothaker, R. D., G. M. Sundaresan, J. P. Hunt, T. J. Goehl, K. S. Rotenberg, V. K. Prasad, W. A. Craig, and P. G. Welling, *J. Pharm. Sci.*, *71*:573 (1982).
178. Kawasaki, T., M. Maeda, and A. Tsuji, *J. Chromatogr.*, *232*:1 (1982).
179. Kawasaki, T., M. Maeda, and A. Tsuji, *J. Chromatogr.*, *233*:61 (1982).
180. Kawasaki, T., M. Maeda, and A. Tsuji, *J. Chromatogr.*, *272*:261 (1983).
181. Takeda, M., M. Maeda, and A. Tsuji, *J. Chromatogr.*, *244*:347 (1982).
182. Alpenfels, W. F., R. A. Mathews, D. E. Madden, and A. E. Newson, *J. Liq. Chromatogr.*, *5*:1711 (1982).
183. Mopper, K., and L. Johnson, *J. Chromatogr.*, *256*:27 (1983).
184. Mizutani, S., Y. Wakuri, N. Yoshida, T. Nakajima, and Z. Tamura, *Chem. Pharm. Bull.*, *17*:2340 (1969).
185. Hirata, T., M. Kai, K. Kohashi, and Y. Ohkura, *J. Chromatogr.*, *226*:25 (1981).
186. Gübitz, G., R. Winersteiger, and R. W. Frei, *J. Liq. Chromatogr.*, *7*:839 (1984).
187. Nakamura, H., and Z. Tamura, *Anal. Chem.*, *50*:2047 (1978).
188. Nakamura, H., and Z. Tamura, *J. Chromatogr.*, *168*:481 (1979).
189. Nakamura, H., and Z. Tamura, *Anal. Chem.*, *51*:1679 (1979).
190. Kanaoka, Y., M. Machida, K. Ando, and T. Sekine, *Biochim. Biophys. Acta*, *207*:269 (1970).
191. Wu, C.-W., L. R. Yarbrough, and F. Y.-H. Wu, *Biochemistry*, *15*:2863 (1976).
192. Jarrot, B., A. Anderson, R. Hooper, and W. J. Louis, *J. Pharm. Sci.*, *70*:665 (1981).
193. Kågedal, B., and M. Källberg, *J. Chromatogr.*, *229*:409 (1982).
194. Takahashi, H., T. Yoshida, and H. Meguro, *Bunseki Kagaku*, *30*:339 (1980).
195. Anzai, N., T. Kimura, S. Chida, T. Tanaka, H. Takahashi, and H. Meguro, *Yakugaku Zasshi*, *101*:1002 (1981).

196. Kanaoka, Y., M. Machida, M. I. Machida, and T. Sekine, *Biochim. Biophys. Acta*, *317*:563 (1973).
197. Machida, M., N. Ushijima, M. I. Machida, and Y. Kanaoka, *Chem. Pharm. Bull.*, *23*:1385 (1975).
198. Miners, J. O., I. Faernley, K. J. Smith, and D. J. Birkett, *J. Chromatogr.*, *275*:89 (1983).
199. Lankmayr, F., K. W. Budna, K. Mueller, F. Nachtmann, and F. Rainer, *J. Chromatogr.*, *222*:249 (1981).
200. Murayama, K., and T. Kinoshita, *J. Chromatogr.*, *205*:349 (1981).
201. Murayama, K., and T. Kinoshita, *Anal. Lett.*, *15*:123 (1982).
202. Toyo'oka, T., and K. Imai, *J. Chromatogr.*, *282*:495 (1983).
203. Imai, K., T. Toyo'oka, and Y. Watanabe, *Anal. Biochem.*, *128*:471 (1983).
204. Nakamura, H., and Z. Tamura, *Anal. Chem.*, *54*:1951 (1982).
205. Goto, J., N. Goto, F. Shamsa, M. Saito, S. Komatsu, K. Suzaki, and T. Nambara, *Anal. Chim. Acta*, *147*:397 (1983).
206. Goto, J., M. Saito, T. Chikai, N. Goto, and T. Nambara, *J. Chromatogr.*, *276*:289 (1983).
207. Goto, J., T. Chikai, and T. Nambara, *J. Chromatogr.*, *415*: 45 (1987).
208. Kubo, I., and S. Komatsu, *J. Chromatogr.*, *362*:61 (1986).
209. Goto, J., S. Komatsu, M. Inada, and T. Nambara, *Anal. Sci.*, *2*:585 (1986).
210. Goto, J., S. Komatsu, N. Goto, and T. Nambara, *Chem. Pharm. Bull.*, *29*:899 (1981).
211. Goto, J., F. Shamsa, N. Goto, and T. Nambara, *J. Pharm. Biomed. Anal.*, *1*:83 (1983).
212. Goto, J., F. Shamsa, and T. Nambara, *J. Liq. Chromatogr.*, *6*:1977 (1983).
213. Takadate, A., M. Irikura, T. Suehiro, H. Fujino, and S. Goya, *Chem. Pharm. Bull.*, *33*:1164 (1985).
214. Mitsui, A., H. Ohta, and A. Ohkura, *J. Chromatogr.*, *344*: 61 (1985).
215. Hojo, T., H. Nakamura, and Z. Tamura, *J. Chromatogr.*, *247*:157 (1982).
216. Kai, M., T. Miyazaki, M. Yamaguchi, and Y. Ohkura, *J. Chromatogr.*, *268*:417 (1983).
217. Hung, Y.-L., M. Kai, H. Nohta, and Y. Ohkura, *J. Chromatogr.*, *305*:281 (1984).
218. Hiraga, Y., and T. Kinoshita, *J. Chromatogr.*, *226*:43 (1981).
219. Hiraga, Y., and T. Kinoshita, *J. Chromatogr.*, *342*:269 (1985).
220. Hara, S., Y. Takemori, M. Yamaguchi, M. Nakamura, and Y. Ohkura, *J. Chromatogr.*, *344*:33 (1985).
221. Mopper, K., W. L. Stahovec, and L. Johnson, *J. Chromatogr.*, *256*:243 (1983).

222. Maiorino, R. M., G. L. Weber, and H. V. Aposhian, *J. Chromatogr.*, *374*:297 (1986).
223. Wintersteiger, R., G. Wenninger-Weinzierl, and W. Pacha, *J. Chromatogr.*, *237*:399 (1982).
224. Hamada, C., M. Iwasaki, N. Kuroda, and Y. Ohkura, *J. Chromatogr.*, *341*:426 (1985).
225. Matsuoka, C., H. Nohta, N. Kuroda, and Y. Ohkura, *J. Chromatogr.*, *341*:342 (1985).
226. Honda, S., T. Konishi, S. Suzuki, K. Kakehi, and S. Ganno, *J. Chromatogr.*, *281*:340 (1983).
227. Honda, S., Y. Matsuda, M. Takahashi, K. Kakehi, and S. Ganno, *Anal. Chem.*, *52*:1079 (1980).
228. Honda, S., M. Takahashi, K. Kakehi, and S. Ganno, *Anal. Biochem.*, *113*:130 (1981).
229. Honda, S., M. Takahashi, Y. Araki, and K. Kakehi, *J. Chromatogr.*, *274*:45 (1983).
230. Honda, S., S. Suzuki, M. Takahashi, K. Kakehi, and S. Ganno, *Anal. Biochem.*, *134*:34 (1983).
231. Bayne, W. F., T. East, and D. Dye, *J. Pharm. Sci.*, *70*:458 (1981).
232. Nelson, P. E., S. L. Nolan, and K. R. Bedford, *J. Chromatogr.*, *234*:407 (1982).

# 10

# Chemiluminescence Derivatization Reactions in Liquid Chromatography

KAZUHIRO IMAI *University of Tokyo, Bunkyo-ku, Tokyo, Japan*

## INTRODUCTION

In fluorescence the energy necessary for the excitation of substances to a light-emitting excited state originates from a light source. This means that there exists a signal (S) from the fluorescence light as well as a background signal (N) caused by the light source (Rayleigh scattering and Raman scattering) and the dark current of the light-detecting photomultiplier or photon counter. Therefore, fluctuation of the light source would cause fluctuations in S as well as in N, indicating that amplification of S does not always afford the enlargement of the S/N ratio. As a consequence it is difficult to get a high S/N ratio in fluorescence analysis. To overcome this problem, the techniques of the time-resolved fluorescence measurements [1] and two-photon excitation fluorescence measurements [2] have been introduced.

In chemiluminescence (CL) the energy for excitation of compounds to a light-emitting excited state is produced by chemical reactions, which means in principle that no light other than the emitted light exists and that the light emerges from a black background. In reality, the background noise N originates from the dark current produced by the employed photomultiplier or photon counter. Since the origin of the CL light is different from that of the dark current, amplification of S by the detector is not necessarily accompanied by amplification of N. Thus, a sensitive detection (a high S/N ratio) as well as a wide dynamic range of detection can be obtained.

Table 10.1 Chemiluminescence Reagents Applied to Liquid Chromatography

| Chemiluminescence reagents | Analytes | Detection limits | Ref. |
|---|---|---|---|
| *Prechromatographic Reactions* | | | |
| *N*-(4-Aminobutyl)-*N*-ethylisoluminol | Toluidines | | 106 |
| | Butylamines, fatty acids, bile acids (cholic acid) | 20 fmol | |
| *Postchromatographic Reactions* | | | |
| Lucigenin | Ascorbic acid | 0.5 mg/liter | 108 |
| | Dehydroascorbic acid | 10 mg/liter | |
| | Heparin | 90 μg | 109 |
| Luminol | Copper(II) | (sub) μm | 107 |
| | Chromium(II) | (sub) μm | 107 |
| | Corticosteroids<br>Carboxylic acids | 500 fmol<br>500 fmol | 110 |
| Peroxyoxalate | Dns-amino acids | 10 fmol | 111 |
| | Dns-amino acids | 0.5 fmol | 112 |
| | Dns-amino acids (microbore LC) | 0.3 fmol | 113 |
| | Dns-amino acids (gradient elution) | 2-5 fmol | 114<br>115 |
| | Dns-amines | 1-15 fmol | 116 |
| | NBD-amines | 20-270 fmol | |
| | OPA-amines | 95-580 fmol | |
| | Dns-hydrazones | 170 fmol | 117 |
| | Flur-norepinephrine, flur-dopamine | 25 fmol | 118 |
| | Polycyclic aromatic hydrocarbons | 0.8 pg<br>0.6-20 pg | 119<br>121 |
| | 3-Aminofluoranthene | 0.1 pg | 120 |
| | Aminopyrene | 60 amol | 122 |
| (Quenching) | Anilines | 2.5-30 ng | 123 |

Table 10.1 (continued)

| | | | |
|---|---|---|---|
| Ozone-induced | Nitrogen and sulfur compounds<br>Alkanes<br>Fluorescent compounds | μg-pg range | 128 |
| Singlet oxygen-induced | Chlorophyls | | 129 |
| Electrogenerated | Aromatic compounds | 50 ng | 131 |
| Thermal energy analyzer | *N*-Nitroso-*N*,*N*-dimethylpiperidinium iodide (microbore LC) | (sub) ng | 132 |
| | Glycerol trinitrate | 0.1 ng | 133 |
| | Pentaerylthritol tetranitrate | 0.1 ng | |
| | Isosorbide dinitrate | 0.2 ng | |

Abbreviations: Dns, 5-*N*,*N*-dimethylaminonaphthalene sulfonyl-labeled; NBD, 7-nitrobenzo-2-oxa-1,3-diazole-4-yl-labeled; OPA, orthophthalaldehyde-labeled; Flur, fluorescamine-labeled.

Recently, much attention has focused on the use of bioluminescence and CL reactions for the sensitive detection of biologically important substances [3-16]. The appearance of a new journal [17] and textbooks [18-20] may also reflect the need for luminescence analysis methods in the field of life sciences. This chapter focuses on the description of CL reactions and/or CL reagents for detection in liquid chromatography (LC). Most of these CL reactions and reagents are summarized in Table 10.1.

## TYPES OF CHEMILUMINESCENCE REACTIONS AND CHEMILUMINESCENCE REAGENTS

Two types of CL reactions can be distinguished. In the first one an energy-rich intermediate, produced by the CL reaction, emits light. Typical examples of this type are CL reactions with luminol (5-amino-2,3-dihydrophthalazine-1,4-dione) [21] and isoluminol (6-amino-2,3-dihydrophthalazine-1,4-dione) [22] (Fig. 10.1), lucigenin (*N*,*N*-dimethyl-9,9-diacridinium dinitrate) [22-24] (Fig. 10.2), gallic acid [25,26] (Fig. 10.3), humic acid [27], and lophine [28,29] (Fig. 10.3) as well as the electrogenerated CL of aromatic compounds [30-34]. In these cases, a catalyst such as hemin, metal ions, or hydrogen peroxide or a reducing organic compound is usually necessary

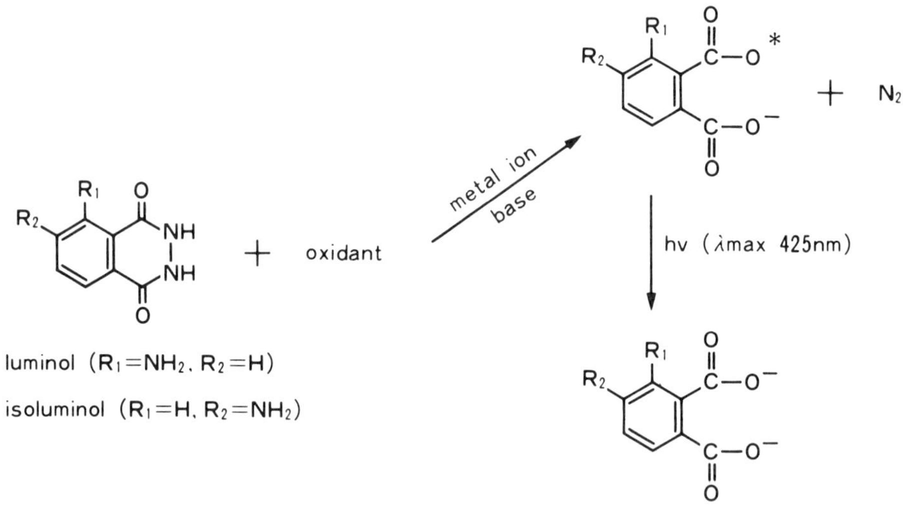


Figure 10.1 Chemiluminescence reaction of luminol and isoluminol.

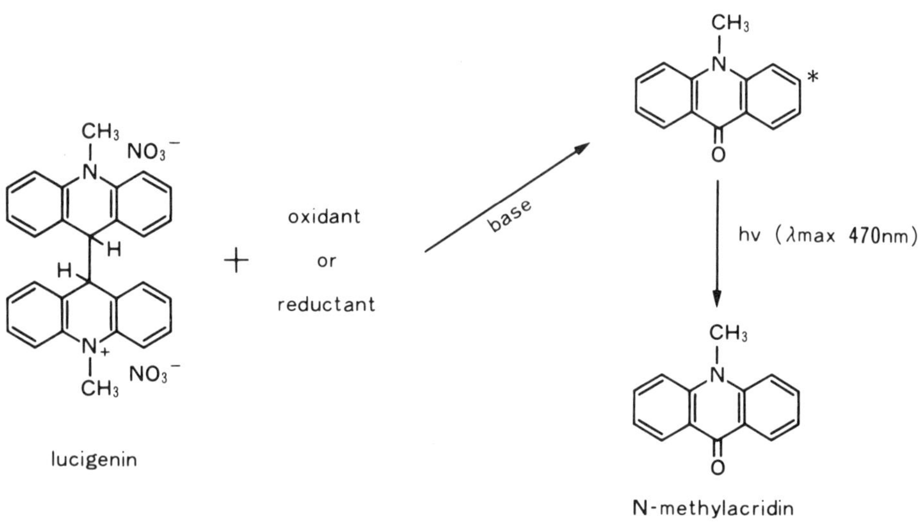


Figure 10.2 Chemiluminescence reaction of lucigenin.

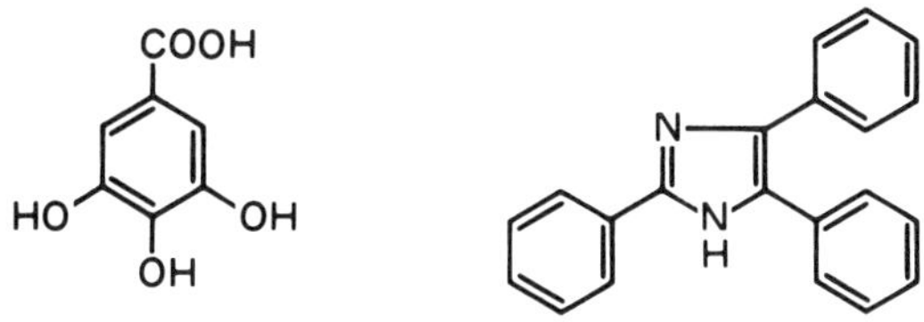


gallic acid lophine

Figure 10.3 Structure of gallic acid and lophine.

and, consequently, this can be measured by the use of the above reactions.

In the second type of CL reaction the energy-rich intermediate which is produced in the early stages of the reaction excites other substances, which in turn emit light. Typical examples are the peroxyoxalate-hydrogen peroxide CL reactions [35,36] (Fig. 10.4), biphenyl peroxide [37,38] (Fig. 10.5), 1,4-diphenyl-2,3-benzodioxine [37,38] (Fig. 10.5), 1-phenylethyl peracetate [37,38] (Fig. 10.5), 3,3,4-trimethyl-1,2-dioxetane-4-one [37,38] (Fig. 10.5), 3,3,4-trimethyl-1,2-dioxetane [39] (Fig. 10.5), and ozone or singlet oxygen [40,41] CL reactions. Usually in these reactions, various kinds of fluorescent substances including aromatic polycyclic hydrocarbons labeled compounds can be assayed.

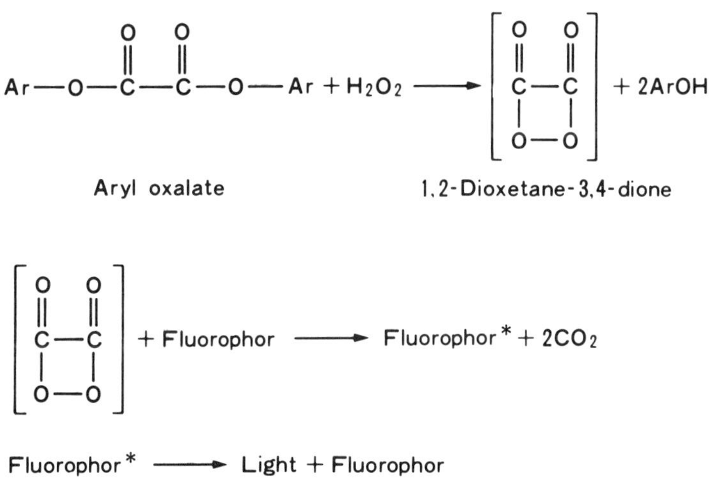


Figure 10.4 Peroxyoxalate chemiluminescence reaction (fluorophor, fluorescent compound).

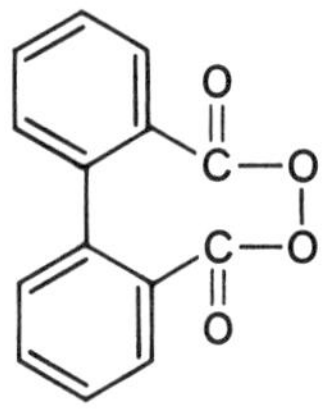

diphenoylperoxide

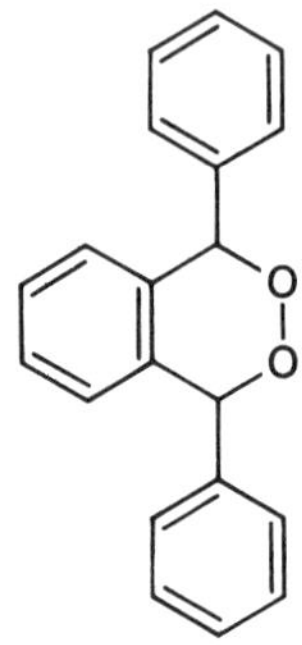

1,4-diphenyl-2,3-benzodioxine

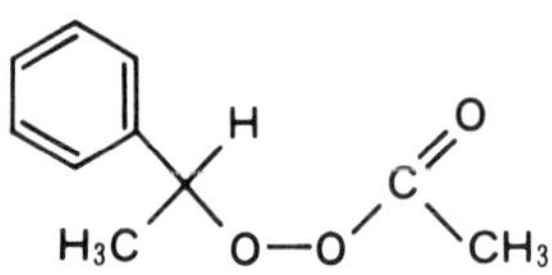

1-phenylethyl peracetate

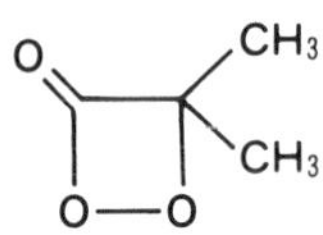

3,3-dimethyl-1,2-dioxetan-4-one

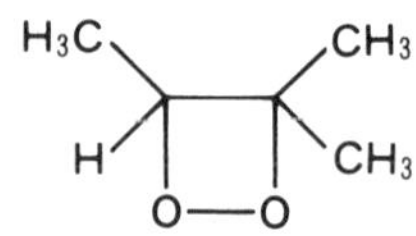

3,3,4-trimethyl-1,2-dioxetane

Figure 10.5 Structure of other type of chemiluminescence reagents.

### Luminol and Isoluminol Chemiluminescence Reactions

Luminol and isoluminol react with hydrogen peroxide or other oxidants in the presence of metal ions in strongly alkaline conditions, resulting in light emission at 470 nm (Fig. 10.1).

Cations such as iron(II), cobalt(II), nickel(II), copper(II), zinc(II), and cadmium(II) were determined by luminol CL reactions [42-45]. On the other hand, suppression of CL by metal-chelating agents [46] enable the detection of chelators such as amino acids [47,48] and proteins [49,50]. For instance, the flow injection analysis based on this suppression permits the detection of amino acids down to 20 nmol [51].

Hemoproteins such as protohemin and myoglobin enhance the luminol CL reaction, creating the possibility of myoglobin determination [52].

Hydrogen peroxide produced by enzyme reactions can be determined using luminol. Human serum glucose was oxidized by mobilized [53] or immobilized [54] glucose oxidase, and the generated hydrogen peroxide could be determined in the nanomole concentration range. Flow injection analysis also permitted the detection of serum glucose with a standard deviation of less than 2% and a sampling frequency of 75 samples/hr [55]. Comparison of the CL method with a spectrophotometric method for the determination of blood glucose was executed [56] and turned out to be far more sensitive [56]. Sucrose was also converted to β-D-glucose by a combination of glucose oxidase, invertase, and mutarotase, and detected with the above-mentioned luminol detection system [57]. Total polyamine and diamine levels were determined using, respectively, amine oxidase and diamine oxidase in the presence of luminol [58,59].

Microperoxidase and lipoxygenase accelerate CL reactions [60, 61]. Hence, the activity of lipoxygenase was measured using the CL reaction with the luminol or the isoluminol analogs mentioned below [61].

In enzymatic immunoassay, the hydrogen peroxide generated by enzymes which is covalently attached to antigens or antibodies can be sensitively detected by luminol CL reactions. Glucose oxidase bound anti-α-fetoprotein, insulin, and 17-α-hydroxyprogesterone were subjected to immune reactions in serum samples, and the residual glucose oxidase was subsequently incubated with glucose to generate hydrogen peroxide, which in turn was detected by the use of luminol and potassium hexacyanoferrate(III) [62].

Isoluminol has about 10% of the efficiency of luminol, but alkylation at the aromatic amino groups leads to a several-fold increase in the CL intensity. With this in mind the analogs *N*-(4-aminobutyl)-*N*-ethylisoluminol (ABEI) and ABEI hemisuccinimide (ABEI-H) were developed [22]. The ABEI- and ABEI-H-labeled analytes reacted with hydrogen peroxide after the addition of microperoxidase and the generated CL was measured for the determination of the analytes. For example, ABEI-labeled 17-α-methyltestosterone (MT), 19-nortestosterone (NT), and zeranol (Z) were covalently bound to anti-MT, -NT, and -Z. After the immune reaction, they were determined by CL produced with hydrogen peroxide and microperoxidase in alkaline medium [63]. Solid phase CL immunoassay of MT [64] and estradiol [65] was also achieved using the respective ABEI-labeled steroids.

Isoluminolamides produce less CL (around 0.1%) than isoluminol itself. *tert*-Butoxycarbonyl (BOC) or carbobenzoxy (Z)-Ala-Ala-Phe-isoluminolamide was developed for the assay of α-chymotrypsin activity by measuring the liberated isoluminol [66].

### Lucigenin Chemiluminescence Reactions

Lucigenin in strong alkaline medium reacts with oxidants such as hydrogen peroxide and sodium periodate and with organic reducing compounds such as ascorbic acid, glucose, creatinine, uric acid [23, 24], conjugated glucuronic acid [67], α-hydroxycarbonyls [68,69], and hydrolyzed products of heparin [70] (Fig. 10.2), whereby CL emission occurs. The detection limits are 0.17, 4.7, 0.64, 1.0, 9.1, 5.2, and 21 mg/liter for ascorbic acid, creatinine, uric acid, glutathione, glucuronic acid, lactose, and glucose, respectively [68]. The 1,2-enediolic structure is said to be responsible for the occurring CL [71].

Acidic compounds such as steroids and bile acid sulfates are able to complexate with the lucigenin in the flow stream and after subsequent extraction by an organic solvent, and using a PTFE porous membrane, they are detected by lucigenin CL [72].

Various acridinium esters can be stimulated to produce light in an alkaline medium [73]. Thus, acridinium ester-labeled antibodies [74] turned out to be efficient tools for immunochemiluminometric assays for α-1-fetoprotein [75], thyroxin [76], and ferritin [77].

### Peroxyoxalate Chemiluminescence Reactions

In peroxyoxalate CL reactions aryloxalate and hydrogen peroxide react in the presence of a fluorescent compound (Fig. 10.4). At the first stage of the process the aryloxalate reacts with hydrogen peroxide to generate an active intermediate, 1,2-dioxetane-3,4-dione, with an energy of around 70 kcal/mol. The generated intermediate interacts with a fluorescent compound and transfers its energy to excite the latter to the singlet excited state, which in turn returns to the ground state under the emission of light. The process of excitation and return to the ground state continues to occur until the production of the active intermediate terminates. As a consequence, one molecule of a fluorescent compound could generate more than one photon equivalent during a number of cycles. This is the reason why this type of CL reaction can be used as a highly sensitive detection system for fluorescent compounds. Fluorescent compounds with excitation energies of less than about 70 kcal/mol and possessing easily oxidizable structures can be effectively excited in these systems [78] (Fig. 10.6). Fluorescein and fluorescein-labeled thyroxin could be detected in the picomole concentration range in flow injection analysis [79]. Some other fluorescent compounds were detected without any separation step at all [80,81]. Recently, another mechanism for the CL reaction besides the one in Fig. 10.4 has been postulated, not including the formation of the active intermediate [82,83] (Fig. 10.7). At least two different intermediates for the generation of light were to be generated [84].

Among the oxalate esters, bis(2,4,6-trichlorophenyl)oxalate (TCPO) is the most popular reagent due to its stability in organic solvents such as ethyl acetate, acetone, and acetonitrile [85]. The most effective CL reaction was attained with bis(2,4-dinitrophenyl)-oxalate (DNPO) [86,87]. Evaluation of the oxalate esters have been

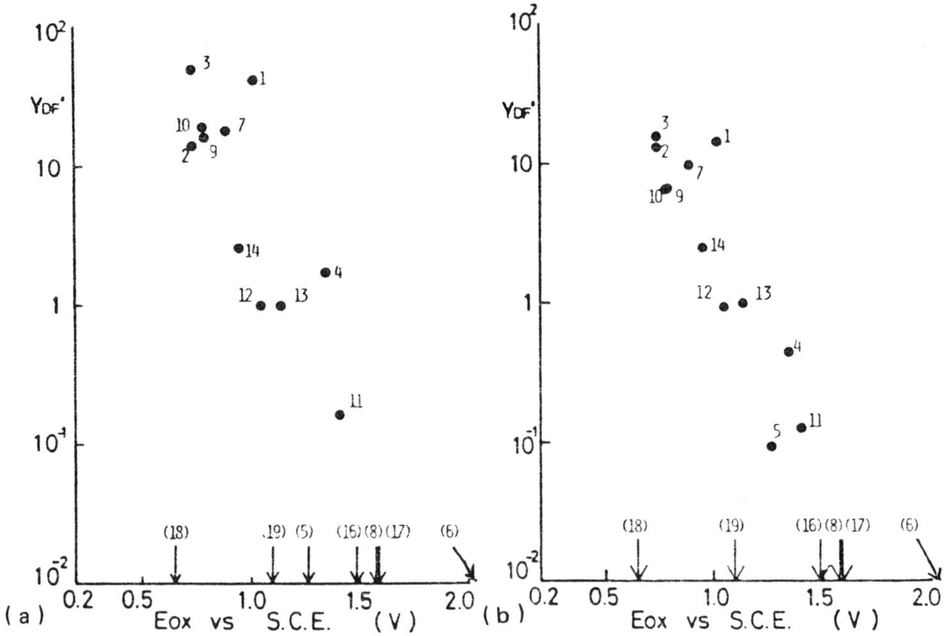


Figure 10.6 Relationship between oxidation potential (Eox) of fluorescent compounds and relative excitation yield (Y' df) [78]. A, bis(2,4-dinitrophenyl)oxalate (DNPO) system; B, bis(2,4,6-trichlorophenyl)oxalate (TCPO) system. Compounds: 1, rhodamine B; 2, rose Bengal; 3, rubrene; 4, 7-nitrobenzo-2-oxa-1,3-diazole-4-yl-(NBD)-proline; 5, fluorescein; 6, riboflavin; 7, perylene; 8, 7-sulfonylbenzo-2-oxa-1,3-diazole-4-yl-(SBD)-mercaptoethanol; 9, 5-*N*,*N*-dimethylaminonaphthalene sulfonyl-(Dns)-glycine; 10, Dns-alanine; 11, 9,10-dibromoanthracene; 12, benzo(a)pyrene; 13, 9,10-diphenylanthracene; 14, 3-methylcholanthrene; 16, umbelliferone; 17, 2-pyridone; 18, α-tocopherol; 19, pyridoxine hydrochloride. (Reprinted with permission from *Anal. Chim. Acta*, *177*:111 (1985).)

achieved [88]. Recently, bis[4-nitro-2-(3,6,9-trioxadecyloxycarbonyl)phenyl]oxalate, which is more soluble in aprotic solvents such as acetone and acetonitrile, was synthesized [89] as a reagent.

Hydrogen peroxide is the only efficient reactant in the peroxyoxalate CL reaction [90]. Thus, hydrogen peroxide [91], the substrates of oxidases such as uric acid [92], glucose [93,94], lactose

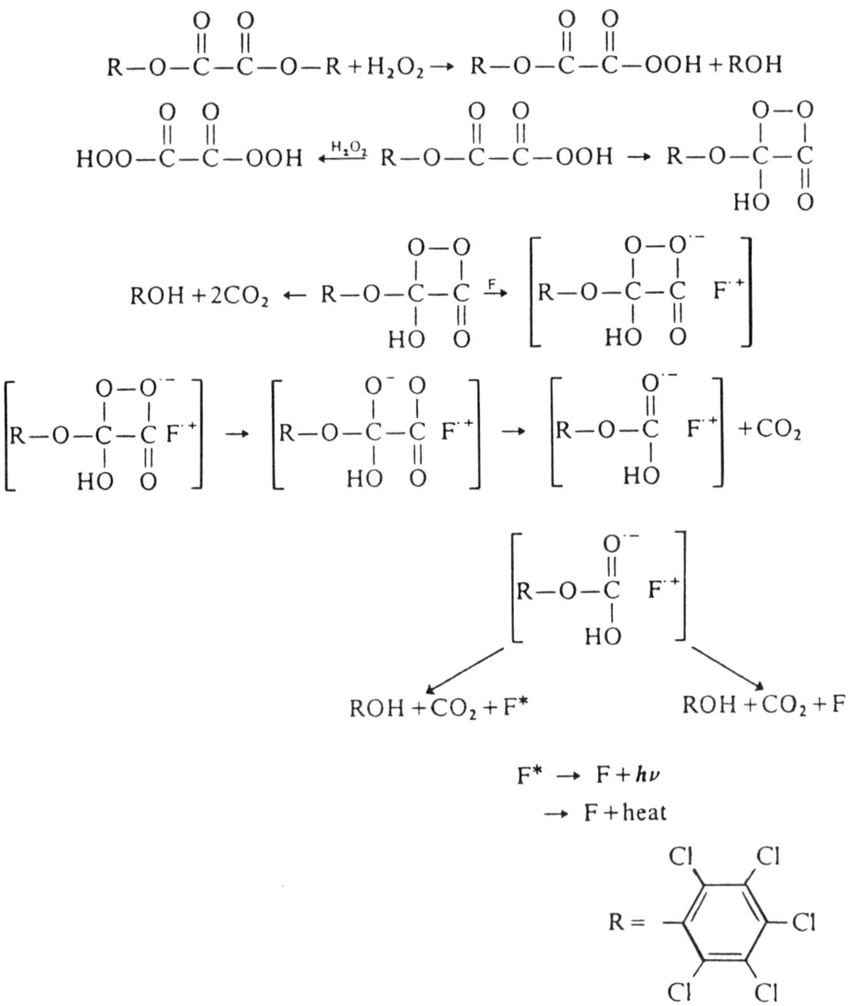

Figure 10.7 A newly proposed mechanism for peroxyoxalate chemiluminescence reaction [83]. (Reprinted with permission from *J. Chem. Soc., Faraday Trans.*, *2*:837 (1984).)

[94], cholesterol and its esters [95], L-amino acids [96], and formaldehyde [97], and the activity of hydrogen peroxide-generating enzymes such as glucose oxidase can be sensitively determined [98]. Immunoassay of 17-α-hydroxyprogesterone [99] and thyroxin [100] was performed using glucose oxidase, TCPO, and 8-anilinonaphthalenesulfonic acid (ANS). The reduced form of nicotinamide adenine

dinucleotide (NADH) reduces oxygen to hydrogen peroxide in the presence of methylene blue. This reaction could be employed to measure NADH produced by lactate dehydrogenase (LDH), and LDH activity in serum could also be determined in an automatic system [101].

### Other Chemiluminescence Reactions

Flow injection analysis of the cobalt(II) cation using gallic acid CL allowed the detection of about 0.6 pg after a 10-μl sample injection [102].

Hypochlorite and hydrogen peroxide react at pH 10 to generate singlet oxygen [103]. Subsequently, this generates CL when it reacts with an acceptor. The overall reaction [104] can be seen in Eqs. 1 and 2.

$$\begin{array}{llll} OCl^- + HOOH & \longrightarrow & H_2O + ClOO^- & \\ ClOO^- & \longrightarrow & Cl^- + O_2(^1\Delta_g) & \\ O_2(^1\Delta_g) + & \longrightarrow & O_2(^3\Sigma_g) + A: & \longrightarrow A + h\nu \end{array} \qquad (1)$$

or

$$O_2(^1\Delta_g) + A \longrightarrow \text{product*} \longrightarrow \text{product} + h\nu \qquad (2)$$

Aromatic compounds in a conducting solution can be stimulated to generate light when a potential is applied between two electrodes [105]. The mechanism is depicted in Eq. 3.

$$\begin{array}{lll} R & \longrightarrow & R^+ + e^- \\ R + e^- & \longrightarrow & R^- \\ R^+ + R^- & \longrightarrow & IR^* + R \\ IR^* & \longrightarrow & R + h\nu \end{array} \qquad (3)$$

## CHEMILUMINESCENCE REACTIONS ADOPTED FOR LIQUID CHROMATOGRAPHY

The following sections review the application of CL reactions or CL reagents in LC detection for the determination of analytes. Isoluminol derivatization followed by LC separation and CL detection of organic substances has been presented. Lucigenin was also used for the CL detection of reducing substances and *p*-nitrophenyl esters of carboxylic acids. Biphenoyl peroxide and the other CL reactions mentioned above have to be performed at elevated temperatures. This is why they have not yet been used for detection

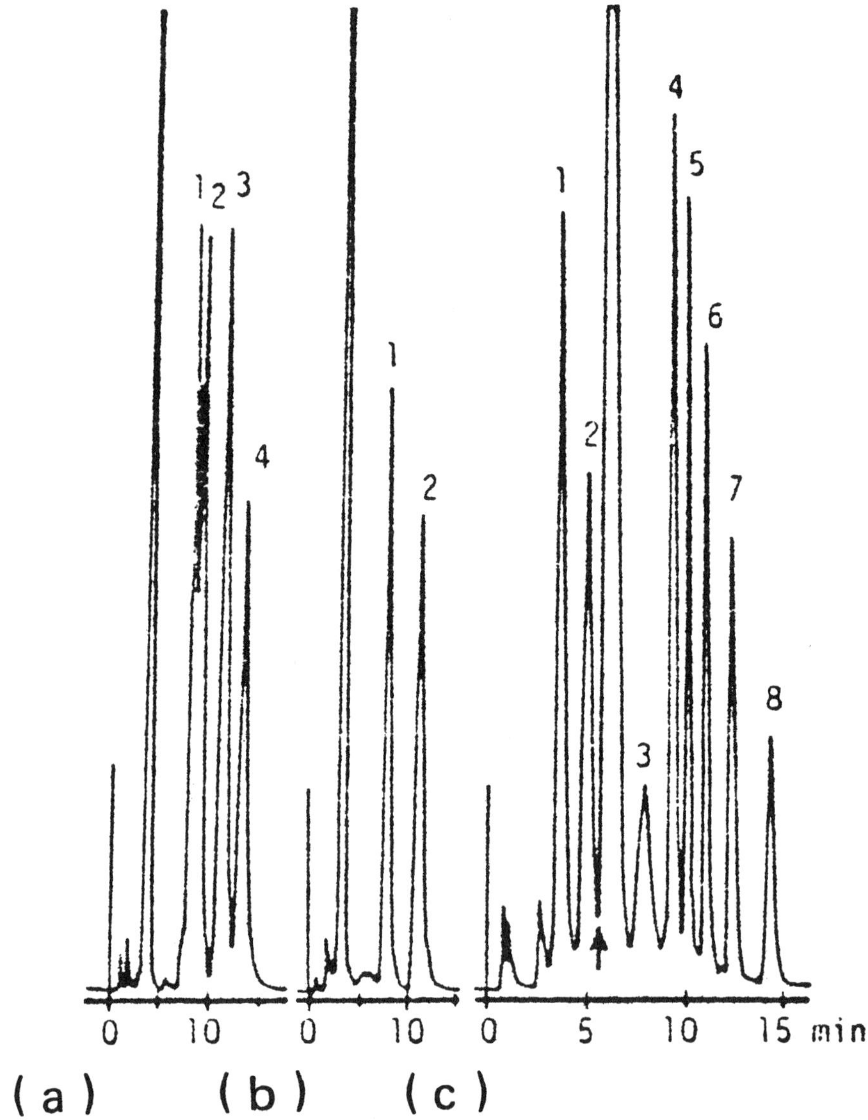


Figure 10.8 Chromatograms obtained from amines labeled with *N*-(4-aminobutyl)-*N*-ethylisoluminol (ABEI) followed by liquid chromatography and chemiluminescence reaction detection [106]. (a) aryl-

purposes in LC. The peroxyoxalate CL reaction has been used as postchromatographic detection system after LC of fluorescent compounds, whereas hydrogen peroxide generated by oxidases could be also employed. Singlet oxygen and electrogenerated CL reactions were also used in postchromatographic reaction detectors.

## Isoluminol in Liquid Chromatography

The prechromatographic labeling of amines such as toluidine isomers and butylamine isomers with ABEI was carried out in methanol after addition of the premixed solution of ABEI and *N,N*-disuccinimidyl carbonate (DSC) in acetonitrile. The mixture was allowed to stand at room temperature for 2 hr [106]. An aliquot of the reaction mixture was subjected to LC, postchromatographically reacted with hydrogen peroxide and potassium hexacyanoferrate(III) in a sodium hydroxide solution, and the generated CL was detected (Fig. 10.8). The sensitivity of this method was good.

For labeling of carboxylic acids such as acetic, propionic, and bile acids, these compounds were mixed with ABEI in methanol and evaporated to dryness under a stream of dry nitrogen gas. Then 2-chloro-1-methylpyridinium iodide (CMPI) and 3,4-dihydro(2H)-pyrido-(1,2-a)pyrimidine-2-one (DPP) in acetonitrile were added and the mixture was heated at 333 K for 2 hr. An aliquot was subjected to LC with the above-mentioned postchromatographic reaction system. Detection limits were to be about 20 fmol. Note that optimum conditions vary according to the flow rate of the column effluent, the reagent solution, the concentration of the reagents, etc. Therefore, the reaction conditions should be optimized for each individual reaction.

Postchromatographic reaction of the column effluent with luminol and hydrogen peroxide permits the detection of transition metals at the low micromole level [107].

## Lucigenin in Liquid Chromatography

LC of ascorbic acid, dehydroascorbic acid, glucose, glucuronic acid, and creatinine [108], heparin [109], and steroids having α-hydroxycarbonyl groups such as cortisol and cortisone [110] followed by

---

substituted amines; 1, aniline; 1, benzylamine; 3, α-phenylethylamine; 4, β-phenylethylamine. (b) aromatic amines; 1, aniline; 2, *N*-methylaniline. (c) aliphatic amines; 1, methylamine; 2, ethylamine; 3, *n*-propylamine; 4, *n*-butylamine; 5, *n*-amylamine; 6, *n*-hexylamine; 7, *n*-heptylamine; 8, *n*-octylamine. (Reprinted with permission from *J. Chromatogr.*, *328*:121 (1985).)

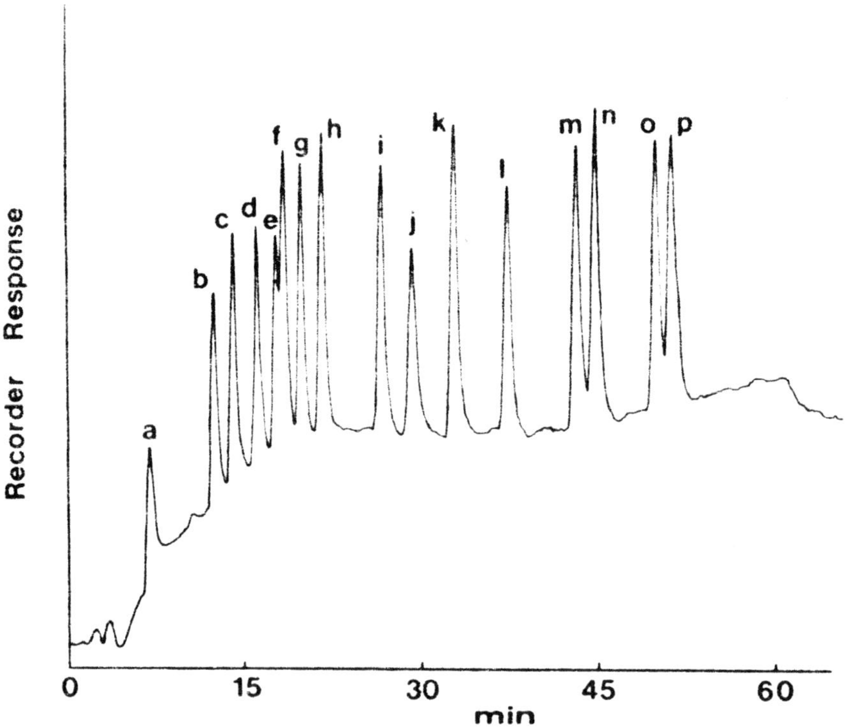


Figure 10.9 Chromatogram obtained by gradient elution of 5-*N*,*N*-dimethylaminonaphthalene sulfonyl–labeled amino acids (20 fmol each) and detection with peroxyoxalate chemiluminescence reaction [115]. a, asp; b, asn; c, gln; d, ser; e, arg; f, thr; g, gly; h, ala; i, pro; j, lys; k, val; l, met; m, ile; n, leu; o, trp; p, phe. (Reprinted with permission from *J. Chromatogr.*, *352*:255 (1986).)

postchromatographic reaction with lucigenin in alkaline medium results in the sensitive CL detection of these analytes.

Derivatization of carboxylic acids with *p*-nitrophenacyl bromide in the presence of sodium ethoxide followed by refluxing for 2 hr results in the *p*-nitrophenacyl esters, which can be subjected to LC with a lucigenin CL reaction detection system [111]. The detection limits were about 500 fmol.

## Peroxyoxalate Chemiluminescence Detection Systems in Liquid Chromatography

Fluorescent compounds or fluorescence-labeled compounds are separated on a column, the mixture of the aryl oxalate and hydrogen peroxide is added to the effluent, and the generated CL is detected by a photomultiplier or photon counter. Thus, a sensitive and selective quantitation of fluorescent compounds or fluorescence-labeled compounds can be achieved.

5-*N*,*N*-dimethylaminonaphthalenesulfonyl (Dns)-labeled amino acids [111-115], Dns-labeled amines [112], Dns-labeled catecholamines [116], Dns-labeled steroids [117], fluorescamine-labeled catecholamines [118], 7-nitrobenzo-2-oxa-1,3-diazole (NBD)-labeled amino acids [116], orthophthaldialdehyde (OPA)-labeled amino acids [118], polycyclic aromatic hydrocarbons [119-121], and aminopyrene [122] have been sensitively detected with the system.

Quenching phenomena of peroxyoxalate CL by sulfite, nitrite, anilines, and organosulfur compounds enable the detection of these compounds by using solid TCPO and immobilized fluorescent compounds set in front of the photomultiplier [123]. Application of the system to microbore column LC [113,121] and gradient solution LC (Fig. 10.9) [114,115] were successful.

It is not easy to mix two or three solutions in the flow stream. So a new mixing device, a so-called rotating flow mixing device, has been presented (Fig. 10.10) [124] in which the two solutions enter a cylinder at the bottom end from opposite directions and mix under rotation while pushing up the cylinder. In this way an efficient mixing of the solutions can be achieved. Optimization of the CL system was studied using DNPO with a large flow cell of 0.5 ml [12] and a high-voltage photomultiplier tube [126].

A column filled with a mixture of immobilized acetylcholine hydrolase and choline oxidase was connected to the analytical column for the separation of cholines, and the hydrogen peroxide generated from acetylcholine was detected using TCPO and perylene at postchromatographic reagents at the low picomole level [127].

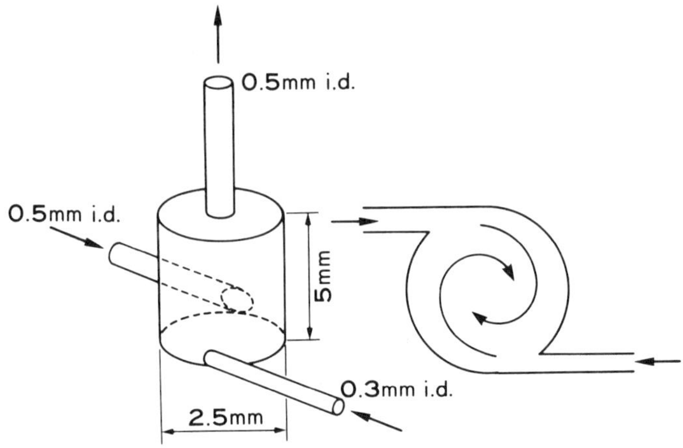


Figure 10.10 A rotating flow mixing device [124].

### Other Chemiluminescence Reactions in Liquid Chromatography

An aerosol spray detector for LC based on ozone-induced and singlet oxygen-induced CL was developed [128,129]. Fluorescent compounds, olefins, divalent sulfur compounds, and certain nitrogen-containing compounds such as hydrazines, azides, nitrogen heterocyclics [128] and chlorophyls [129] were detected.

A cell design for electrogenerated CL was also presented [130]. With this system aromatic compounds such as naphthalene, acenaphthene, fluoranthene, etc., were separated on a reversed phase column and detected at submicrogram levels [131].

LC equipped with a thermal energy analyzer CL detector permitted the detection of a few nanograms of several ionic nitrosamines [132]. Glycerol trinitrate, pentaerythritol tetranitrate, and isosorbide dinitrate were also detected at the subnanogram concentration level [133]. The applications of these CL reactions and reagents are summarized in Table 10.1.

## REFERENCES

1. Furuta, N., and A. Otuki, *Anal. Chem.*, *55*:2407 (1983).
2. Pfeffer, W. D., and E. S. Yeung, *Anal. Chem.*, *58*:2103 (1986).
3. Glick, D., *Clin. Chem.*, *23*:1465 (1977).
4. Seitz, W. R., in *Methods in Enzymology*, Vol. 57, Academic Press, New York, 1978, p. 445.
5. *Clin. Chem.*, *25*(9) (1979).

6. *Photochem., Photobiochem., 30*:(1) (1979).
7. Wood, W. G., H. Fricke, J. Haritz, A. Godow, H.-S. Krausz, B. Tode, C. J. Strasburger, and P. C. Scriba, *J. Clin. Chem. Clin. Biochem., 22*:349 (1984).
8. Imai, K., and R. Weinberger, *Trends Anal. Chem., 4*:170 (1985).
9. Mottola, H. A., and H. B. Mark, *Anal. Chem., 54*:62R (1982).
10. Wehry, E. L., *Anal. Chem., 54*:131R (1982).
11. Kricka, L. J., and G. H. G. Thorpe, *Analyst, 108*:1274 (1983).
12. Wehry, E. L., *Anal. Chem., 56*:156R (1984).
13. Mottola, H. A., and H. B. Mark, *Anal. Chem., 56*:96R (1984).
14. Majors, R. E., H. G. Barth, and C. H. Lochmuller, *Anal. Chem., 56*:300R (1984).
15. Wehry, E. L., *Anal. Chem., 58*:13R (1986).
16. Miller, J. N., *Analyst, 109*:191 (1984).
17. *J. Bio. Chemiluminescence*, Wiley, New York.
18. DeLuca, M. A., and W. D. McElroy (eds.), *Bioluminescence and Chemiluminescence: Basic Chemistry and Analytical Applications*, Academic Press, New York, 1981.
19. Knox Van Dyke (ed.), *Bioluminescence and Chemiluminescence: Instruments and Applications*, Vols. 1,2, CRC Press, Boca Raton, 1985.
20. Collins, W. P. (ed.), *Alternative Immunoassay*, John Wiley and Sons, New York, 1985.
21. White, E. H., O. Zafriou, H. H. Kagi, and J. H. M. Hill, *J. Am. Chem. Soc., 86*:940 (1964).
22. Schroeder, H. R., and F. M. Yeager, *Anal. Chem., 50*:1114 (1978).
23. Gleu, K., and W. Petsch, *Angew. Chem., 48*:57 (1935).
24. Totter, J. R., *Photochem. Photobiol., 22*:203 (1975).
25. Slawinsa, D., and J. Slawinski, *J. Anal. Chem., 47*:2101 (1975).
26. Stieg, S., and T. A. Nieman, *Anal. Chem., 49*:1322 (1977).
27. Marino, D. F., and J. D. Ingle, Jr., *Anal. Chim. Acta, 124*:23 (1981).
28. Marino, D. F., F. Wolff, and J. D. Ingle, Jr., *Anal. Chem., 51*:2051 (1979).
29. MacDonald, A., K. W. Chan, and T. A. Nieman, *Anal. Chem., 51*:2077 (1979).
30. Chandross, E. A., and F. I. Sonntag, *J. Am. Chem. Soc., 86*:3179 (1964).
31. Park, S., and D. A. Tryk, *Rev. Chem. Intermed., 4*:43 (1980).
32. Haapakka, K. E., and J. J. Kankare, *Anal. Chim. Acta, 130*:415 (1981).
33. Haapakka, K. E., and J. J. Kankare, *Anal. Chim. Acta, 138*:263 (1982).
34. Haapakka, K. E., *Anal. Chim. Acta, 139*:229 (1982).

35. Rauhut, M. M., L. J. Bollyky, B. G. Roberts, R. H. Whitehouse, A. V. Iannotta, A. M. Samsel, and R. A. Clarke, *J. Am. Chem. Soc.*, *89*:6515 (1967).
36. Rauhut, M. M., *Acc. Chem. Res.*, *2*:80 (1969).
37. Schuster, G. B., *Acc. Chem. Res.*, *112*:366 (1979).
38. Schuster, G. B., B. Dixon, J.-Y. Koo, S. P. Schmidt, and J. P. Smith, *Photochem. Photobiol.*, *30*:17 (1979).
39. Kopecky, K. R., and C. Mumford, *Can. J. Chem.*, *47*:709 (1969).
40. Bruening, W., and F. J. M. Conche, *J. Chromatogr.*, *142*:191 (1977).
41. Marino, D. F., and J. D. Ingle, Jr., *Anal. Chim. Acta*, *123*:247 (1981).
42. Hartkopf, A. V., and R. Delumyea, *Anal. Lett.*, *7*:79 (1974).
43. Neary, M. P., W. R. Seitz, and D. M. Hercules, *Anal. Lett.*, *7*:583 (1974).
44. Delumyea, R., and A. V. Hartkopf, *Anal. Chem.*, *48*:1402 (1976).
45. Burguera, J. L., M. Burguera, and A. Townshend, *Anal. Chim. Acta*, *127*:199 (1981).
46. Burdo, T. G., and W. R. Seitz, *Anal. Chem.*, *47*:1639 (1975).
47. Pentel, S., and H. Weisz, *Anal. Chim. Acta*, *74*:275 (1975).
48. Pabivanets, B. I., A. K. Ryabov, and E. M. Shnaiderman, *Z. Anal. Khim.*, *30*:2439 (1975).
49. Hara, T., M. Toriyama, and K. Tsukagoshi, *Bull. Chem. Soc. Jpn.*, *56*:1382 (1983).
50. Hara, T., M. Toriyama, and T. Ebuchi, *Bull. Chem. Soc. Jpn.*, *58*:109 (1985).
51. MacDonald, A., and T. A. Nieman, *Anal. Chem.*, *57*:936 (1985).
52. Olsson, T., K. Bergstroem, and A. Thore, *Clin. Chim. Acta*, *138*:31 (1984).
53. Auses, J. P., S. L. Cook, and J. T. Maloy, *Anal. Chem.*, *47*:244 (1975).
54. Bostick, D. T., and D. M. Hercules, *Anal. Chem.*, *47*:447 (1975).
55. Ridder, C., E. H. Hansen, and J. Ruzicka, *Anal. Lett.*, *15*:1751 (1982).
56. Worsfold, P. J., J. Farrelly, and M. S. Matharu, *Anal. Chim. Acta*, *164*:103 (1984).
57. Koerner, C. A., and T. A. Nieman, *Anal. Chem.*, *58*:116 (1986).
58. Fagerstroem, R., P. Seppanen, and J. Janne, *Clin. Chim. Acta*, *143*:45 (1984).
59. Bachrach, U., and Y. M. Plesser, *Anal. Biochem.*, *152*:423 (1986).
60. Olsson, B., *Anal. Chim. Acta*, *136*:113 (1982).
61. Lilius, E.-M., and S. Laakso, *Anal. Biochem.*, *119*:135 (1982).

62. Maeda, M., and A. Tsuji, *Anal. Chim. Acta, 167*:241 (1985).
63. Jansen, E. H. J. M., R. H. van den Berg, G. Zomer, R. Both-Miedema, C. Enkelaar-Willemsen, and R. W. Stephany, *Anal. Chim. Acta, 170*:21 (1985).
64. Jansen, E. H. J. M., C. A. Laan, R. H. van den Berg, R. W. Stephany, and G. Zomer, *Anal. Chim. Acta, 170*:29 (1985).
65. Boever, J. de, F. Kohen, M. Dhont, D. VandeKerckhove, and G. van Maele, *Anal. Chim. Acta, 170*:117 (1985).
66. Branchini, B. R., J. D. Hermes, F. G. Salituro, N. J. Post, and G. Claeson, *Anal. Biochem., 111*:87 (1981).
67. Klopf, L. L., and T. A. Nieman, *Anal. Chem., 57*:46 (1985).
68. Veazey, R. L., H. Nekimkin, and T. A. Nieman, *Anal. Chem., 51*:2092 (1979).
69. Maeda, M., and A. Tsuji, *Anal. Sci., 2*:183 (1986).
70. Steen, R. A., and T. A. Nieman, *Anal. Chim. Acta, 155*:123 (1983).
71. Veazey, R. L., H. Nekimkin, and T. A. Nieman, *Talanta, 31*:603 (1984).
72. Maeda, M., and A. Tsuji, *Analyst, 110*:665 (1985).
73. McCapra, F., *Acc. Chem. Res., 9*:201 (1976).
74. Weeks, I., I. Beheshti, F. McCapra, A. K. Campbell, and J. S. Woodhead, *Clin. Chem., 29*:1474 (1983).
75. Weeks, I., A. K. Campbell, and J. S. Woodhead, *Clin. Chem., 29*:1480 (1983).
76. Weeks, I., M. Sturgess, K. Siddle, M. K. Jones, and J. S. Woodhead, *Clin. Endocrinol., 20*:489 (1984).
77. Weeks, I., and J. S. Woodhead, *Clin. Chim. Acta, 141*:275 (1984).
78. Honda, K., K. Miyaguchi, and K. Imai, *Anal. Chim. Acta, 177*:111 (1985).
79. Mahant, V. K., J. N. Miller, and H. Thakrar, *Anal. Chim. Acta, 145*:203 (1983).
80. Grayeski, M. L., and W. R. Seitz, *Anal. Biochem., 136*:277 (1984).
81. Sherman, P. A., J. Holzbecher, and D. E. Ryan, *Anal. Chim. Acta, 97*:21 (1978).
82. Catherall, C. L. R., T. F. Palmer, and R. B. Cundall, *J. Chem. Soc., Faraday Trans. 2, 80*:823 (1984).
83. Catherall, C. L. R., T. F. Palmer, and R. B. Cundall, *J. Chem. Soc., Faraday Trans. 2, 80*:837 (1984).
84. Alvarez, F. J., N. J. Parekh, B. Matuszewski, R. S. Givens, T. Higuchi, and R. L. Ahowen, *J. Am. Chem. Soc., 108*:6435 (1986).
85. Mohan, A. G., and N. J. Turo, *J. Chem. Educ., 51*:528 (1974).
86. Rauhut, M. M., *Acc. Chem. Res., 2*:80 (1969).
87. Honda, K., J. Sekino, and K. Imai, *Anal. Chem., 55*:940 (1983).

88. Honda, K., K. Miyaguchi, and K. Imai, *Anal. Chim. Acta, 177*:103 (1985).
89. Imai, K., H. Nawa, H. Tanaka, and H. Ogata, *Analyst, 111*:209 (1986).
90. Rauhut, M. M., L. J. Bollyky, B. G. Roberts, M. Loy, R. H. Whitman, A. V. Iannotta, A. M. Semsel, and R. A. Clarke, *J. Am. Chem. Soc., 89*:6515 (1967).
91. Zoonen, P. van, D. A. Kamminga, G. Gooijer, N. H. Velthorst, and R. W. Frei, *Anal. Chim. Acta, 167*:249 (1985).
92. Scott, G., W. R. Seitz, and J. Ambrose, *Anal. Chim. Acta, 115*:221 (1980).
93. Williams, D. C., III, G. F. Huff, and W. R. Seitz, *Anal. Chem., 48*:1003 (1976).
94. Riggin, V. I., *J. Anal. Chem. USSR, 34*:619 (1979).
95. Riggin, V. I., *J. Anal. Chem. USSR, 33*:1265 (1983).
96. Riggin, V. I., *J. Anal. Chem. USSR, 38*:1328 (1983).
97. Riggin, V. I., *J. Anal. Chem. USSR, 36*:1111 (1981).
98. Williams, D. C., III, G. F. Huff, and W. R. Seitz, *Clin. Chem., 22*:372 (1976).
99. Arakawa, H., M. Maeda, and A. Tsuiji, *Chem. Pharm. Bull., 30*:3036 (1982).
100. Arakawa, H., M. Maeda, and A. Tsuji, *Clin. Chem., 31*:430 (1985).
101. Williams, D. C., III, and R. W. Seitz, *Anal. Chem., 48*:1478 (1976).
102. Yamada, M., T. Komatsu, S. Nakahara, and S. Suzuki, *Anal. Chim. Acta, 155*:259 (1983).
103. Khan, A. U., and M. Kasha, *J. Am. Chem. Soc., 92*:3293 (1970).
104. Kautsky, H., *Biochem. Z., 291*:271 (1937).
105. Park, S., and D. A. Tryk, *Rev. Chem. Intermed., 4*:43 (1980).
106. Kawasaki, T., M. Maeda, and A. Tsuji, *J. Chromatogr., 328*:121 (1985).
107. Neary, M. P., R. W. Seitz, and D. M. Hercules, *Anal. Lett., 7*:583 (1974).
108. Veazey, R. L., and T. A. Nieman, *J. Chromatogr., 200*:153 (1980).
109. Steen, R. A., and T. A. Nieman, *Anal. Chim. Acta, 155*:123 (1983).
110. Maeda, M., and A. Tsuji, *J. Chromatogr., 352*:213 (1986).
111. Kobayashi, S., and K. Imai, *Anal. Chem., 52*:424 (1980).
112. Mellbin, G., *J. Liq. Chromatogr., 6*:1603 (1983).
113. Miyaguchi, K., K. Honda, and K. Imai, *J. Chromatogr., 316*:501 (1984).
114. Miyaguchi, K., K. Honda, and K. Imai, *J. Chromatogr., 303*:173 (1984).

115. Miyaguchi, K., K. Honda, T. Toyo'oka, and K. Imai, *J. Chromatogr.*, *352*:255 (1986).
116. Mellbin, G., and B. E. F. Smith, *J. Chromatogr.*, *312*:203 (1984).
117. Koziol, T., M. L. Grayeski, and R. Weinberger, *J. Chromatogr.*, *317*:355 (1984).
118. Kobayashi, K., J. Sekino, K. Honda, and K. Imai, *Anal. Biochem.*, *112*:99 (1981).
119. Sigvardson, K. W., and J. W. Birks, *Anal. Chem.*, *55*:432 (1983).
120. Sigvardson, K. W., J. M. Kennish, and J. W. Birks, *Anal. Chem.*, *56*:1096 (1984).
121. Grayeski, M. L., and A. J. Weber, *Anal. Lett.*, *17*:1539 (1984).
122. Weinberger, R., C. A. Mannan, M. Cerchio, and M. L. Grayeski, *J. Chromatogr.*, *288*:445 (1984).
123. Zoonen, P. van, D. A. Kamminga, C. Gooijer, N. H. Velthorst, and R. W. Frei, *Anal. Chem.*, *58*:1248 (1986).
124. Kobayashi, S., and K. Imai, *Anal. Chem.*, *52*:1548 (1980).
125. Jong, G. J. de, N. Lammers, F. J. Spruit, U. A. Th. Brinkman, and R. W. Frei, *Chromatographia*, *18*:129 (1984).
126. Weinberger, R., C. A. Mannan, M. Cerchio, and M. L. Grayeski, *J. Chromatogr.*, *288*:445 (1984).
127. Honda, K., K. Miyaguchi, H. Nishino, H. Tanaka, T. Yao, and K. Imai, *Anal. Biochem.*, *153*:50 (1986).
128. Birks, J. W., and M. C. Kuge, *Anal. Chem.*, *52*:897 (1980).
129. Shoemaker, B., and J. W. Birks, *J. Chromatogr.*, *209*:251 (1981).
130. Schaper, H., *J. Electroanal. Chem.*, *129*:335 (1981).
131. Blatchford, C., and D. J. Malcolme-Lawes, *J. Chromatogr.*, *321*:227 (1985).
132. Massey, R. C., C. Crews, D. J. McWeeny, and M. E. Knowles, *J. Chromatogr.*, *236*:527 (1982).
133. Yu, W. C., and E. U. Goff, *Anal. Chem.*, *55*:29 (1983).

# Index